Medienkulturen im digitalen Zeitalter

Series Editors

Kornelia Hahn, Salzburg, Austria

Rainer Winter, Klagenfurt, Austria

Fortgeschrittene Medienkulturen im 21. Jahrhundert zeichnen sich dadurch aus, dass alle Kommunikation durch Erfahrungen mit „neuer", digitaler Medientechnologie beeinflusst ist. Es kommt nicht nur zu vielfältigen Transformationen von Praktiken und Identitäten. Überdies entstehen neue Identifikationen und Gebrauchsweisen. Auch die Medien selbst werden verändert, weil Inhalte leichter verfügbar sind, sich Plattformen und Produzenten vervielfältigen und multiple Konvergenzen herausbilden. Die Verknüpfung von traditionellen und neuen Medien führt immer mehr zur Entfaltung komplexer und intensiver Medienkulturen, die unser Leben maßgeblich prägen. Dabei ist Medienkommunikation immer bereits in spezifische Kulturen eingebettet und wird eigensinnig implementiert. Die Reihe enthält empirische und theoretische Beiträge, die gegenwärtige Medienkulturen als spezifische Facette des sozialen Wandels fokussieren. Die damit verbundenen medialen Transformationen sind gleichzeitig Untersuchungskontext als auch Gegenstand der kritischen Reflexion. Da Medien in fast allen sozialen Situationen präsent sind, gehen wir nicht von einem Gegensatz zwischen Medienkultur und Nicht-Medienkultur aus, sondern eher von einem Kontinuum bzw. einem Spektrum an Veränderungen. Während bisher die Erforschung der medienbasierten Fernkommunikation überwiegt, gibt die Reihe auch der face-to-face oder kopräsenten Kommunikation und Interaktion in Medienkulturen ein Forum. Die Beiträge basieren damit auf Untersuchungskonzeptionen, in deren Zentrum die soziologische Analyse von Medienkulturen steht.

More information about this series at http://www.springer.com/series/11768

Douglas Kellner

Technology and Democracy: Toward A Critical Theory of Digital Technologies, Technopolitics, and Technocapitalism

Douglas Kellner
Graduate School of Education and Information
Studies, UCLA, Los Angeles, USA

ISSN 2570-4087   ISSN 2570-4095 (electronic)
Medienkulturen im digitalen Zeitalter
ISBN 978-3-658-31789-8   ISBN 978-3-658-31790-4 (eBook)
https://doi.org/10.1007/978-3-658-31790-4

Lektorat: Cori A. Mackrodt
This Springer VS imprint is published by the registered company Springer Fachmedien Wiesbaden
GmbH part of Springer Nature.
The registered company address is: Abraham-Lincoln-Str. 46, 65189 Wiesbaden, Germany

Contents

Introduction: Technology and the Demands of Democracy

Abstract

To meet the challenges of a new technological society and culture, we must develop a critical theory of technology which criticizes uses or types of technology as tools of domination, rejects the hype and pretensions of new media and technologies, and that calls for appropriate restructuring of technology to democratize politics, society, and everyday life. A critical theory of media and technology sees how new technologies can be used, and perhaps redesigned and restructured, for positive purposes such as enhancing education, democracy, and overcoming the divide between haves and have nots, while enabling individuals to democratically and creatively participate in a new economy, society, and culture. The following essays collect Douglas Kellner's articles on these topics and provide a comprehensive overview of his critical theory of technology for the first time.

Keywords

Critical theory • Technology • Democracy • Capitalism

As we enter a new millennium, it is clear that we are in the midst of one of the most dramatic technological revolutions in history that is changing everything from the ways that we work, communicate, participate in politics, and spend our leisure time. The technological revolution centers on computer, information, communication, and multimedia technologies, is often interpreted as the beginnings of a knowledge or information society, and therefore ascribes technologies

D. Kellner, *Technology and Democracy: Toward A Critical Theory of Digital Technologies, Technopolitics, and Technocapitalism*, Medienkulturen im digitalen Zeitalter, https://doi.org/10.1007/978-3-658-31790-4_1

a central role in everyday life. This Great Transformation poses tremendous challenges to critical social theorists, citizens, and educators to rethink their basic tenets, to deploy the proliferation of digital technology and media in creative and productive ways, and to restructure the workplace, social institutions, schooling, and our democratic institutions to respond constructively and progressively to the technological and social changes that we are now experiencing.

To meet the challenges of an always expanding and transformative technological society and culture, we must develop a **critical theory of technology** which criticizes uses or types of technology as tools of domination, rejects the hype and pretensions of digital media and technologies, and that calls for appropriate restructuring of technology to democratize politics, society, and everyday life. A critical theory of media and technology sees how digital technologies can be used, and perhaps redesigned and restructured, for positive purposes such as enhancing education, democracy, and overcoming the divide between haves and have nots, while enabling individuals to democratically and creatively participate in a new economy, society, and culture. Yet it is also concerned with how digital technologies are used for surveillance, and threats to privacy with megacorporations amassing tremendous data bases that they can use to surveil and influence consumers, voters, and social groups and individuals. Digital technologies give government and megacorporations tremendous power, and concerns over privacy, surveillance, and unregulated social media have produced debates and calls for government regulatory action as I will discuss in this book.

At the same time that we are undergoing technological revolution, important demographic and socio-political changes are occurring in the United States and throughout the world. Emigration patterns have brought an explosion of new peoples into the U.S. and elsewhere in recent decades, and Western democracies are now more racially and ethnically diverse, more multicultural, than ever before. The demands and promise of U.S. democracy must meet the challenge of providing people from diverse races, classes, and backgrounds with the tools and competencies to enable them to succeed and participate in an ever more complex and changing world.[1] The current explosion of digital technologies and social

---

[1] Studies at the beginning of the twenty-first century reveal that women, minorities, and immigrants constitute roughly 85 percent of the growth in the labor force, while these groups represent about 60 percent of all workers; see Duderstadt 1999–2000, p. 38, and "Workplace 2000: Workers and Workplace in the 2000s" at https://files.eric.ed.gov/fulltext/ED290887.pdf (accessed December 10, 2020). In the past decades, the number of Hispanics in the United States increased by 35 percent and Asians by more than 40 percent, and by 2020 more than half of the high school students are African-American or Latino; see "Racial/Ethnic Enrollment in Public Schools" at https://nces.ed.gov/programs/coe/pdf/coe_cge.pdf (accessed on December

media have ignited furious debates over their substance, trajectory, and effects which pose major challenges to critical social theory and a radical democratic politics: first, how to theorize the dramatic changes in every aspect of life that digital technologies and social media are producing; and, secondly, how to utilize digital technologies to promote progressive social change to create a more egalitarian and democratic society in an era marked by rampant technological development and the seeming victory of market capitalism over its historical opponents, amidst global crises of democracy as authoritarian regimes, such as the just terminated Trump presidency, challenge democracy, threaten our environment and health, and provide a demand for a democracy that meets citizens' needs, protects the environment, and helps produce global peace, stability, and well-being.

In this book, I will suggest some ways to theorize the current technological revolution without falling into either technological or economic determinism, as well as the modes of technophilia or technophobia. I will argue that one needs to theorize the diffusion of evolving digital technologies and social media within the context of a series of transformations that advanced industrial and technological societies are undergoing: (1) in the context of the current stage of capitalist development, as a crucial part of the global restructuring of capitalism, and thus to think together the current development and imbrication of technology and capitalism; and (2) as embodying a set of technologies and practices that themselves can be restructured and reconstituted to carry out individual and group projects aiming at a democratic transformation of society, merging theory and practice.

In carrying out this project, one needs to avoid the extremes of either exaggerating or downplaying the autonomous role of technology in this process, as if technology were either the demiurge of the contemporary world, or an unimportant epiphenomenon of a much greater force, such as capitalism or human self-development. In addition, one must avoid two extremes which would either denigrate and demonize technology in the mode of technophobia, or celebrate and deify it in the mode of technophilia. Instead, a critical theory of technology attempts to develop a dialectical optic that avoids one-sided approaches in theorizing and evaluating the genesis of relatively new digital technologies and their often contradictory effects.

I propose to develop democratic and activist perspectives on digital technologies and social media, suggesting some ways that they might be used for such things as self-valorisation and empowerment, democratization, and progressive

---

10, 2020). Meanwhile, a "tidal wave" of children of baby boomers entered college in the 2000s; see Atkinson 1999–2000, pp. 49–50.

social transformation, in opposition to their roles of strengthening the forces of corporate and state domination, or narcissistic possessive individualism. Yet I do not want to fall into the utopianism of the boosters of digital technologies and social media, nor the pessimism and defeatism of those who merely see evolving information and communication technologies as an instrument of capital and the state, which are destructive and wholly negative social and cultural forces. In addition, I will take on the issue of theorizing the information society and the so-called "Information Superhighway" and will argue that the "information society" is the now dominant ideology of technocapitalism that identifies technological progress and human well-being with new technologies and the market, while presenting the state as an obsolete force of domination and bureaucratic state apparatus that is seen as an impediment to progress, freedom, and other positive values. In this view, it is the market and individual entrepreneurship that has made possible the dramatic technological revolution of the present and any state regulation is seen as an impediment for further progress. Such an ideology is used to dismantle the welfare state and must thus be put in question and subjected to political critique.

For a Critical Theory of Digital Technology

In studying the exploding array of discourses which characterize the digital technologies and social media, I am bemused by the extent to whether they expose either a technophilic discourse which presents digital technologies as our salvation, that will solve all our problems, or they embody a technophobic discourse that sees digital technologies and social media as our damnation, demonizing them as the major source of all our problems. It appears that similarly one-sided and contrasting discourses greeted the introduction of other new technologies in the twentieth century, often hysterically. To some extent, this was historically the case with film, radio, TV, and now computers, the Internet, social media, and the technoculture.[2] Film, for instance, was celebrated by some of its early theorists as providing new documentary depiction of reality, even redemption of reality, generating a new art form and new modes of mass education and entertainment.

Film was also demonized from the beginning for promoting sexual promiscuity, juvenile delinquency and crime, violence, and copious other forms of immorality. Its demonization led in the United States to a Production Code that rigorously regulated the content of Hollywood film from 1934 until the 1950s and

---

[2] On the demonization of new technologies in past eras when they first appeared such as radio, film, and television, see Barnouw (1990).

1960s—no open mouthed kissing was permitted, crime could not pay, drug use or attacks on religion could not be portrayed, and a censorship office rigorously surveyed all films to make sure that no subversive or illicit content emerged (Schatz 2010).

Similar extreme hopes and fears were projected onto radio, television, and now computers. It seems that whenever there are new technologies, people project all sorts of fantasies, fears, hopes, and dreams onto them, and I believe that this has been happening since the 1990s with computers and digital technologies. It is indeed striking that if one looks at the earlier literature on new technologies—and especially computers—it was either highly celebratory and technophilic, or sharply derogatory and technophobic. For technophilia, one can open any issue of *Wired*, or popular magazines like *Newsweek*, one can read Bill Gates' book *The Road Ahead* (1995), or some of the academic boosters of digital technologies like Nichols Negroponte, Kevin Kelley, or the earlier works of Sherry Turkle. These folks are sometimes referred to as digerati: intellectuals who boost new digital technologies and they also include Alvin Toffler, George Gilder, David Gelernter, (incidentally, one of the Unabomber's victims), and countless wannabees who write for the media, specialist journals, and other publications who in the 1990s got on the digital bandwagon and urged everyone to extract whatever joys and cultural capital it would yield.[3]

Boosters of the information society promised more jobs, new economic opportunities, better education, a bountiful harvest of information and entertainment, and expanding prosperity in an info-topia that would make Adam Smith proud. With powerful economic interests behind the emergent digital technologies, one expects the technological revolution to be hyped and promoted in extravagant forms. And obviously there is academic capital to be gained through promoting new technologies and the latest Big New Thing, so it is not surprising that numerous individuals, promoted, and still extoll, digital technologies and social media, often in an uncritical fashion.

Mainstream media too took up the cause of championing new digital technologies with major newspapers like the *New York Times* and the *Los Angeles Times* devoting entire sections to touting the proliferating gadgets and practices of the new cyberculture. Business sections of print publications hyped "the new economy" and magazines like *The Red Herring* and *Fast Company* puffed up every "new new thing" involved with the technological "revolution" and the spectacularly proliferating technoculture. The moment of the new cyberculture arrived

---

[3] For accounts of the promotion of computers, digital technologies, and the Internet as emancipatory forces when they emerged in the 1990s, see Turner (2008).

in the mid-1990s or so with the media, politicians, technophilacs of all strips, and many of our academic colleagues celebrating information and communication technology (ICT) as the key to the present and hope of the future.

Certain advocates of postmodern theory and cultural studies also celebrated a "technological sublime" (Tabbi 1996), which postulates a radically novel realm of experience and forms of culture and identity which break with allegedly moribund modern forms and practices. Following Lyotard's (1984) equation of the postmodern aesthetic with the sublime (as opposed to the modern promotion of the beautiful), many postmodern and other theorists have viewed technology itself as constituting a realm of the sublime that is revolutionizing art, everyday life, and human subjectivity, providing exciting aesthetic forms and higher dimensions to human experience (i.e. cyberspace, technoculture, virtual reality, multimedia hypertext, and so on; See Best and Kellner 2001). Similar celebrations of the technoculture abound within the field of cultural studies and the emerging field of cyberstudies,[4] which often assume an uncritical and technophilic posture toward ICTs.

Technophilic politicians in the 1990s boom era included Bill Clinton and Al Gore and Newt Gingrich in the United States and Tony Blair and his New Labor cohort in England. These promoters of the information society promised more jobs, exciting economic opportunities, more leisure, better education, enhanced democracy, a bountiful harvest of information and entertainment, and growing prosperity for all. With powerful economic interests behind the emergent technologies, one expects the technological revolution to be hyped. And obviously there is academic capital to be gained through boosting ICTs, so it is not surprising that many championed digital technologies, often in an uncritical fashion. What is perhaps more surprising, however, is the extent of wholly negative discourses on computers and information technologies. Since their emergence in the 1980s and 1990s as a global phenomenon central to everyday life, in the past decades, a large number of books on computers, the Internet, and cyberspace have appeared by a wide range of writers whose discourse is strikingly technophobic.

---

[4] A Google search for technoculture and cultural studies reveals that there is a "Technocultural Studies Undergraduate Curriculum" Program at the University of California—Davis; "*Technoculture* | a peer reviewed scholarly journal is seeking critical essays and creative works from a broad range of academic disciplines that focus on cultural studies of ..."; and there are countless books and articles on technoculture, some of which I engage in the following studies. See https://www.google.com/search?biw=1067&bih=487&ei=PNcWYP m5EbbL0PEP2JWOsA4&q=technoculture+cultural+studies&oq=technoculture+cultural+ studies&gs (accessed January 31, 2021).

One strand of this vast technophobic literature now aimed at computers, digital technologies and social media goes back to 1960s and earlier criticism of technology by Theodor Rozack (1969, 1994); Neil Postman (1985, 1992); Jerry Mander (1978, 1996) and other long time critics of media culture and technology, who began to aim their anti-technology jeremiads at the technoculture. The same arguments these writers previously used against technology in general, they used against the evolving digital information and communication technologies, so there was recycling of a lot of technophobic positions in the emergent critical discourses on technoculture.

I will now lay out some of these earlier critiques that might not be familiar to all contemporary audiences to depict the origin and genesis of some of the technophobic literature, although the amount is so vast by now that others may find their own preferred examples. Critiques emerged from the philosophical community, including Albert Borgmann's *Across the Postmodern Divide* (1994) which claims that new technologies are taking us into the sphere of hyperreality, a term he borrows from Baudrillard, and that we are losing touch with our bodies, with nature, with other people and with focal things and practices—an argument developed in more popular form by Slouka (1995).[5]

Lorenzo Simpson's book on technology and modernity (1994) provides another technophobic polemic against how technology is alienating and oppressing us. Postmodern theorists Arthur Kroker and Michael Weinstein have written a book *Data Crash* (1995)—a highly demonizing and technophobic text which suggests that our culture has crashed, imploded, into hyperreality, and that we've lost touch with reality altogether, that we are ruled by a new virtual class, that we have entered a new stage of virtual capitalism, which comes to a great surprise to those still laboring in sweatshops or factories. Yet perhaps the once most famous technophobe was the Unabomber whose *Manifesto* is a compendium of anti-technological, technophobic discourses, condemning industrial-technological society in its totality (Kaczynski 1995). The Unabomber was once the FBI's most

---

[5] I provide more detailed critiques of Albert Borgman'swork in "Crossing the Postmodern Divide with Borgmann, or Adventures in Cyberspace," in Eric Higgs, et al., editors, *Technology and the Good Life?* Chicago: University of Chicago Press, 2000: 234–255. I provide critical engagement with the works of Lorenzo Simpson and Andrew Feenberg who I discuss next and later in the book in "Lorenzo Simpson's Conversations with Technology, Modernity, and Postmodernity: Some Critical Reflections," *Research in Philosophy and Technology*, Volume 18 (1999): 227–245, and "Andrew Feenberg, Critical Theory, and the Critique of Technology." In *Critical Theory and the Thought of Andrew Feenberg*, edited by Darrell P. Arnold and Andreas Michel. Switzerland: Palgrave Macmillan, 2017, 263–284.

wanted for sending bombs through the post to individuals who worked in the technoculture (See Wiehl and Pulitzer 2020). His "Manifesto" echoes countercultural writers and theorists like Marcuse, Ellul, and other critics of the technological society who condemned its dehumanizing features and its tendencies toward massification, robbing individuals of power and freedom. Putting his ideas into practice, the Unabomber sent bombs to representatives of the industrial-economic order, maiming and killing many victims, before being apprehended and tried in 1997–1998.

Other technophobic missives include Clifford Stoll's *Silicon Snake Oil: Second Thoughts on the Information Highway* (1995), which provides a fascinating contrast with Gates book, attacking everything that Gates affirms, providing positive–negative mirror images of each other, both of which are highly one-sided and demonstrate the need for dialectical perspectives. Some comrades on the Left also enrolled in the ranks of the technoculture forces, including Kevin Robins and Frank Webster who advocate a neo-Luddism (1986 and 2000). Leftist critics often fail to note any progressive aspects to the emergent technologies which they interpret primarily as capitalist tools, used by capital to ensure its hegemony and to alternately dominate and overpower or seduce the working class into virtual dreams and technofetishism. Robins and Webster see the ICTs as ushering in an age of "Slaves without Athens," downplaying the democratizing potential of the technologies. Thus, while Robins and Webster are aware of the magnitude of the restructuring of capital and of the importance of technologies in this restructuring, they primarily maintain a gloomy pessimism, believing that ICTs are simply tools of capital hegemony and not also forces of resistance and democratization.

Likewise, David Noble has been publishing sharp and historically-grounded critiques of technology for decades (see 1977, 1984, 1995, 1997), and in an often-published critique of 1990s University initiatives to require faculty to create Websites for their courses, Noble insists that this is a form of unpaid labor that does not really promote a quality education (1998). In his 1997 jeremiad *The Republic of Technology*, Noble argues that from the beginning, major scientists, inventors, and ideologues of science and technology perceived technologies as vehicles of salvation, of redemption of fallen humans who would be restored to a godlike state through the marvels of technoscience, and disregarded human needs and limitations.

It is indeed curious that technology has become for many a religion and center of ultimate concern for growing numbers in the technoculture, while at the same time it is a focus of technophobic attack upon which any number of social anxieties are projected. Responding to the one-sidedness of dominant perspectives, a new discourse of "technorealism" appeared in 1998 in response to much media hoopla (see https://www.technorealism.org). Yet, like much of the digerati discourse of the tech.boom period, its' advocates lack adequate theorizing of the emergent technologies and robust critique, as the technorealists for the most part failed to theorize the technologies within the framework of their imbrication of a restructuring of global capitalism and in addition do not articulate an adequate standpoint of critique.

Against one-sided technophilic or technophobic approaches, I would argue that we need to develop a critical theory of technology in order to sort out positive and negative features, the upside and downside, the benefits and the losses in the development and trajectory of the new technologies. It is necessary, I believe, to counter promises of technological utopia, that computers will solve all our problems, produce jobs for everyone, generate a wealth of information, entertainment, and education, connect everyone, and overcome boundaries of gender, race, class—claims that we hear from promoters of the high tech industry like Bill Gates, or politicians like Al Gore and others. Yet we also need to counter technological dystopia, that digital technologies are our damnation, that they are vehicles of alienation, mere tools of capital, the state, and domination.

Both approaches are one-sided and reveal the need for a dialectical theory that plays off extremes against each other to generate a more inclusive position, indicating how technology can be used as instruments of domination and emancipation, as tools of both dominant societal powers and of individuals struggling for democratization and empowerment. A critical theory of technology requires a substantive vision of what technology is, what it does and what it could do, as well as a normative vision that delineates positive and negative uses. This requires articulation of a standpoint of critique, from which one can make distinctions between positive and negative uses of technology.

The critical theory of the Frankfurt School, which I draw upon in my studies, criticizes existing institutions, social relations, and phenomena from a normative standpoint through which existing realities can be judged deficient and oppressive (Kellner 1989a). I would suggest that those forms and uses of technology that enhance positive values such as democracy, community, freedom, self-development, and the like should be deemed life-enhancing and meritorious, while those forms and uses of technology which promote domination and oppressive forces like the alienation of labor, massification, or militarism, while undermining democracy, community, freedom, creativity, and other positive values, should be criticized as blameworthy. Of course, often one cannot make such a clear distinction, there can be unintended consequences of introducing new technologies, and technologies are often highly contradictory and ambivalent, combining positive and negative functions and effects.

Most contemporary critiques of technology operate with dualistic and usually ontologized categorical distinctions between such things as technique and being (Heidegger), technical action and social interaction (Habermas), devices versus focal things and practices (Borgmann), and instrumentality and meaning (Simpson), in which the former is devalued as modes of technological domination and alienation, whereas the latter is valorized as the authentic sphere of human meaning and value. This mode of critique thus ontologizes technology and excludes it a priori from the higher forms of human being and activity. Such approaches separate technology from culture and reify a notion of technical or instrumental action in which all action that involves technical imperatives follows a logic of things, of instrumentality, abstracted from human purposes and meaning. They therefore fail to see how technology can involve meanings and contribute to human development and enhance human life and well-being.

Dominant currents in the philosophy of technology thus essentialize technology, decontextualize it, and abstract it from social relations, culture, and human meaning, thus failing to see how deeply embedded technology is, and has long been, in the very fabric of everyday life. Such essentialist conceptions should be distinguished from a critical theory of technology that sees technology as central to human life, socially constructed and constituted, and deeply involved with human beings in a specific society at a particular point in history. Critical theory also engages specific technologies, like the Internet and social media, and criticizes their uses in specific socio-historical contexts; it promotes the reconstruction and refunctioning of technology to serve positive values like democracy or human development, and new alternative technologies to promote human and social liberation and well-being.

Technology can either be extremely domineering and destructive, or creative and life-enhancing depending on the technology in question, its specific uses in particular contexts, and the values that are being pursued in concrete situations. Yet it should also be seen that technology is highly ambivalent, that its' positive and negative aspects are often interconnected, and that it is thus often extremely difficult to appraise and evaluate, like Facebook which on one hand allows families to communicate and share their lives and, on the other hand, can help promote and elect authoritarians like Donald Trump and rightwing populists throughout the world, as has happened in the 2016 U.S. presidential election and during the Trump era (see Kellner 2017; Nance 2018; Corn and Ishikoff 2018; Hettena 2018; Stelter 2020).

The ambiguity of technologies in part derives from the centrality of technology in human life, its deep embeddedness in every integral dimension of human life ranging from the economy, to the polity, to social and everyday life, and culture and human subjectivity itself. Human beings are technical beings, technologies are extensions of human faculties which in turn come to shape human thought, behaviour, and interaction, as Marshall McLuhan has argued (1964). Technology is pivotally embedded in the human adventure, and is thus bound up with the nature of the very beings that we are. For this reason, social constructivist conceptions of technology might miss the depth and pathos of technology, its centrality in the human adventure, and the extent to which it influences the organization of human society and culture in all known historical periods.

To be sure, on one hand, technology is socially constructed, specific societal biases and interests are encoded in technology, and the social relations in which technology is produced and used will help determine its nature and uses. Yet a critical theory of technology is concerned to articulate the potentials and downsides of technology, to develop a substantive vision of the role of technology in human life, and to project ways that technology can serve human self-valorisation and the promotion of strong democracy.

A critical theory of technology must thus articulate critical perspectives that can distinguish between technologies that are ecologically appropriate, life-enhancing, and that increase social and civic engagement, contrasted to technologies that are ecologically destructive, serve as instruments of domination and oppression, and strengthen authoritarian societies, while undermining democracy. A critical theory of technology will thus critique the oppressive and authoritarian forms and uses of technology and sketch ways in which the restructuring and refunctioning of technology can promote progressive social change and the creation of a strong democracy, a good life, and a good society.

Hence, a critical theory of technology is driven by philosophical vision of normative conceptions of democracy, and a Good Life and Society, judging technology according to ethical, social, and political norms, and seeing the construction and reconstruction of technology as fundamental to the human society and everyday life. In this book, I will engage the dialectic of Democracy and Technology, and delineate ways in which technology does or does not enhance or undermine democracy and human life. I shall argue that technology can both strengthen or undermine democracy and that it is up to citizens and the polity to protect and enhance democracy by increasing the number of voters, protecting the vote, and working to develop a stronger, more secure, and more participatory system of representative democracy where strong voting rights are secured and enhanced.

In contemporary democratic societies, like the U.S., there are integral connections between technology and voting as voting machines are used throughout the country. Yet this makes it extremely important that voting machines are secured, monitored, and used to ensure a fair counting of the voting. There have been claims by both Democrats and Republicans in the twenty-first century that voting machines have been misused to distort the results of elections, making it all the more important that attention is paid to improving the technology of voting and that it enhances the system of checks and balances and maximizes voting and citizen participation which are foundation stones of contemporary U.S. democracy.

Consequently, a dialectical normative optic to develop a critical theory of technology that spells out its positive and negative—or ambivalent—aspects shows how technology, for instance, can be used to strengthen or undermine democracy and supports developing technologies that will enhance voter participation and civic engagement, willing ensuring fair elections and the rights of all citizens to vote. This is not to reject radical critiques of technology, or of specific technologies, out of hand, for often the critiques are valid and important. It is a mistake, however, to dismiss technology per se as merely a mode of domination and oppression, though it may be so in many cases and threaten positive values and human life, like nuclear weapons. Yet technologies, like the computer, for instance, were initially used and developed by big government, corporations, and the military as a centralized instrument of social control and power. At computers became more powerful and were linked with military destruction, they were, with much justice, criticized in the 1960s for contributing to big institutional domination, the dehumanizing and disempowerment of humans, and the proliferation of destructive and life-threatening weapons systems. Yet in the 1980s and 1990s computers were recreated, made "personal," and have been significantly different in their constitution and effects than their earlier incarnations, although over the past decades they have become increasingly ambiguous and contested, as we shall see in later chapters.

A critical theory of technology thus creates a historically specific and normative critique of technology. It not only attacks life-negating and oppressive aspects of technology, but valorises empowering, democratizing, and positive forms and uses. Crucially, it attempts to discover and invent ways that technology can serve the interests of human emancipation and well-being, while aspiring to delineate emancipatory functions and uses for technology—which may require the reconstruction of existing technology and the creation of what Herbert Marcuse called a "new technology" that synthesized art and technology (1964, p. 227 f.; see the discussion in Kellner 1984 and development in Feenberg 1991, 1995, 2002, 2017).

As for the standpoint of critique and the normative criteria that differentiate emancipatory from oppressive constructions and uses of technology, they themselves are historical, evolving, and subject to change and development. Conceptions of democracy, freedom, and human well-being are constantly shifting and so one's normative standards are historical, subject to the vicissitudes of history. The Frankfurt School, for instance, shifted in the 1930s and 1940s from socialist conceptions of critique, that would evaluate phenomena from the standpoint as to whether they promoted or retarded the growth of socialism and/or promoted capitalism, fascism, and domination, to what they called "immanent critique," which took the norms of existing liberal democratic societies in the 1930s fascist era as yardsticks to measure and criticize failures in specific societies to protect, or realize these norms (Kellner 1989a).

Thus, the Frankfurt School in the 1930s assumed the validity of the norms of enlightenment, democracy, human rights, individualism, freedom, and other positive Enlightenment ideals to criticize the suppression of these norms in existing fascist, communist, and capitalist societies. In *Dialectic of Enlightenment*, Horkheimer and Adorno (1972) argued that these once-progressive Enlightenment values had turned into their opposite, that enlightenment, rationality, culture, and other bourgeois ideals had shifted from a form of emancipation and progress to forms of oppression and domination, as science, technology, industry, and instrumental rationality created machinery of war, death camps, and nuclear annihilation, as well as systems of social control, massification, and oppression. Henceforth, the Frankfurt theorists attempted to develop new strategies of critique and opposition to the new forms of technological domination and power.

My critical theory of technology will also deploy strategies of immanent critique, but will develop stronger conceptions of democracy, freedom, and the good society than the classical Frankfurt School which eschewed presenting normative visions or conceptions of a good society, focusing on critique of contemporary capitalist societies in the era of fascism, capitalist crises, and the Cold War. I will carry out critiques of ideological notions of democracy, empowerment, and freedom being promoted by the avatars of new computer and multimedia technologies, as well as conservatives and others who appropriate these concepts for their ends. For instance, I would reject liberal notions of democracy that relegate it to voting and separation of powers alone, although coming out of the Trump era we should recognize the importance of voting and having a series of checks and balances to wield against an authoritarian and would-be-dictator like Trump.

Liberal democracy is thus part of a normative concept of democracy but I would include participatory democracy, and the need to create a system that allows and educates individuals to participate in the construction of their own society as in the strong democracy advocated by Benjamin Barber (1984). I would also connect participatory democracy to new social movements (see Laclau and Mouffe 1985), and throughout this book will describe how progressive social movements have used technology to strengthen democracy and will deploy a concept of technopolitics aimed at showing how digital technologies, social media, and other technologies can be used in social struggle to promote progressive social change and a strong democracy.

I shall also describe impediments to democracy and develop analyses of technology and politics that will engage theorizing digital technologies and social media in the context of threats to democracy in the current global restructuring of capitalism and surge of right-wing authoritarian politics that we have experienced in the Trump era, and the ways that technology is both central to this process and yet provides both potentially progressive and regressive uses and effects. What is at stake, therefore, is theorizing at once how digital technologies can be used as instruments of domination by corporations, the state, and authoritarians, and how they can also be used for democratization, for creating a more egalitarian society, and for empowering individuals and groups who are currently disenfranchised and without power with the ends of promoting social justice and a strong democracy– a task that I will undertake in the following chapters of this book.

One also needs to distinguish between technology as part of a societal system, as a force of production that inscribes but is also shaped by a system of relations of production contrasted to technology as a set of specific instruments and practices used by particular individuals with their own ends and goals in sight. This involves theorizing connections between technology and the economic, political, cultural, and social dimensions of contemporary society, and seeing how technology can be used differently by varying groups and individuals in specific contexts. It also involves analysing the social construction of technology and how certain societal biases, interests, and values are embedded in current forms of technology, ranging from supercomputers and our personal digital devices to nuclear energy and weapons.

In the current mode of social organization, technology plays such a major role, however, that there has been an explosion of theories of technological determinism which make an autonomous technology the organizing principle of contemporary society, thus occluding the force and power of economic and political dimensions, and erasing the efficacy of human practice. Theories of technological autonomy and determinism often use the discourse of post-industrial, or postmodern, society to describe current developments. This discourse deploys an ideal-type distinction between a previous mode of industrial production characterized by heavy industry, mass production and consumption, bureaucratic organization, and social conformity, contrasted to the new post-industrial society characterized by "flexible production," or "post Fordism," in which new technologies serve as the demiurge to a new postmodernity (Harvey 1989; Kellner and Best 1991, 1997, 2001). For postmodern theorists such as Baudrillard (1993), technologies of information and social reproduction (e.g. simulation) have permeated every aspect of society and have created a new social environment that has left behind reality and the world of modernity, as we enter a technoculture and undergo an implosion of technology and the human, while we mutate into a new species (see Baudrillard 1993 and the analyses in Kellner 1989b, 1995). For other less extravagant theorists of the technological revolution, we are evolving into a new post-industrial techno-society, culture, and condition where technology, knowledge, and information are the axial or organizing principles (Bell 1976; Masuda 1980; Kovalchuk 2021).

Theorizing Technocapitalism and the Information Society

The postindustrial society is sometimes referred to as the "knowledge society," or "information society," in which knowledge and information are given roles more prominent than earlier days (see the survey in Webster 1995). It is now well-documented that the knowledge and information sectors are increasingly important domains of our contemporary moment and, as many have noted, the theories of Daniel Bell and other postindustrial theorists are not as ideological and far off the mark as many of us once argued. Yet in order to avoid the technological determinism and idealism of many forms of this theory, one should theorize the information or knowledge "revolution" as part and parcel of a new form of technocapitalism marked by a synthesis of the information and entertainment industries and producing a new form of "infotainment society." The limitations of earlier theories of the "knowledge society," or "postindustrial society," as well as current forms of the "information society," devolve around the extent to which they exaggerate the role of knowledge and information and advance an idealist vision that excessively privileges the role of knowledge and information in the economy, in politics and society, and in everyday life, downplaying the role of capitalist relations of production, corporate ownership and control, and hegemonic configurations of corporate and state power.

Yet while perceiving the continuities between previous forms of industrial society and the new modes of society and culture described by discourses of the "post," we should also grasp the novelties and discontinuities (Best and Kellner 1997). Webster (1995, p. 5, passim) wants to draw a line between "those who endorse the idea of an information society" and "writers who place emphasis on continuities." Although he puts my writings in the camp of those who emphasize continuities (188), I would argue that we need to see both continuities and discontinuities in the current societal transformation we are undergoing, that we deploy a both/and logic in this case and not an either/or logic. In other words, we need both to theorize the novelties and differences in the current social restructuring, as well as the continuities with the previous mode of societal organization. Such a dialectical optic is, I believe, consistent with the mode of vision of Marx and neo-Marxists such as those in the Frankfurt School.

In any case, the concept of the information society and information superhighway became part of a new dominant ideology of contemporary technocapitalism. The notion of the information society goes back to post-industrial society theorists such as Daniel Bell in the 1970s (Bell 1973, 1976; Webster 1995), though the information superhighway concept followed by some decades, promoted by the Clinton-Gore administration report *The National Infrastructure: An Agenda for Action* issued in 1993,[6] followed by Vice-President Al Gore's popularization of the concept of the "Information Superhighway" in a March 1994 speech at the World Telecommunication Development conference.[7] These conceptions proliferated in the 1990s, with the Singapore government publishing their Vision of an Intelligent Island in 1993, conceiving of Singapore as an information island –[8] a concept recently elaborated in government promotion of Singapore as a high-tech city in 2020.[9]

In 1994, Japan followed the Clinton-Gore report with its publication *Reforms Towards the Intellectually Creative Society of the twenty-first Century*; the UK produced a report *Creating the Superhighways of the Future*; Norway published its plan *National Information Network*; Sweden released a report *Wings to*

---

[6] See "The National Information Infrastructure: Agenda for Action," Clinton White House Archives at https://clintonwhitehouse6.archives.gov/1993/09/1993-09-15-the-national-inf ormation-infrastructure-agenda-for-action.html (accessed January 31, 2021).

[7] See Al Gore's March 1994 speech at the World Telecommunication Development conference archived in the Clinton Presidential Library at https://clinton.presidentiallibraries. us/items/browse?advanced%5B0%5D%5Belement_id%5D=39&advanced%5B0%5D%5Bt ype%5D=is+exactly&advanced%5B0%5D%5Bterms%5D=History+of+the+Office+of+ the+Vice+President&sort_field=added&sort_dir=a&output=omeka-xml (accessed January 31, 2021).

[8] National Computer Board, Singapore (1992) "A Vision of an intelligent island: the IT2000 report." Singapore: SNP Publishers at https://www.nlb.gov.sg/biblio/6317434 (accessed January 31, 2021).

[9] See Sharmishta Sivaramakrishnan, (2019) "*3 reasons why Singapore is the smartest city in the world,*" the World Economic Forum report, November 14 2019 at https://www.weforum. org/agenda/2019/11/singapore-smart-city/ (accessed January 31, 2021).

Human Ability; and Denmark issued a report *Info-Society 2000*. European Union and G7 reports followed, generating a vast media and academic literature on the information society and information superhighway.[10]

These documents for the most part advocate economic liberalization, deregulation, and resurgent market forces as the best route to develop a healthy and robust information infrastructure, and thus promote the still hegemonic neo-liberal market ideology championed by Thatcherism and Reaganism in the 1980s. Hence, whereas Daniel Bell (1976) did not perceive the emerging post-industrial society as relying primary on market mechanisms and in fact advocated technocratic-inspired control mechanisms, the dominant discourses of the following decades relentlessly advocate a deregulated capitalist market system as the road to the information society. These discourses and theories tend toward a form of technological determinism which sees the emergent digital technologies producing the information society as inevitable and beneficial.

Al Gore, Bill Gates and others celebrate the information superhighway as the route to more jobs, better education, economic prosperity, and a global culture and networked world, and provide the most extravagant celebrations of new technologies and their beneficial social impact. Yet a broad array of political reports and discourses also embody technophiliac academic and popular celebrations of the new technologies and information superhighway, discussed above, and a dominant media discourse is equally as enthusiastic (the latter is not surprising since, as I document below, mergers between the giant media conglomerates and the key institutions of the information and computer industries are proceeding at a dramatic pace). What, then, is being occluded or mystified in the discourses of the information society and information superhighway and what are the limitations of this discourse?

As noted, the official information society discourses tends to be highly techophilic and uncritical, promising a wealth of bounties from the evolving digital technologies, though there is a technophobic academic discourse that equally one-sidedly bemoans the proliferation of the new technologies and the negative consequences for the human species that will be displaced as master of the universe and lose control over its destiny. Secondly, the technophilic discourse tends to be deterministic, as if it were fated that digital technologies propelled by inexorable market forces would dramatically proliferate, and that all anyone could do was to get on the bandwagon, to be wired and connected, and to participate in the

---

[10] See the analysis of these national programs of technology development in Bill Martin, "Information society revisited: from vision to reality," *Journal of Information Science*, February 1, 2005 at https://doi.org/10.1177/0165551505049254 (accessed January 31, 2021).

joys and benefits of the Digital Technology Revolution. Such discourses down-play the costs and dislocations of digital technologies, the powerful economic and political interests that will reap the substantial portion of the benefits, and the fact that capital is promoting these technologies as the essential ingredient in a global restructuring that itself is fraught with peril, uncertainty, and suffering for much of the world, as it intensifies inequalities, has a negative impact on the earth, and that propels us into an uncertain future.

The discourse of the information society thus occludes the economic forces and dynamics behind the proliferation of ever-expanding technological revolutions, the costs as well as the benefits, the extent of the social and economic changes, and the political issues at stake in the debate; i.e. the dismantling of the welfare state, the endangering of democracy, the threat of autocracy, destruction of the environment from overdevelopment, and widening distances between the haves and the have nots. After several decades now of the ever-cascading growth of digital technologies and social media, by the 2020s, we see this "revolution" has produce increased power for capital, the state, the media, and high-tech and social media conglomerates; has enabled autocrats to emerge throughout the world, including enabling the fateful and destructive election of Donald Trump in 2016, and has produced ever-spiralling inequality between the superrich and the poor and eve-ryone else.[11] Furthermore, in the course of these studies we will see that the implementation of digital technologies and their promotion are part of a global restructuring of capitalism and that the mergers of the information and entertainment industries and new syntheses of information and entertainment are producing a new infotainment society. Consequently, understanding these developments and developing a politics to promote a democratization of digital technologies and social media is of crucial importance in the contemporary moment.

Metaphors and Ideologies for the Technological Society

As noted, the "information superhighway" emerged in the 1990s as the key ideological discourse that legitimates the development of technocapitalism and the concept of the information society. The discourse of the "information superhighway" was early on a dominant ideology of the emerging technoculture,

---

[11] See Drew DeSilver, "U.S. income inequality, on rise for decades, is now highest since 1928," *Fact-Tank*, December 5, 2020 at https://www.pewresearch.org/fact-tank/2013/12/05/u-s-income-inequality-on-rise-for-decades-is-now-highest-since-1928/ (accessed on December 10, 2020).

followed by the "infotainment society" as a primary component of the con-temporary technocapitalist society. Information and communication technology, broadcast media and social media have been hyped to the maximum by the corporate PR/propaganda industry in the U.S. media because mega-tech and media corporations are major players in technocapitalism, as these megacorporations that own big media are continually merging with computer and information industries, and thus digital technologies are both a source of profit and of social power and prestige (see McChesney 2013). Thus, while one could envisage competition between the established media institutions and new institutions of the information and computer society, their mergers have created a situation where the media are cheerleaders and promoters of cascading digital technologies and social media as they roar across the information superhighway and relentlessly expand the infotainment society.

Moreover, the libertarian individualism and "free market" entrepreneurialism originally associated with the discourses of the new digital technologies and information superhighway are still today part of the ideological arsenal against the Welfare State which associates Big Government with harmful regulation, excessive taxes, and wasteful welfare spending. The prestige and power of the always-expanding digital technology and media culture thus feeds into "free market" ideologies and mitigates against social democratic and pro-welfare discourses.

The discourses of the "empowering" digital technologies (Bill Gates) articulate with ideologies of individualism which have long been crucial for capitalism. "The individual" continues to be both the structure and part of the ideology of the information superhighway and very texture of the information society. The concept of the individual emerged in early modern society in the Renaissance and Enlightenment as the source of knowledge, discovery, and creativity, as well as a valuable political unit whose rights and freedoms must be protected.

Likewise, in the U.S., the pioneering individual who was conquering the wilderness of the frontier to create a new life, while "civilizing" the space colonized, was celebrated as the New Adam and American Individual Hero. This ideology overlooks the nasty fact that the original native inhabitants were often put in reservations, enslaved or murdered, while slaves kidnapped from Africa worked plantations in the South. Yet as capitalist society developed, and as threats to individual freedom and well-being intensified, a romanticization of the individual continued, even as massification and increased social domination robbed actually existing individuals of the freedom and creativity once maintained as the basis of bourgeois society. This romantic individualism, this celebration of the sovereign subject, and denigration of those powers and institutions that would threaten it, have metamorphosed into a dominant ideology in today's technoculture.

The Silicon Valley techno-pioneers assumed the same role as the pioneers who settled the nation, extolling themselves as explorers and entrepreneurial heroes of the new digital frontier. This time, they invited everyone to join them, simply by buying a computer, going online, and participating in the New Frontier of the Internet and Silicon Valley Techno-Utopia. The digerati constantly extol the wonders of the Brave New Technoworld, the pleasure of fast connectivity, with all of human knowledge at each person's disposal, attended by the joys and pleasures of becoming digital and exploring the exciting new frontiers and lands of cyberspace (Negroponte 1995).

As a discourse, the individual and information superhighway ideologies are rich with connotations and seductive images and concepts. The notion of the information superhighway and discourse of "surfing" or "cruising" the 'web or 'net carries connotations of fast travelling, of adventure, and of individual excitement and adventure—connotations enhanced by the discourse of the "electronic frontier" with the connotations of exploration, the establishment of new communal spaces, and of being on the cutting-edge of the new. The metaphors of the 'net and 'web also point to connectedness, rhizomatic and multilayered levels of experience and texture, that naturalize and domesticate the highly artificial and complex technological worlds of the new computer networks.

In fact, the "natural" discourses of the information superhighway (i.e. surfing, cruising, the net, the web, connectivity, etc.) transform nature into culture and make the dramatic development of the information society a force of nature, a natural event that cannot be stopped. Indeed, the discourse appropriates both biological/natural metaphors and the figure of evolution to make it appear that the development of the new technologies and resultant social transformation is a natural process that in addition is a force of human progress, of development to higher spheres of social evolution. Such metaphors of nature and progress cover over the social constructedness of digital technologies, the corporate interests behind the project of technocapitalism and the infotainment society, and the social struggles over its future.[12]

In addition, Bill Gates' notion of a "friction-free" capitalism (1995) also covers over the messiness, conflictedness, and suffering created from the reorganization of capitalism in which there are necessarily winners and losers, and tremendous pain from dislocation, downsizing, and economic downward mobility, uncertainty and anxiety. In general, there can be no friction-free capitalism as capitalism

---

[12] I discuss further "Metaphors of Cyberspace and Digital Technologies" in Chap. 2.

itself depends on competition, antagonisms, and what Joseph Schumpeter called "creative destruction."[13] It is an ideological illusion and fantasy to believe that capitalism could eliminate friction, conflict, and suffering, especially through the market-mechanism alone which is predicated on self-interest and a Darwinian logic of the survival of the fittest.

I would also argue that current conceptions of the information society and emphasis on information technology as its demiurge are by now too limited; the digital technologies and social media are **modes of information and entertainment**,[14] and it is becoming harder and harder to separate them as they coalesce into a **corporate information-entertainment post-industrial complex**. Indeed, as I have been suggesting, digital technologies and social media are much more than solely information technology, but are also technologies of entertainment, communication, and play, encompassing and restructuring both labor and leisure. In addition, they embody modes of power and domination and a contested terrain of social and political struggle.

Furthermore, the infotainment society transforms information into entertainment and present information in the forms of compelling narrative, titillating scandal, and sometimes dazzling spectacle. The highly competitive industries of news and information use all the tricks of technology and media spectacle to get individuals to tune into their programs, or click on their news sites. The infotainment society also divides society into tribes who live in a bubble or liberal-radical or conservative-hardright media and infotainment sources that provides a fertile ground for disinformation and conspiracy theories. Once one has bought into a tribe, the followers believe the leaders and denigrate the opinions and views of the other tribe, whose discourses are denigrated as "lies" or "hoaxes." This has certainly happened to the media of news and information in the Trump era (Stelter 2020), and reveals in often-ugly forms the dark sides of the infotainment society.

This process of mushrooming infotainment and increased spectacle has been going on for some time. Previous forms of entertainment were rapidly absorbed within the Internet, and the computer and subsequent digital devices and social media became increasingly major forms of information, entertainment, play, communication, and connection with the outside world. As clues to the enormity of the transformation that has been going on for decades now, and as indicators of the syntheses of information and entertainment in the infotainment society, I

---

[13] On Schumpeter's concept of capitalism as "creative destruction," see Schumpeter 2009, and the sources and readings on Springer Link at https://link.springer.com/referenceworkentry/10.1007%2F978-1-4614-3858-8_407 (December 10, 2020).

[14] See Mark Poster (1991) on the concept of the mode of information.

would suggest reflections on the massive mergers of the major information and entertainment conglomerates that have taken place in the United States during the past decades which have seen the most extensive concentration and conglomeration of information and entertainment industries in history, including earlier examples such as:

CBS and Westinghouse: $5.5 billion
MCA and Seagrams: $5.6 billion
Time Warner and Turner: $ 7.5 billion
Disney/Capital Cities/ ABC $19 billion
NBC and Microsoft/ megabillions[15]

These mergers bring together corporations involved in TV, film, magazines, newspapers, books, information data bases, computers, and other media, suggesting an implosion of media and computer culture, of entertainment and information in a **new corporate info-entertainment postindustrial complex**. There have also been massive mergers in the telecommunications industry (in the U.S. between Southwest Bell and California Bell and New York and Atlantic Bell, with a merger between AT&T and major regional systems almost occurring, and with MCI negotiating a $37 billion merger with WorldCom, which topped British Telecommunications and GTE offers), and mega-mergers continue throughout the high-tech world.[16] The corporate media, communications, and information industries are frantically scrambling to provide delivery for the wealth of information, entertainment, and other services that will include increased internet access, cellular telephones and satellite personal communication devices, and video, film, and information on demand, as well as Internet shopping and more unsavory services like pornography and gambling.

Consequently, the mergers between the immense information, computer, and entertainment conglomerates disclose a synergy between digital technologies and media, which combine entertainment and information, undermining such a distinction. These mergers call for an expansion of the concept of information revolution, or information society, into concepts of the **infotainment society** and

---

[15] For the latest statistics on media conglomerates and mergers see the entry on in Wikipedia at https://en.wikipedia.org/wiki/Concentration_of_media_ownership (accessed on December 10, 2020).

[16] See "The Latest Mergers and Acquisitions in Telecommunications and Technology," *Cloudscene,* February 26, 2020 at https://cloudscene.com/news/2020/02/latest-mergers-and-acquisitions-telecommunications-and-technology/ (accessed December 11, 2020).

info-entertainment postindustrial complex, in order to highlight the imbricati-ons of information and entertainment in the evolving media, communications, and info-technologies of the past decades, continuing into the present and anticipating an ever-evolving future. Together, these corporate mergers and the products and services that they are producing constitute a new form of **technocapitalism** and new infotainment society that it is our challenge to theorize and attempt to shape to more humane and democratic purposes than the accumulation of capital and corporate/state hegemony.

Technocapitalism, the Infotainment Society and the Info-Entertainment Postindustrial Complex

I thus want to argue that the synthesis of entertainment and information in the technological and information revolution is part of the creation of an always-proliferating infotainment society and Info-Entertainment Postindustrial Complex that itself is part and parcel of a global restructuring of capitalism. Few theories of the information revolution and evolving digital technologies contextualize the structuring, implementation, marketing, and use of digital technologies and social media in the context of the vicissitudes of contemporary capitalism.[17] The ideo-logues of the information society act as if technology were an autonomous force and either neglect to theorize the interconnections of capital and technology, or use the advancements of technology to legitimate market capitalism (i.e. Gates 1995). More critical theorists of the momentous changes in the economy and society often fail to theorize the ways that the restructuring of capital are connec-ted with technological revolution and the emergence of a dominant technoculture. Offe (1985) and Lash (1987), for instance, see important changes in the economy, polity, culture, and society, but see this as a disorganization of capitalism, as its unravelling, rather than as the reorganization of capital with a global economic and technological infrastructure.

---

[17] To be sure, theorizing evolving digital technologies and the Internet in the context of the hegemony of corporate capitalism has been the project of Robert McChesney for many decades (see McChesney 1993, 2000, 2004 and 2013). While McChesney's optic is political economy, I will approach digital technologies from the approach of critical social theory and will focus on the use of digital technologies for progressive politics and education, as well as critiquing ideologies of the information society and the downside of technological revolution. Yet the studies by McChesney, Herbert Schiller, Noam Chomsky, Edward Hermann, and others of the political economy of the media are an essential component of the project, though my focus is on the social theory and cultural aspects of developing a critical theory of technology which I have been working on from this optic for decades.

While most of the prophets and promoters of the information society tend to be technological determinists, many of the (neo)Marxists who criticize its ideologies and practices tend to be economic determinists. Both economic and technological determinisms, however, often neglect the role of continuing conflict and struggle, the possibilities of intervention and transformation, and the ability of individuals and groups to remake society to serve their own needs and purposes. In all determinist conceptions, technology and society are conceived as matrixes of power and domination, while humans are seen as passive objects of manipulation and empowering uses of technology are not considered. With Lewis Mumford (1934), however, we should insist that humans can take command of their social circumstances and technology, shape their social environment to enhance their lives, and use technology to empower themselves and democratize society. Technologies are instruments that can be actively deployed by human beings. Although they are shaped by social forces to serve specific ends, they can be reconfigured, reshaped, and deployed against the purposes for which they are designed. This is close to what autonomous Marxists call self-valorization, as opposed to capital-valorization, using the forces of production and communication against capitalist relations of production and values (see Negri 1989; Cleaver 2000).

Yet to avoid the romanticism of voluntarism and humanism, we need to be clear concerning the major economic, social, political, cultural, and technological forces that have been restructuring every aspect of life and develop strategies of democratic social transformation based on this knowledge. I introduced the term "technocapitalism" to describe the synthesis of capital and technology in the current organization of society (Kellner 1989a) and will expand on this concept throughout the book. Unlike theories of postmodernity (i.e. Baudrillard) which often argue that technology is the new organizing principle of society, and not the economic relations, I propose the term technocapitalism to point to both the increasingly important role of technology and continued primacy of capitalist relations of production. I would argue that contemporary societies continue to be organized around production and capital accumulation, and that capitalist imperatives continue to dominate production, distribution, and consumption, as well as other cultural, social and political domains. Workers continue to be exploited by capitalists and capital continues to be the hegemonic force—more so than ever after the collapse of communism.

The term technocapitalism points to a configuration of capitalist society in which technical and scientific knowledge, automation, computers, and high-tech play a role in the process of production analogous to the role of human labor power, mechanization of the labor process, and machines in an earlier era of capitalism, while producing as well new modes of societal organization and forms of culture and everyday life. In short, technocapitalism refers to a dialectic of capital and technology in which each are mutually shaping and continually reconstituting each other in the current configuration of technocapital, in which technology may be a primary and structuring force in some situations and capital in others, yet for the most part the dialectic of capital and technology is the motor and structuring force of the current forms of the economy, society, polity, and everyday life.

We are in a parallel situation, I believe, to the Frankfurt school in the 1930s which theorized the new configurations of economy, polity, technology, society and culture brought about by the transition from market to state monopoly capitalism which was producing new forms of social and economic organization, technology, and culture with the rise of giant corporations and cartels, a capitalist state to help organize capitalism whether in a fascist or a state capitalist form, and with culture industries and mass culture serving as new modes of social control, new forms of socialization, and a new configuration of culture and everyday life (Kellner 1989a). My thesis is that today media culture (Kellner 2020) and digital technologies are vitally transforming every aspect of social life in a process that is creating novel forms of society, sometimes described as postmodern society, the information society, cybersociety, global postFordism, and various other terms which I perceive as part of the configurations of technocapitalism.

The concept of technocapitalism thus points to syntheses of technology and capital and attempts to avoid technological or economic determinism. Further, the restructuring of capital is producing a very specific social configuration that I propose calling "the infotainment society" and "the Info-Entertainment Postindustrial Complex" in order to point to the mergers of information and media industries and to the significance of technologies of information, entertainment, and social reproduction. In terms of political economy, the new postindustrial form of technocapitalism is characterized by a decline of the state and increased power of the market, accompanied by the growing power of globalized transnational corporations and global governmental and corporate bodies and the decline of the nation-state and its institutions. To paraphrase Max Horkheimer, whoever wants to talk about capitalism, must talk about globalization, and it is impossible to theorize globalization without talking about the restructuring of capitalism (see Cvetkovitch and Kellner 1996; Kellner 1998a, 2014).

While knowledge, information, and education are playing a more important role than ever in the organization of contemporary society, this is because, I would argue, capital is restructuring itself through the implementation of digital technologies and information into every sphere of life, and is motored by scientific and technological knowledge. The dangers are that corporate control of knowledge, information, entertainment, and technology will provide a tremendous concentration of corporate power without any countervailing forces. The ideologues of the technological revolution and information society are forever arguing that education is the key to future prosperity, that education must be made available to all, and that it is thus the top social priority. This would be fine if education were to be expanded and made accessible to more individuals and if it were able to augment the realm of knowledge and literacies needed to participate in today's society and culture, rather than just to serve as a sophisticated enhancement of job training, focusing on transmitting the skills and knowledge that capital needs to expand and multiply—issues that I take up in Chap. 10.

Yet it is clear that digital technologies are revolutionizing not only labor, production, and leisure, but also education and schooling. The past years have seen major implementation of digital technologies and on-line teaching in the educational process, generating fierce debates over how to deploy digital technologies in education, how to make them accessible for everyone, and whether they are enhancing or harming education. Whether digital technologies will ultimately enhance or diminish and harm education is not yet decidable, but it is clear that individuals need to develop intensified computer literacies, as well as print literary and, I would add, critical media literacies, social and cultural literacies, digital literacies and ecoliteracy (see Kellner and Share 2019 and Chap. 10 below). As we approach an increasingly complex new world, we need to greatly expand and rethink education and literacy and to devise strategies to use technology to strengthen and democratize education.

The dangers are, first, that existing inequalities will be reproduced by the increased importance of computers and digital literacies, which will privilege existing elites at the expense of others. Secondly, there is a danger that the values and cultural forms of the infotainment society will permeate education, as well as every sphere of culture and everyday life, rendering education more and more a form of entertainment, of multimedia interaction, in which consumption of media material will replace active study, practice, and experimentation. This need not be the case, of course, since interaction with multimedia can be as active and as creative as with book and print material, and the modes of popular entertainment can to some extent serve valuable educational purposes if combined with the cultivation of critical media and digital literacies.

Technopolitics and New Public Spheres

Since media and digital technologies are in any case dramatically transforming every sphere of life, the key challenge is how to theorize this great transformation and how to devise strategies to make productive use of the emergent technologies. Obviously, radical critiques of dehumanizing, exploitative, and oppressive uses of diverse technologies in the workplace, schooling, public sphere, and everyday life are more necessary than ever, but so are strategies that use emergent technologies to rebuild our cities, schools, economy, and society. I want to focus, therefore, in the remainder of this section on how diverse technologies can be used for increasing democratization and empowering individuals. In previous articles (Kellner 1995, 1996, 1998a), I have argued that new technologies are creating a new public sphere, a new realm of cyberdemocracy, and are thus challenging public intellectuals to gain technoliteracy and to make use of the new technologies for promoting progressive causes and social transformation—themes that I develop in later chapters of this book.

Given the extent to which capital and its logic of commodification have colonized ever more areas of everyday life in recent years, it is somewhat astonishing that cyberspace is by and large decommodified for large numbers of people—at least in the overdeveloped countries like the United States. In the U.S., government and educational institutions, and some businesses, provide free Internet access and in some cases free computers, or at least workplace access. With flat-rate monthly phone bills (which I know do not exist in much of the world), or connections to a business, University, or organization that provides free computer use, one can thus have access to a cornucopia of information and entertainment on the Internet for free, one of the few decommodified spaces in the ultracommodified world of technocapitalism.

Obviously, large sections of the world do not even have telephone service, much less computers, and there are vast inequalities in terms of who has access to computers and who participates in the technological revolution and cyberdemocracy today. Critics of digital technologies and cyberspace repeat incessantly that it is by and large young, white, middle or upper class males who are the dominant players in the cyberspaces of the present, and while this was once true, statistics and surveys indicate that many more women, people of color, seniors, and other minority categories are becoming increasingly active.[18] Moreover, it appears that computers and a diversity of digital devices are becoming part of

---

[18] See PEW Research, "Internet/Broadband Fact Sheet," June 19, 2019 at https://www.pew research.org/internet/fact-sheet/internet-broadband/ (accessed December 11, 2020).

the standard household consumer package and are as common as television sets in the contemporary moment, and certainly more important for work, social life, and education than the TV set. In addition, there are plans afoot to wire the entire world with satellites that would make the Internet and communication revolution accessible to people who do not now even have telephones, televisions, or even electricity.

However widespread and common—or not—computers and digital technologies become, it is clear that they are of essential importance for labor, politics, education, and social life, and that people who want to participate in the public and cultural life of the future will need to have computer access and digital literacies. Moreover, although there is the threat and real danger that the computerization of society will increase the current inequalities and inequities in the configurations of class, race, and gender power, there is the possibility that a democratized and digitized public sphere might provide opportunities to overcome these inequities. I will accordingly address below and in following chapters some of the ways that oppressed and disempowered groups are using the digital technologies and social media to advance their interests and progressive political agendas. Yet first I want to dispose of another frequent criticism of the Internet and computer activism.

Critics of the Internet and cyberdemocracy frequently point to the corporate origins of the 'net and its central role in the practices of dominant corporate and state powers. Yet it is amazing that the Internet for large numbers is decommodified and is becoming increasingly decentralized, becoming open to more voices and groups. Thus, cyberdemocracy and the Internet should be seen as a site of struggle, as a contested terrain, and progressives should look to its possibilities for resistance and circulation of struggle. Dominant corporate and state powers, as well as conservative and rightist groups, have been making serious use of media and digital technologies to advance their agendas and if progressives want to become players in the political battles of the future they must devise ways to use these technologies to advance a progressive agenda and the interests of the oppressed and forces of resistance and struggle.

There are by now copious examples of how the Internet, social media, and cyberdemocracy have been used in progressive political struggles. A large number of insurgent intellectuals are already making use of these technologies and public spheres in their political projects. The peasants and guerrilla armies struggling in Chiapas, Mexico from the beginning used computer data bases, guerrilla radio, and other forms of media to circulate their struggles and ideas. Every manifesto, text, and bulletin produced by the Zapatista Army of National Liberation who occupied land in the southern Mexican state of Chiapas in 1994 was immediately circulated through the world via computer networks. In January 1995, the Mexican government moved against the movement and computer networks were used to inform and mobilize individuals and groups throughout the world to support the Zapatistas struggles against repressive Mexican government action.[19] There were many demonstrations in support of the rebels throughout the world, prominent journalists, human rights observers, and delegations travelled to Chiapas in solidarity and to report on the uprising, and the Mexican and U.S. governments were bombarded with messages arguing for negotiations rather than repression; the Mexican government accordingly backed off their repression of the insurgents and as of this writing in January 2021, they have continued to negotiate with them.

Moreover, a series of struggles around gender, sex, and race are also mediated by digital communications technologies, including in the last years of the Trump administration the Dreamer, #Me Too, Black Lives Matter, and Trump Resistance movements. Earlier, after the 1991 Clarence Thomas Hearings in the United States on his fitness to be Supreme Court Justice, Thomas's assault on claims of sexual harassment by Anita Hill and others, and the failure of the almost all male US Senate to disqualify the obviously unqualified Thomas (see Hill 1998), prompted women to use computer and other technologies to attack male privilege in the political system in the United States and to rally women to support women candidates. The result in the 1992 election was the election of more women candidates than in any previous election and a general rejection of conservative rule, and eventually several women Supreme Court Justices appointed to the court.

Many feminists have established websites, mailing lists, and other forms of cybercommunication to circulate their struggles.[20] Younger women, once deploying the concept of "riotgrrrls," have created electronically-mediated 'zines, web sites, and discussion groups to promote their ideas and to discuss their problems and struggles. African-American women, Latinas, and other groups of women have been developing web sites and discussion lists to advance their interests. And AIDS and other health activists have used digital technologies and social media to disseminate and discuss medical information and to activate their constituencies for courses of political action and struggle—an effort especially relevant

---

[19] See the collection of essays with an Introduction by Harry Cleaver in *Zapatista: Neoliberalism, the Chiapas Uprising & Cyberspace*. Galmuri Publishing House, Seoul, Korea, 1998, on = line at https://la.utexas.edu/users/hcleaver/bookoutline.htm (accessed December 27, 2020).

[20] See the Duke University Press site "Feminism(s) and Tech" at https://sites.duke.edu/womenandadvertising/exhibits/tech-ads-and-women/feminist-movements-technology-and-advertising/ (accessed on December 11, 2020).

during the 2020–2021 Global COVID-19 pandemic which has disproportionally hit communities of color.

Likewise, African-American insurgent intellectuals have made use of broadcast and computer technologies to promote their struggles in movements from the 1980s through Black Lives Matter (Kellner and Satchel 2019). John Fiske (1994) has described some African-American radio projects in the "technostruggles" of the present age and the central role of the media in struggles around race and gender. African-American "knowledge warriors" have been using radio, computer networks, and other media to circulate their ideas and counter-knowledge on a variety of issues, contesting the mainstream and offering alternative views and politics. In addition, activists in communities of color—like Oakland, Harlem, and Los Angeles—have been setting up community computer and media centers to teach the skills necessary to survive the onslaught of the mediazation of culture and computerization of society to people in their communities.

Obviously, rightwing and reactionary groups can and have used the digital technologies to promote their political agendas as well. In a short time, one can easily access an exotic witch's brew of ultraright websites maintained by the Ku Klux Klan, myriad neo-Nazi groups including Aryan Nation and various Patriot militia groups, which have become all-too-visible during the Trump era in which miscreants such as the Proud Boys, QAnon conspiracy nuts, and other far-right groups of the Trump Storm Troopers who have gained media access, support, and a dark legitimacy through their embrace by Trump and his followers on the right. Internet discussion lists also promote these views and the far right is extremely active on many Internet forums, as well as having their radio programs and stations, public access television programs, video, and even rock music production. These groups are hardly harmless, having promoted terrorism of various sorts ranging from church burnings to the bombings of public buildings, and in 2021 included the invasion of Congress. Adopting quasi-Leninist discourse and tactics for ultraright causes, these extremist groups have been successful in recruiting working class members devastated by the developments of global capitalism which have resulted in widespread unemployment for traditional forms of industrial, agricultural, and unskilled labor.

In the 2016 U.S. Presidential election, social media played a major role in disseminating misinformation about Hillary Clinton and electing Donald Trump, while throughout Trump's reign his shock troops were spurred to acts of violence by Trump's daily incendiary Twitter feeds, a danger that continues as Trump is driven out of office after his defeat by Joe Biden, as he refuses to concede the election while claiming it was fraudulent, keeps his storm troopers riled up and ready for disruption, leading to a White Riot and occupation of the Capital on January 6, 2021, driving the Democrats to impeach Trump for the second time.[21]

The Internet and technoculture is thus a contested terrain, used by Left, Right, and Center to promote their own agendas and interests. The political battles of the future may well be fought in the streets, factories, parliaments, and other sites of past struggle, but political struggle today is already mediated by media, computer, and information technologies and will increasingly be so in the future. Those interested in the politics and culture of the future should therefore be clear on the important role of the new public spheres and intervene accordingly.

Some Concluding Remarks

In the light of the projects of technocapitalism and rightwing politicians and regimes to dismantle the Welfare State, it is up to citizens to create new public spheres, new politics, and to use the digital technologies and social media to discuss what kinds of society we want and to oppose the society we don't want, to demand more education, health care, welfare, and benefits from the state, and to struggle to create a more democratic and egalitarian society. Yet one cannot expect that generous corporations and a beneficent state are going to make available to citizens the bounties and benefits of the evolving info-technological economy. Rather, it is up to individuals and groups to promote democratization and progressive social change.

Thus, to globalization from above of corporate capitalism, one could support a globalization from below, from individuals and groups in struggle using the digital technologies and social media to create a more egalitarian and democratic society (see Chap. 6). Individuals and groups all over the world are using digital technologies and social media to advance progressive goals and the new public spheres of cyberspace are more open to cultural and intellectual intervention than the media spaces controlled by the giant corporations. Social struggles

---

[21] Donald Trump's election and Nightmare Reign of Horror was in part facilitated by the intervention in Facebook and other social media by Russia and other interested parties (see Jamieson 2017; Kellner 2017; Nance 2018; and Isokoff and Corn 2017, as well as the 2018 PBS Documentary "The Facebook Dilemma"). As I edit this book in January 2021, Trump has been permanently banned from Twitter after his encouragement of his far-right Storm Troopers to invade Washington and to storm the Capital on January 6, 2021 in the last days of his presidency in a failed attempt to overthrow the Electoral College tally of the election which decisively choose Biden as the winner; see https://blog.twitter.com/en_us/topics/company/2020/suspension.html (accessed January 9, 2021).

ranging from native peoples in the Mexican state of Chiapas, to dockworkers in London, to oppressed peoples of North Africa, to anti-corporate campaigns world-wide against McDonald's and Nike, to recent struggles by Dreamers, #MeToo, and Black Lives Matter have used digital technologies against the dominant corporate powers. Moreover, groups like African-Americans, Latinos, gays and lesbians, and others excluded from the democratic dialogue are using digital technologies and social media to promote democratization and advance their interests.

Of course, the digital technologies might exacerbate existing inequalities in the current class, gender, race, and regional configurations of power and give major and statist and corporate forces powerful tools to advance their interests, as well as providing hostile states weapons to engage in asymmetrical warfare against the U.S. and other democratic countries. In this situation, it is up to the people, to us, to devise strategies to use digital technologies and social media to promote democratization and progressive social change. For as digital technologies become ever more central to every domain of everyday life, developing a progressive technopolitics in the digital public spheres will become more and more important.

Changes are certainly happening, we are undergoing a Great Transformation, but we are, I believe, too early in this adventure to determine its structure and the ways that it is transforming social relations, cultural forms, and effecting everyday life, and "creative destruction," innovation, and dramatic changes of the techno-culture have characterized it from the beginning and no doubt will continue to do so. It is clear, however, that a technological revolution has been going on, that it has already had massive effects, and that it is a great challenge to us concerning how we will theorize and actually use digital technologies—or whether they and the forces that control them will themselves use us in their projects and we will uncritically surrender to the technoculture and the objects it keeps providing us.

Thus, it is not only a challenge to social theorists to theorize the always expanding digital technologies and their effects, and to activists to devise stra-tegies for using technology and social media to promote progressive political change, but it is a challenge to each individual to determine how they will live with digital technologies and cyberspaces, how they will themselves deploy them, and whether digital technologies in their lives will ultimately be empowering or disempowering, and democratizing or de-democratizing. For as long as human beings have vision, goals, and autonomy, we can design, shape, and restructure our technologies, as well as being shaped and constrained by them.

In this book, I will accordingly address these issues starting in Chap. 2 with a study of "Metaphors of Cyberspace and New Technologies," that will attempt to conceptually map the field of the technoculture by exploring the meanings of dominant metaphors. In Chap. 3, I engage "Technology and Alienation," exploring the downsides and potential alienating effects of immersion in a high-tech world, as well as possibilities for dis-alienation and empowerment, followed in Chap. 4 with analysis of "The Media, Democracy, and Spectacle," which will begin exploring the notion of technopolitics and how we can use digital technologies and social media for social transformation and to meet the demands of democracy, a theme that will run throughout the book. In Chap. 5, I discuss "Intellectuals and Digital Technologies in a New Era of Struggle," focusing on the political challenges of a high-tech world and how individuals can be active participants in the great adventures of the digital revolution.

Following this line of inquiry in Chap. 6 on "Globalization, Technopolitics and Revolution," I analyze in more detail the global economy, politics, and culture from the standpoint of technology and a networked world. In Chap. 7, I present French theorist and technology critic Paul Virilio in an analysis of "Virilio, War, and Technology," which explores destructive and unsettling aspects of digital technologies. Chap. 8 continues this motif in a study of "Vicissitudes of High-Tech War," following by analysis of filmmaker Stanley "Kubrick's *2001* and the Dangers of Techno-Dystopia." I conclude with a study of "Technological Transformation, Digital Literacies and Democracy: Toward a Reconstruction of Education," which explores how technology can be used for a democratic reconstruction of education and society in the spirit of John Dewey, Paulo Freire, and Herbert Marcuse.

I am writing this study from a U.S. perspective, having taught courses on technology at the University of Texas and UCLA since the 1980s, but would suggest that my arguments have broader reference in an increasingly globalized society and world marked by a networked economy, increasing migration and multiculturalism, and a proliferating Internet-based cyberculture. There is by now a tremendous amount of books and articles on the global economy, technological revolution, new cultural spaces, and the implications for every aspect of life from education to war.[22] I have drawn on this literature to develop critical theories of technology, technocapitalism, technopolitics, and social transformation that I present in the following chapters. These studies suggest how we can

---

[22] See, for example, the monumental studies by Castells (1996, 1997, 1998, 2001), and the analyses of the restructuring of capital, technological revolution, and the postmodern turn in Best and Kellner (2001), both of which I draw on, but update, in these studies. While Castells is an excellent historian and sociologist of the Internet, I find him generally too positive and celebratory of the Internet and technoculture, though will draw on his work in various chapters of the book.

use digital technologies and social media today to meet the demands of democracy which will require future struggles to create a more egalitarian, ecological, and just democratic society which it is our challenge to defend and nourish if we do not want to regress into the barbarism that we have witnessed aspects of in the twenty-first century and that continues to threaten our lives, health, environment, well-being, and prospects of a more advanced and secure democratic society and polis.

Metaphors of Cyberspace and Digital Technologies

2

> *"Metaphor is never innocent. It orients research and fixes results"* (Derrida 1978, p. 17).
>
> *"Metaphors are serious things. They affect one's practice"* (Hall 1992, p. 282)

Abstract

Chapter 2 develops a study of "Metaphors of Cyberspace and New Technologies," that conceptually maps the field of the technoculture by exploring the meanings of dominant metaphors. The "information superhighway" emerged as a key ideological discourse that legitimates the development of technocapitalism and the concept of the information and/or knowledge society, followed by other ideologies that developed with the emergence of new digital technologies and social media that I explore in this and following chapters. I would indeed go so far as to claim that the "information superhighway" is the dominant ideology and the infotainment society is the primary project of the contemporary technocapitalist society. The infotainment society is hyped to the maximum by the U.S. media because these corporations are major players

---

This study is unpublished but was presented to many audiences who I thank for useful discussion, including Popular Culture Association, San Diego, March 1999; Symposium on Post-Humanities, April 1999 at the Department of Information Studies, UCLA; Conference on Media-Body-Imagination, Freie Universitat, Berlin, Germany, September 2006; and McGill University, Montreal, October 2006.

in this project, and since the same corporations that own big media are merging with computer and information industries, digital technologies and social media are both a source of profit and of social power and prestige. Thus, while one could envisage competition between the established media institutions and new institutions of the information and high-tech society, their mergers have created a situation where the media are cheerleaders and promoters of the always-expanding digital technologies, social media, and the devices that are part of the infrastructure of a high-tech industrial-entertainment complex.

Keywords

Metaphors of Cyberspace and New Technologies information superhighway •
Technocapitalist society • Infotainment society • Digital technologies • Social
media • High-tech industrial-entertainment complex

In the Introduction, I noted that the "information superhighway" emerged as a key ideological discourse that legitimates the development of technocapitalism and the concept of the information and/or knowledge society, followed by other ideologies that developed with the emergence of new digital technologies and social media that I explore in this and following chapters. I would indeed go so far as to claim that the "information superhighway" is the dominant ideology and the infotainment society is the primary project of the contemporary technocapitalist society. The infotainment society is hyped to the maximum by the U.S. media because these corporations are major players in this project, and since the same corporations that own big media are merging with computer and information industries, digital technologies and social media are both a source of profit and of social power and prestige. Thus, while one could envisage competition between the established media institutions and new institutions of the information and high-tech society, their mergers have created a situation where the media are cheerleaders and promoters of the always-expanding digital technologies, social media, and the devices that are part of the infrastructure of a high-tech industrial-entertainment complex.

As I argued in the Introduction, we are currently undergoing a dynamic technological revolution which is changing everything from work to politics to sex and war. This revolution is creating novel forms of experience, culture, and identity that require new language and concepts to describe the emergent cultural and social spaces and forms generated by proliferating information and communication technologies and social media.

Metaphors traditionally function as figures which organize and make sense of experience, help us represent novelties, and help generate concepts to map and orient us in our transactions with the natural and social world. In this chapter, I will sort out the dominant metaphors for the technological revolution and digital society such as the information superhighway, cyberspace, the net and the web, the electronic frontier, social media, Facebook, Microsoft, Apple, Twitter, YouTube, and other dominant metaphors and metaphorical names for the always evolving digital technologies and socio-cultural discourses, focusing this chapter in providing readings of the dominant metaphors in discourses describing and celebrating digital technologies and social media, including metaphors of desktop computer icons and sites accessed by the personal devices that for some sectors of youth have supplanted the desktop computer that previous generations grew up with in order to interrogate the language of work and social interaction that illuminates what we do with home computers and digital devices. I show that the dominant metaphors draw from the human body, everyday life, home and business, nature, travel, technology, and the military and space travel. I analyze both how past ideological tropes are used to render familiar digital and virtual technologies and cultural spaces familiar, and in some cases to market, celebrate, and legitimate them. Yet the evolving metaphorical space of the technological revolution is full of tensions and contradictions which open possibilities for the reconstruction of technology and everyday life.[1]

I begin by reflection on what metaphors we have used over the last decades to describe the emergence of information and communication technology and the domain of cyberspace, the Internet, and the whole world of computers and digital devices. I started this research in the 1990s when one of my fundamental research interests was to theorize the then new technologies and just emergent cyberspace of the Internet, and I began carefully examining the language used to describe the evolving technologies, cultural spaces, experiences, and subjects produced. While in the beginning, cyberculture and information and communication technologies (ICTs) were quite novel, by now they are obviously normalized and a taken for granted part of everyday life.

---

[1] I should note that some scientists are against metaphor as such, including Allen Newell who writes: "It is clearly wrong to treat science as metaphor, for the more metaphorical, the less scientific" (cited in Bardini 2000, p. 239). I disagree with this notion in relation to philosophy, the social sciences, and theorizing digital technologies and agree with Bardini who demonstrates that the whole history of computers and digital technology can be imagined as a "dance of metaphors" (op. cit. p. 34). In this chapter, I join Bardini and others in exploring this dance and discussing how it metaphorically evolved.

Aristotle defined metaphor as the act of "giving the thing a name that belongs to something else."[2] I am interested in this analysis in delineating the spheres from which the names or metaphors for familiar computer-mediated technologies and activity have been drawn and the ideological functions of these metaphors. At stake is the difference that exists between the "thing" and the "something else" and the adequacy and appropriateness of the metaphors we use to describe our thing—which is the new digital technologies, social media, cultural spaces, and activities that the high-tech industry and digerati name and help structure.

One problem with metaphors of emergent technologies and spaces is that they equate the often novel and heterogeneous phenomenon with previously familiar phenomena, and thus occlude the difference and radicalism of the novelty or otherness of new technologies. It seems when confronting something new and different we—and the people who construct these metaphors, often for commercial purposes—use previous phenomena for reference points, to assimilate the new, to blunt the unfamiliar and different.

Another problem is that our dominant metaphors naturalize and provide ideological legitimation for the emergent and always mutating ICTs. Consequently, I want to interrogate some of the dominant metaphors for the now ubiquitous technologies and field of electronic culture and communication and sort out:

- The origins and sources of the dominant metaphors; where they come from and what clusters of meanings they generate;
- criteria for judging metaphors;
- appropriateness of metaphors in the sense of fitness;
- connotations and effects of metaphors; i.e. what do the dominant metaphors connote and imply and what are their social functions and effects?

Finally, I'll raise the question if the metaphors drawn from previous modes of experience are appropriate to the phenomena in question or whether we need new concepts and language to grasp the radicalism of the difference and novelty of, say, information and communication technologies, social media, and the novel spaces, experiences, and social relations they open. I'm assuming that language is important and that our metaphors are crucial in shaping how we see and understand the world, as the opening quotes by Derrida and Hall suggest.

---

[2] See the analysis and sources assembled by Jan Garrett "Aristotle on Metaphor," March 28, 2007 at https://people.wku.edu/jan.garrett/401s07/arismeta.htm (accessed December 12, 2020).

I'll suggest that the metaphors we use are often very revealing, and that we need to sort out and reflect upon the language of our technoculture to help construct more appropriate and critical language to describe the now dominant field of electronic culture and communication, or at least to reflect on the connotations of the metaphors and language we use to describe a primary mode of our experience and labor so as to critically dissect what sort of meanings and values are embedded in our metaphors.

In a dominant and evolving field of culture and communication, the language is up for grabs and subject to contestation and reconstruction, and so I hope eventually to contribute to help reconstruct the language we use to describe information and communication technologies, social media, and their multiple and often very important effects—although this should ultimately be the task of subsequent generations, so I am simply presenting a provisional analysis of metaphors of cyberspace based on existing technologies, social media, and cyberculture in the beginnings of the second decade of the twenty-first century.

Anthropomorphic Metaphors, the Self, and the Body

Not surprisingly, many early and key metaphors for ICTs derive from key anthropomorphic activities of the human mind and body. Metaphor, as McLuhan writes (1964), derives from the Greek *metapherin*, to bridge or bear a transfer in meaning from one figure to another, and operates by selecting some common and familiar feature or activity to provide a bridge for understanding something new—and what could be more familiar than our bodies, ourselves?

Initially, computers performed the important anthropological function of counting and computing. In the nineteenth century, a computer described someone who computes, or a group working together to calculate numbers, referring to individuals who calculated, i.e. computed, mathematical sums and that was indeed at first the function of early computers which were computing machines. The first computers in the 1800s used wooden punch cards to automate the design of woven fabrics and similar punch cards were used to calculate numbers. In 1822, English mathematician Charles Babbage invented a steam-driven calculating machine that was able to compute tables of numbers, called a difference machine, while in the U.S. in 1890, inventor Herman Hollerith designed the punch card system to calculate the 1880 U.S. census and later established a company that would become IBM. Other important moments in the history of computing occurred in the 1930s when Alan Turing developed an idea for a universal machine, which he would call the Turing machine, that would be able to compute anything that is computable, followed by Bill Hewlett and David Packard in 1939 building Hewlett-Packard computer. These developments helped lead to creation of the Electronic Numerical Integrator and Calculator (ENIAC) in 1943, and then code-breaking machines and computers that directed aircraft and munitions in World War II.[3]

After the war, the UNIVAC computer was developed that would become the first commercial computer for business and government use and since then IBM, and companies throughout the world developed computers that were initially mysterious mega-machines controlled, often secretly, by business and government. Computers, the machines that is, were thus initially the province of the military, government and intelligence services, and big business. Yet when corporations began to sell computers to the public in the late 1970s and 1980s, they began to be marketed as consumer items for the home and office which required a more anthropomorphic and friendly image.

Initially, one of the first important metaphorical interventions in the contemporary computer age was evolving the term "personal computer," or "PC." Important steps in personal computing include the 1973 Xerox Alto, developed at Xerox's Palo Alto Research Center (PARC) which had a graphical user interface (GUI) that later served as inspiration for Apple's Macintosh, and Microsoft's Windows operating system. Major computer companies began relentlessly promoting PCs in the 1980s, using concepts like Xerox PARC's metaphor of computers as "personal dynamic media," a marketing term still in use today. This refiguring computers as personal, dynamic, and an interactive medium reconceptualized computers from giant calculating machines standing alone in storage rooms to working and interacting tools that could be used by anyone, anywhere, for a variety of individual and social purposes (see Bardini 2000).

Picking up on this, IBM in the early 1980s began marketing the "PC" by stressing how personal computers were different from the big mainframe computers that IBM had been associated with, and they and other firms promoted PCs as home or desk personal computers. Henceforth, our computers became personal, they were seen as mine and an extension of one's self, that empowered the owner and user, and thus the PC was not a big impersonal instrument of a corporation or big organization but something anyone could own and use.

Indeed, early in the PC era, people did come to see and use computers as extensions of themselves and interpret them in familiar analogical and anthropomorphic metaphorical terms. Sherry Turkle in her first major book on computers,

---

[3] See Mara Calvello, A Complete History of Computers: From the 1800s to Now, July 8, 2019 at https://learn.g2.com/history-of-computers (accessed December 12, 2020) for a useful overview of the history of computers, and Isaacson (2015) for a lively history of key personalities and events in the history of computing.

The Second Self (1984) saw computers as a second self, as an externalization and enhancement of essential human powers, of who we are, of what we generally do (i.e. write, communicate, research and store, play and perhaps create images, ideas, music, and new technologies and cultural space). The computer, claims Turkle, produces more powerful selves, enhances human capacities and thus functions as a technology that is an extension of the human, intensifying basic human powers.

In her second major book, *Life on Screen* (1995) Turkle argued that the difference between modernity and postmodernity is the difference between giant mainframe computers, symbolic of giant organizations, and personal computers, more individualized, flexible, malleable, and so on that symbolize the difference between modern selves that are masculine, rationalistic, abstract, and relatively fixed contrasted to more aestheticized, feminized, malleable, and multidimensional postmodern selves.

Computer selves are also bodily and material. Consider to what extent the term digitization derives from digit, a finger root, as well as number, and consider that early computer commands involved point and click, enter, delete, and other physical activities that altered the materiality of the text or artifact produced.

The root "face" also early on entered computer metaphors as in the term "interface" which often uses faces literally to greet computer users, and as Turkle notes, many people personalize their computers with names, or humanoid icons or facial images, just as people use smiley face icons in emails, or maybe frowny ones, if you are depressed or angry. As Bardini (2000) points out, "The earliest notion of an interface comes from the Greek prosopon, a 'face facing another fact' (Heim 1993, p. 78)." This prosopon embodies the notion of a computer interface, in which one face (the user) interacts with another face (the computer), as if it were alive and as if they were directly communicating and interacting (Badrini, op. cit).

The computerized self, moreover, was conceived as a privileged self, with magical new forces ready to serve and expand the self. Alan Kay (1993) used the term "delegation" to describe what computers do; just as some people have maids who are delegated certain activities if you are middle-class, or perhaps you have butlers or other servants who are delegated certain activities if you are upper class, so too can you, with your personal computer, delegate certain tasks to computers—calculating, doing searches, merging texts and references in writing, or other functions.

Others use terms like "agents" to describe what the computers do for us and call for more "intelligent agents" that will run our homes, cars, and personal lives, just as our computers tend to enable our work lives. This is the major thrust of

Nicholas Negroponte's *Being Digital* (1996) and other MIT scientists and theorists who envisage the computerization of everything, including the soul (this is a story for another day or more metaphysical text, although I'll have some metaphysical comments in the following pages since we are arguably dealing with new modes of being).

The boosters of the new technologies want the computer agents to do all sorts of things: to store and access information; to serve as a filter to eliminate spam or viruses or unwelcome intruders; and to engage in an ever-expanding range of activities. Indeed, as Bardini argues (2000), some early computer visionaries like Douglas Engelbart foresaw the computer as augmenting the human being, creating more intelligent and productive humans whose work with computers would create a higher species or Nietzschean Ubermensch, although Engelbart did not use quite such extravagant language.

The computer user will perhaps use both push and pull technology to retrieve, store, organize, and disseminate information. They will engage in business activity and help make money or organize finances, as well as enable new modes of social interaction and play. They will operate in new research terrains and write in a highly more advanced form than those in which previous writers scribbled with pens or knocked off texts with typewriters.

Continuing this line of metaphor, you have a menu that tells what your computer has to offer, i.e. what it can do for you and what uses and functions you can consume and deploy. You probably have a server, at least at school or on a work network, that holds and serves up to you all sorts of information and data, serving your need for information, connectivity, productivity, and functionality. Earlier digital natives were "served" by information services like Genie, America On-Line, or Compuserve, but these were often replaced by University or business servers for those in the University or employed by business or government with a high-tech profile or dimension—which is becoming more and more of us. I've been served for decades at UCLA by Bruin-On-Line and servers maintained by the Department of Education and Information Studies, and do not know where my data is backed-up but hope it is secure as I have more than forty years of work stored in some server (which I'm told is safe on a Cloud).

Note that servers incorporate magic (like Genie), or iconic totem names (like America-on-line or Bruin-on-line), which are enhancement metaphors of different sorts and make you part of something really super like America or the UCLA Bruins, thus making the user a superior human being who is served by magical technology, covering over that many are computer addicts, while others are forced by work requirements to spend hours of on-line slavery every day.

Connected to my bodily personal self are, perhaps above all, home and work and, not surprisingly, a wide range of computer metaphors are drawn from familiar domains of everyday life, in particular everyday domestic life and the workplace. As noted, our technologies are very personal and have become very homey: we have our own Home page, either to begin browsing, or to present ourselves to the world, implying that we are living and rooted in a whole new space—which is, however, configured and presented to us within dominant corporate parameters.

Initially, a personalized MySpace became the thing for young people and today it is replaced by newer social media sites used for multiple purposes like Facebook. Some of my students and colleagues trade baby pictures, movie suggestions, or videos on YouTube which once seemed the Big New Thing of the day. Previously, P2P, another personalized metaphor, was the thing as people traded movies, TV shows, music, sometimes pirated, which lead to titanic copyright and political battles. These were largely replaced by digital social media sites like Facebook, Twitter, and Instagram, which have community building, dating, news and information dissemination, and other services.

"Facebook" seems a deceptive and manipulative metaphor, like the rapacious corporation that owns it since it is not really a book, but a selectively-curated set of behaviors and posts that can be sold to advertisers. It is also a debased form of connecting people as someone is not necessarily a "friend" because I click on an icon that says he or she is, whereas friendship from Aristotle to Mr. Rogers was something personal and definitive of individuals, not something that one amassed on a corporate forum. And to "unfriend" is nasty and part of a negative culture of insult that has created much hurt, suffering, and worse so Zuck and his creepy colleagues can become as rich as Midas.

Next, turning to computer design and interface we find our computers are coded metaphorically as very familiar parts of everyday life like desktop, laptop, or notebook, highlighting that we can work on our desktops, laps, or carry around and take notebooks anyplace we go and can use them in any site. We have computer screens filled with icons and menus which in turn are figured in terms of very ordinary objects and activities like file cabinets for our texts, providing access and storage. We have a trashcan for disposal of obsolete or waste material. We used to have telephone icons for our modems, and maybe some people still do, and almost everyone has mailboxes for E-mail which has been ubiquitous and highly important since the beginnings of computer culture.

With the monopoly power of Microsoft, many of us have Windows—to look out at into computer world, or, as we'll see, onto multiple worlds. For Sherry Turkle (1995), "Windows" is a key metaphor for the postmodern self—multiple, flexible, decentered, and in the realm of simulation, in which RL (i.e. real life) is just one more window. Each Window opens up to one of our significant everyday worlds and realms of everyday life: our texts, our mail, our research, our entertainment and games, and our various projects and interests.

Key computer metaphors also appropriate the figures and discourse of business, deploying work metaphors, from desktop computers which refer to office as well as home. Other office icons include files, documents, trashcan, and spreadsheets which derive from the workplace, and people used to log-on their computers every day, just as many punch-in at work. Moreover, you can be monitored in your virtual work just as in real work, and surveillance files are like a captain's log, a report on various employees and their activities. Computer surveillance has Big Brother watching over you, as in Orwell's *1984*), and your on-line activity is monitored by a variety of agencies, many of which are unknown to you. So like Kafka's K you are perpetually on trial without knowing what your monitors are going to do with it, and thus the human species says "Goodbye to Privacy," or sees it disappearing with each advance of the Logarithm.

Many of us are intellectual workers and to some extent computers were first experienced and used as glorified typewriters, as more efficient ways to produce texts. Then with e-mail, they became more efficient ways to write letters and memos, and to interact with people professionally or socially. The growth of the Internet and computer databases enhanced research and the initially text-based and then more multimedia World Wide Web dramatically multiplied sources and types of information and texts—which in turn required a whole new set of metaphors as we'll see in the next section.

Metaphors of Activity and Movement

Recalling McLuhan's notion of metaphor as bridge and transfer of meaning, the activities of using a computer—commonly described as surfing, cruising, browsing—are all familiar activities from everyday life—drawn from nature (surfing), car culture (cruising), not to forget book culture (browsing). In fact, McLuhan to the contrary this time, book culture is having a remarkable afterlife, indeed is flourishing, in digital culture and a lot of our metaphors are appropriately drawn from book culture, such as documents, files, and bookmarks. We **read** our files, **write** e-mail, create web **pages**, negotiate hypertexts and the like, pointing to continuities with print culture, putting in question McLuhan's notion of an absolute rupture with print culture (1964) in electronic media and computer culture, although arguably configuring bookmarks is more like setting up your own cable system, or developing a switching system, than marking a book. I would argue, however, that we need to view the continuities **and** discontinuities between book culture and digital culture, and see the increased importance of reading and writing and traditional print literacies at the same time we develop multiple new functions for computers and digital devices.

H.G. Wells' concept of the "world brain" provides a most illuminating metaphor and anticipation of the Internet. In the 1930s, Wells, by then one of the most famous writers and intellectuals in the world, conceived of a "World Brain" embodied in a new, free, comprehensive, authoritative, and evolving "World Encyclopaedia" that could enable individuals throughout the world to make use of information resources and contribute to general enlightenment.[4] As many commentators have noted, the Internet and Wikipedia can be interpreted in this light, although there are fierce battles over whether these sources provide adequate scholarly references, and this debate highlights the importance of critical digital literacies that I will take up in this book (see Chap. 10).

Early home computer were introduced in the 1980s and early home computers served basic needs, they mimicked basic activities of everyday life, some work related, some personal, and some interpersonal like e-mailing, chatting or playing games. In any case, by the 1980s computers were beginning to play a key role in everyday life. In 1982, the computer was the "Machine of the Year" in *Time* magazine, replacing their traditional "Man of the Year" format, and in 1984, Orwell's totemic year, Apple featured an ad with the Mac smashing the totalitarian state envisaged in Orwell's iconic novel *1984*. Hence, by the 1980s, ICTs were taking on a variety of functions, both work and play related, and for some were a key mode of keeping busy, assuaging loneliness or boredom, and thus served deep anthropological needs as they became more and more rooted in everyday life.

Many of these early anthropomorphic metaphors, and metaphors drawn from everyday life, are fairly descriptive and straightforward. They were coined to sell computers to businesses and individuals, and the icons, the interface design, the key metaphors were intended to convince the consumer that they needed the computer to work and engage in a wide range of activities, and that they were not threatening and strange, but could be used at home for play, leisure and social activity, and a variety of other functions. Given the early threatening and negative cultural images of computers, which equated computers with massification and Big Machines and Business (i.e. IBM), the new computer industries of the 1990s

---

[4] Wells 1938. The *Wikipedia* entry on Wells' concept of the "World Brain" documents how this metaphor has long been used by scholars as a metaphor and anticipation of the Internet and Wikipedia; see "World Brain," *Wikipedia* at https://en.wikipedia.org/wiki/World_Brain (accessed on January 24, 2021).

wanted to make their computers friendly and non-threatening and so naturally used anthropomorphic metaphors and figures from everyday life.

Such metaphors were perhaps appropriate for early computers for work and business, but their literality occludes the novelty and radicalism of computers, the extent to which they are transforming every realm of experience, and though the metaphors are straightforward, they are perhaps too banal and too literal for the weirdness and otherness of technoculture, multimedia, the Internet, and so on, so it is not surprising that many metaphors would later derive from technology and technoculture itself. But first an excursion into nature.

Nature and the Naturalizing of Technology

As noted, many, but not all, of the early metaphors for what computers do, or you do with ICTs, are anthropomorphic, centered in the body and one's personal self, and grounded in home and work. Yet **nature** too quickly became more of the metaphor repertoire of computer culture. As Leo Marx pointed out in *The Machine in the Garden* (2000) nature metaphors were used to naturalize and sell technologies, and this was certainly the case with computer technology, as in **Microsoft**. Many people's programs are run by programs or software that are often produced by Microsoft, which is as I'll suggest, another semiotically brilliant marketing concept like PCs.

To provide access to the connotations of "soft" in software and Microsoft, I might cite Paul Levinson's book *The Soft Edge* (1997), subtitled "*a natural history and future of the information revolution.*" The Edge is "soft" because information technology is to be distinguished from the old hard edge industrial technology— a standard distinction of post-industrial or information society technology that one finds in a whole range of writers who distinguish between hard, mechanistic, centralized and now backward industrial civilization contrasted to the emergent, soft, flexible, free-flowing postindustrial culture and economy. This is in some ways an ideological construct, but the difference between a hard mechanical and industrial age and a soft postindustrial one is fairly well-accepted in social theory and common discourse.

One might compare, for instance, software and the soft edge with Charles Babbage's "the analytical engine," or "difference engine," which deployed nineteenth century mechanistic metaphors from the industrial revolution, terms derived from the realm of calculation and mechanism, to describe the computers of an earlier century. Such mechanistic metaphors are hard and cold, compared with the soft revolution of digital technologies (indeed "digital" has a soft connotation of sensuous fingers as well as a harder connotation of numbers).

Hence, the anthropomorphic metaphors of early PC days are supplemented by natural and technical metaphors like Microsoft which forged a brilliant metaphorical synthesis—micro and soft, technology and nature—as its corporate logo, positioning it in the forefront of the personal computer revolution, as well as the software revolution, merging the two key components of the technological revolution—the micro and the soft—into one metaphorical figure.[5]

Continuing our examination of nature metaphors and corporate logos, let's reflect upon Apple. IBM and Microsofts' main competitor in the early personal computer world was **Apple** which also promoted its Macintosh and **Mac** computers (does anyone hear McDonalds here? i.e. a move from fast and easy food to fast and easy computing?). While the Microsoft metaphor connoted touch and a fusion of nature and culture, the Apple metaphors connote nature and an item associated with the pleasant activity of eating. Certainly, an apple is as American as apple pie, an apple a day keeping the doctor away, and an apple for the teacher, but what, really, does an apple have to do with computers? And what's up with the Mac, is this an appropriation of the (dubious) cultural cache of McDonalds, that also connotes eating as well as a successful prototypically American global corporation? And what about a prototypical American nickname like Mac, as in "Hey Mac, whatcha doin' today?" Where does this stuff come from?

Steve Rivkin tells us that in his Steve Jobs biography, Jobs told Walter Isaacson he was 'on one of my fruitarian diets' and had just come back from an apple farm, and thought the name sounded 'fun, spirited and not intimidating'.[6] And so Apple was born and since it was recently named the most valuable corporation in the world,[7] I guess the crunchy brand name worked to attract its users and apparently didn't make them sick from over-eating (i.e. computing), as the Apple

---

[5] To the question: Who came up with the term Microsoft? Bill Gates, Paul Allen, or some unknown PR type? Uncle Google informs us: "'So why combine "micro" and "soft?"' In the 1970s, before Microsoft was a thing, Gates and Allen got their start developing MICROcomputer, or MICROprocessor, SOFTware. And it was actually Allen that originally came up with the name." *Windows Central*, November 3, 2017 at https://www.windowscentral.com/where-did-name-microsoft-come (accessed on November 28, 2020).

[6] See Steve Rivkin, "How Did Apple Computer Get Its Brand Name?" *Branding Strategy Insider*, November 17th, 2011 at https://www.brandingstrategyinsider.com/how-did-apple-computer-get-its-brand-name/#.X-jqYYdYbm4 (accessed November 28, 2020).

[7] Ben Popken, "Apple is now worth $2 trillion, making it the most valuable company in the world," *NBC News*, August 19, 2020 at https://www.nbcnews.com/business/business-news/apple-now-worth-2-trillion-making-it-most-valuable-company-n1237287 (accessed January 24, 2021).

cult continues to this day to tell skeptics that its junk is the best and I suppose this argument will go on until the end of Apple and a new millennium.

In another more strange appropriation from nature, I might note **animal metaphors**, in computer world like the mouse, a bug, and (shudder, shudder) a virus. The mouse was from early on used as part of interface, developed by Doug Engelbart (see Bardini (2000)). Someone once told me that only a male could think of a mouse to navigate a computer, and my earlier Window-phobia was related in some ways to not liking to play with a mouse, a phobia I soon got over when Windows made the mouse necessary and highly efficient and Windows opened to things I really wanted like an Internet full of information, music, daily mail, and other wonders.

And, what about **bugs** which if they got into a computer or its code cause malfunctions? Bill Gates talks of **moths** in the original mainframe computer ENIAC, and indeed at one time moths and bugs literally got into big mainframes that had to be cleaned from time to time, but as time went on bugs and viruses have become metaphors of choice for the horrors that can infect computers.

In fact, viral culture became a metaphor for computer culture as Arthur and Marilouise Kroker in their books (1997) and their journal *C-Theory* promoted the notion of viral culture in the Age of AIDS, computer viruses, and ever present viruses of a variety of manmade and natural origins—indeed I am currently writing during the CoVid-19 pandemic of 2020–2021 which has kept me quarantined at home during most of the year and is giving me now the opportunity to finally finish this book with minimal interruptions. The Krokers were right, already in the 1990s, that the delirious high-tech culture of the digital age was marked by speed, viral defusion, hypertext, and constant change and mobility, and that, indeed, the cyberculture itself was going viral and had effected participants in the carnivalesque frenzy of its technoculture and its never-ending delights and challenges.

Further, an interesting symptom of the implosion between biology and technology, bios and techne, appears in the phenomenon of "**viruses**" in computer systems. From the 1980s to the present, viral panic haunts the worlds of both microcircuitry and the flesh. Hackers have implanted viruses in the Internet and computer systems around the world, such that data on one's hard drive or server could be destroyed. In the computer and the body, viruses can take over their host and clone their carrier genetic codes by instructing the host to make replicas of itself. Software programs bear names like Flushot, Disk Defender, and Antidote, and there are now "vaccinated workshops," "sterilized networks," and the offering of technological "hygiene lessons" (see Ross 1991, p. 111).

Also, curiously, the metaphor of "phishing" has come into use to describe hackers entering into our computers to steal and deal information, identities, and so on, in which the all-American ritual of fishing becomes something sinister and frightening. To be sure, "phishing" derived from phone "phreaking" that emerged in hacker subculture as one of the ways of getting into information systems and in a classic Derridean move changing the "f" to "ph"—constituting a difference visible on the screen but not audible in spoken language (thanks to Jonathan Sterne for this point).

Moving on, one of the key moves of cyberculture digerati is to collapse the distinction between nature and culture. Many key metaphors assimilate computers and new technologies to nature and life forms, like the Web, A.I. (artificial intelligence) or A.L. (artificial life), and metaphors of nature and evolution are often used to describe evolving or developing computer systems and networks.

Indeed, *evolutionary metaphors* are favorites of digerati, ideologues for the new technologies, like Kevin Kelly who talks of networks as "hives," or "swarms," and claims that new technologies are evolving life-forms that undo the distinction between the born and the made—claims taken up by posthumanists like Bruce Sterling, who claim that there is an implosion between nature, technology, and humans in cyberspace.[8]

Bill Gates (1999) in his book *Business @The Speed of Thought* revels in such naturalistic and anthropomorophic metaphors, referring to "digital nervous systems" as the key to happening and successful business. These naturalizing metaphors drawn from nature make what is constructed, artificial, and marketed appear natural, part of nature, just like early ideologues of capitalism appropriated metaphors from Darwin to present capitalism as natural. I might note that the obverse of the naturalization of culture and technology is the culturalization of nature, in which life itself is conceived as an information system, a set of self-reproducing data in which DNA, the very stuff of life is interpreted as information.

More benign and positive nature metaphors include "*surfing*," drawn of course from California beach culture.[9] The metaphor of "surfing" the net is interesting in that a metaphor derived from nature, from riding the waves of the ocean, would

---

[8] For a critique of this discourse, see Steven Best and Douglas Kellner, "Kevin Kelly's Complexity Theory: The Politics and Ideology of Self-Organizing Systems," *Organization and Environment*, Vol. 12, Nr. 2 (1999): 141–162.

[9] On how California slang and culture helped construct the discourse, metaphors, and icons of computer culture, see Richard Barbrook's and Andy Cameron's article "The Californian Ideology" at https://www.alamut.com/subj/ideologies/pessimism/califIdeo_I.html and.

.

be transferred to the technologically mediated spheres of cyberspace. The term surfing connotes riding the crest of a force of nature and of a pleasurable very fast movement from one space to another, precisely the experience of moving from one web site to another in the slower earlier days of the Internet. Indeed, the rhythms of surfing apply in a peculiar way to earlier Net experience: as one moves from one site to another there is a slow motion gliding as one waits for the new site to appear, a crescendo of excitement as the data—and perhaps images and sounds appear—and a return to a slower, less exciting mode of experience as one settles down to access whatever information or images one seeks, as well as the possibility of a crash! For just as every wave can bring your body down, so too could any site crash your system, a frequent experience in my early days with computer culture at the University of Texas.

Surfing also connotes no directionality or goal but to ride the wave where it takes you, staying on the surface and going wherever the wave takes you, which was the early experience of Internet travel, as one clicked from one site to another, often discovering something new and surprising, before it became somewhat naturalized and mapped with favorites, Google search engines and the like.

It's also interesting to ponder whether surfing rules out directed searches and the possibility of reflective and directed action, as some surfers came to be non-reflective and just go for the wave, whereas other surfers seek out specific waves and beaches and may have a reflective strategy. So in this sense the "reflective surfer" like the experienced net traveler, seeking particular sites and pleasures, whereas the non-reflective surfer is like the newbie who delights in clicking and exploring and is not so concerned where he or she may eventually arrive. So suffering is perhaps no longer a particularly relevant metaphor for all Internet explorers though it maybe be a perfectly appropriate metaphor for Internet activity of some—indeed, does anyone surf anymore, or do we just click and consume?

The metaphor of "surfing" has other cultural origins, also evoking TV's couch-bound "channel surfer." Digerati like Steven Johnson (1997) attacks the "terrible injustice" of this metaphor since the conventional image of the TV channeling surfer is, for Johnson, that of a passive, superficial, and bored couch potato, clicking for something of interest to passively consume, whereas the Web surfer for the 1997 Johnson is—in the idealized form of the digerati—actively involved, interactive, and seeking something of value. In fact, however, the metaphorical transference is more apt than Johnson would have it. TV channel-surfers are also often searching for something of interest and the fact of clicking from one TV channel to another can show a more active and discriminating viewer than previous TV watchers who tended to watch whatever was on the channel they were viewing; i.e. early audience surveys showed that the majority of TV watchers stayed with the same channel the entire evening and network programmers assi-duously attempted to provide a line-up, with an attractive opening program, that would keep the audiences hooked for the entire evening. With today's cable TV, and devices like Roku, one can choose from a vast menu of television shows, movies, and news and information, from all over the world and many histori-cal epochs. Indeed, Johnson later more positively valorized television in the era when HBO, Tivo and more sophisticated programming and technology made TV culture cool again for Internet digerati.

Further, web-surfing can be as superficial and uninvolved as TV channel sur-fing, as individuals click from one link to another just to see what is there, a legitimate explanation for much TV channel surfing as well, to pass time, or sim-ply to seek an image, sound, graphic or whatever that engages one's attention. Metaphors change as technology changes but it is striking to me that so many of the earlier computer and cyberspace metaphors are still relevant, and some are even still in use.

As for "the net" and "the web," these metaphors refer to the web of life, ano-ther biological metaphor for a technological artifact, while "the net" refers to networks like, earlier, TV networks and today digital networks. It also refers to human activities like "networking" and indeed the techno-self of technocapita-lism thrives and lives through networking for business advantage, friends, lovers, and advancing one's self interests. Yet these metaphors clash with "information superhighway" or "surfing" and other bodily metaphors, as one doesn't speed down a web or net, which connote traps and getting stuck in something, which Neil Postman and other technophobes think indeed happened to youth in Internet culture.

To be sure, new metaphors emerge such as "cloud" for what has become the largest data storing site and a realm where our computers have a second life, as do social media such as Facebook, Twitter, YouTube, and other popular sites, while entire corporations and governments engage in "cloud-computing" (Carr 2013, p. 236ff). The nature metaphor is striking as is the transcendent site of clouds as part of nature and the sky above the earth, lending it a magical quality. Yet the cloud is also vulnerable to malfunctions as Nicolas Carr describes in an account of an Amazon Web Services cloud-computing complex failing during a violent electrical storm near its facility in Northern Virginia (Carr 2010, p. 235). While metaphors may provide soothing images such as clouds in the sky we must also remember that they are part of a weather complex that could involve violent storms and destruction of electrical systems, a disquieting thought when we reflect that much of our date, including mine, is stored in a cloud.

Travel and the Frontier: Metaphors of Adventure

It is interesting that Bill Gates tended to prefer "**navigating**" and his Internet programs have "navigators" which connote adventure, exploration, and travel. Microsoft indeed privileges the **metaphor of travel** in its corporate culture and hype: where are you going today? was one of its major advertising figures. Or consider Bill Gates' first book *The Road Ahead* with blue skies and a road before us on its cover, assimilating pristine nature to technology but also the open road of adventure.

So in addition to appropriating figures from nature to describe activity in the realm of electronic culture and communication, we also deploy **car and transportation** metaphors: the information superhighway, the **Road Ahead**, and the transportation or routing of data. Data routing and switching systems, which configure and route "packets" of data draws upon earlier railroad analogies: (see Bill Gates 1995, p 104 on metaphors of switches and railroad tracks). Routing packets of information and data are part of transportation metaphors which also evoke the telegraph and the instant traveling of messages and information.

"Cruising" and the "information superhighway" are automobile metaphors, connoting traveling and adventure, and assimilate car culture to computer technology. As a key metaphor, the "information **superhighway**" is rich with connotations and images. "Superhighway" suggests fast traveling, high speed with few stoplights, but also signifies familiar, well-charted, and benign sites and places. It carries metaphorically the prestige of automobiles, of adventure, and technological transcendence of time and space.

Moreover, "information superhighway," with its connotations of free information, open to all, a public space, suggests that it is a gift to us, it's decommodified, and free—which is partly true of privileged computer worlds facilitated by Universities, businesses, or other institutions. As we know, the Internet was initially produced and financed by the government or public institutions like universities, and was initially free to most of us who worked for government or universities.

On the other hand, there are hidden fees, overt in commercial Internet services, and in corporations as well such as those that want to colonize the space who dream of charging microfees for every time it is used. Hence, there remains the danger that always abides that the freeway, the superhighway, will become a tollroad—and perhaps it is a miracle that it's been free for many for so long—although obviously many do pay for its services in different ways.

Returning to **travel and spatial** metaphors, we need to engage the metaphor of **frontier**, which has been popularized by the Electronic Frontier Foundation, and digerati celebrating the frontiers of the new technoculture. The term "frontier" connotes pioneers, exploring new territory, establishing new settlements and communities; i.e. websites and even virtual communities and cities. The concept of frontier also suggests novelty, entering a new cultural terrain and dangers and excitement. For the digerati of the cyberculture, ICTs are constructing and traversing new frontiers, new modes of communication and interaction, new sources of knowledge and creativity, and new sources of danger.

To some extent, the metaphor of the "frontier" and establishment of new communities in unexplored spaces is useful to explain the amazing development of the Internet. Like the Western experience of the United States, in which intrepid pioneers entered spaces where no person had ventured before, so too in the early days of the Internet, bold individuals explored the new spaces, established homesteads (i.e. list serves, bulletin boards, websites, chatrooms, etc.) which sometimes became part of and attracted communities.

Along with the hardy pioneers come missionaries, merchants, and teachers, and pretty soon there were established communities and towns. As in the Wild West, there were wars between various sectors of the pioneer communities, outlaws and bad guys (i.e. flamers, hackers, and just plain old obnoxious individuals). Just as some pioneers welcomed newcomers, others were hostile to them who were dismissed as newbies or clueless interlopers. In some sites, eventually harmonious relations were established as civilizing influences such as rules, protocols, and norms became established to govern communal life, providing models, in some cases, of self-governance and participatory democracy. Of course, powerful individuals often assumed dictatorial powers, bullied and tried to control the communities and spaces, and sometimes there were revolts against such arbitrary authority. And just as big corporations eventually settled in, wielded tremendous power, and attempted to control communal and political life in the West, so too have giant corporations appeared and have attempted to colonize the world of the Internet.

The concept of a frontier thus suggests new lands to explore, a pushing beyond established boundaries into something new and the establishment of new settlements such as web sites, Moos/Muds and various multi-user dimension sites, as well as new social relations and identities. Yet the metaphor of the frontier also evokes danger of colonization and here the worry is that corporations and government will colonize cyberspace, will come to control it either through policies, censorship, or establish toll-roads that will keep some people out and off, while establishing new domains of corporate or state control.

On the wilder side, as in Wild West shows, where men played women, and in the "real" frontier women assumed male roles, or whites played Indians in

Wild West Shows and Hollywood cowboy films, so too is there also gender and race bending in cyberspace (see Turkle, *Life on Screen* (1995) or Stone, *The War of Desire and Technology at the Close of the Mechanical Age [1996]*). Further, while there was transgressive sexuality in the frontier, so too on the Internet is there weird and raunchy sexuality in cyberspace, taboos broken, new pleasures, harassment and hurt—and embarrassment that may cost you a job if you are caught exhibiting your junk on Zoom, or someone sends someone's naked picture out to the Internet.

There is also gambling on the frontier and when you enter new spaces you can engage excessively in tabooed activities and get into addictions like porn or gambling, which have become big problems on the Internet. And just as criminality blossomed in the West and in frontier spaces, so too on the Internet does crime flourish and there is constant danger of violation and bad things happening such as identity theft, cybercrime, and the like.

Hence, metaphors of the frontier point to both positive and negative dimensions of the Internet, and capture its uncharted terrain and novelties, as well as continuities with previous forms of culture and experience. The frontier leads us to another important site of metaphors which will take us to outer space and the realms of warfare.

Military and Space Metaphors

The new frontiers of cyberculture connote traveling where no one has traveled before, and it is indeed a mode of space exploration, in which new spaces are created or discovered all the time, there is colonizing of these spaces, wars over space and sites, battles going on for control and power, as there was on the frontier or in science fiction films. And so the motto of *Star Trek*—Go where no man has gone before—points to connections with space travel and will take us quickly to military metaphors.

A Marshall McLuhan Institute newsletter in the late '90 s argued that Internet discourse is basically militarist, marked by its origins with military metaphors and terms like "flame," "erase," "delete," kill, abort, or cancel. These terms are all somewhat violent, male, macho, and militarist. Likewise, "jacked in" and "booted up" sound vaguely militarist, and certainly "crash" is a crucial military metaphor. The term "spam" could also be military in origins as this term described what World War Two soldiers ate out of tin cans and hated. So perhaps spam was projected by military dudes onto computers as the ultimate in disgusting waste, and certainly this continues to be one of the horrors of the Internet.

The most extreme military metaphors, however, come from French thinkers Jean Baudrillard and Paul Virilio. In general, the metaphors one mobilizes (i.e. military terms) for one's theoretical discourse often carries a theoretical agenda like technophilia or technophobia and in the case of Virilio and Baudrillard their metaphors are extremely technophobic.

Baudrillard (1984) uses nuclear metaphors of implosion, irradiation, and extermination (of meaning in the media, of the body in cyberspace, of the real in the virtual and so on). He also talks of a nuclear orbit, suggesting military satellites in space, where we are ejected from the world and trapped in a hyperreal nuclear orbit of our information world and devices.

Paul Virilio uses even more violent military metaphors, speaking of the information bomb, the "accident of accidents" in which the Internet crashes and the whole global, networked economy and system of communication goes down in an apocalyptic catastrophe.[10]

Computer security geeks also use military metaphors. Attacks on computer systems require computer security experts to fight off the attack and regain control over the systems, protecting data and data flow, and destroying or building a wall against the intruder. The bad guys uses nefarious means to breach security and attack the site, disabling its programs and perhaps destroying some of the data and property, or infecting it with deadly viruses. The good guys protect the site and ward off attacks which they attempt to block from penetrating the site and wreaking havoc.

In terms of a technopolitics which calls for resistance, struggle, and counterhegemony, here too there are military metaphors like strategy and tactics, infowar, and the term technological revolution has been used both as an ideological term to legitimate new ICTs as cool and happening and by radicals who want to use the technology for revolutionary social transformation—a topic I explore later in this book in Chap. 5–7.

Cyberwar is, in fact, a new type of warfare that has been increasingly deadly in recent years. Cyberwar played an important role in the 2016 presidential election and helped elect Donald Trump president (Nance 2016; Jameison 2017; Kellner 2017; Hettenda 2018). The Democratic National Committee email server, and that of Clinton's campaign manager, John Podesta, among others, were hacked by the Russians, and Putin and the Russians used WikiLeaks and various global internet networks to circulate fake news, bots, and anti-Clinton and pro-Trump

---

[10] Virillo's views will be explored in more detail in Chap. 7 and I wrote the first systematic book on Baudrillard in Kellner 1989 if you are interested in further exploration of his metaphors of cyberspace and beyond.

stories.[11] From this perspective, key global—digital networks helped Trump win, which along with the anti-globalization discourse with which Trump conned his followers, makes Election 2016 the first US presidential election where global actors using high-tech cyberwar and computer networks and social media like Facebook played a significant and perhaps decisive role in securing a presidential election.

In early September 2017, it was revealed that the Russian intervention in election 2016 did not only include hacking the Democrats and releasing embarrassing or divisive emails on WikiLeaks, but that the Russians were actively engaging in an infowar against Hillary Clinton and for Donald Trump on Facebook, Twitter, Google, and various social network sites. While Facebook initially denied that its platform was used to spread anti-Clinton propaganda, in response to Congressional inquiries and media reports, they confessed that they had created hundreds of fake accounts that attacked Clinton, or spread divisive disinformation, which were linked to Russian intelligence services, and that Facebook had sold at least $100,000 worth of anti-Clinton ads to Russian sources.[12] Fake news stories and fake postings on Facebook and Twitter had weaponized social media and led President Obama to warn Mark Zuckerberg about Russian penetration and use of Facebook before the election but evidently the warning was not heeded.[13] Twitter was also weaponized by fake individuals linked to Russian intelligence which used bots to send messages to thousands of individuals, constituting en toto one of the most consequential cyberattacks in election history.[14]

---

[11] The Russian hacking is documented in Nance 2016, Kellner 2017, Hettena 2018, and many mainstream media sources, although it is denied in pro-Trump sources from the swamps and whacko-worlds of "alternative facts" which may be the enduring legacy of the Trump presidency. For a concise analysis of how the Russian hacking interfered in the 2016 election and dangers for the future of US democracy, see Massimo Calabresi, "Hacking Democracy. Inside Russia's Social Media War on America." *Time*, May 29, 2017, pp. 30–35.

[12] Editorial, "Russia's Fake Americans," *The New York Times,* September 9, 2017: A22 and Scott Shane, "The Fake Americans Russia Created to Influence the Election," *The New York Times,* September 2, 2017 at https://www.nytimes.com/2017/09/07/us/politics/russia-facebook-twitter-election.html (accessed October 15, 2017).

[13] Adam Entous, Elizabeth Dwoskin and Craig Timber, "Obama tried to give Zuckerberg a wake-up call over fake news on Facebook," *Washington Post*, September 12, 2017 at (accessed October 15, 2017) at https://www.washingtonpost.com/business/economy/obama-tried-to-give-zuckerberg-a-wake-up-call-over-fake-news-on-facebook/2017/09/24/15d19b12-ddac-4ad5-ac6e-ef909e1c1284_story.html (accessed October 15, 2017).

[14] Daisuke Wakabayashi and Scott Shane, "Twitter, With Accounts Linked to Russia, to Face Congress Over Role in Election," *New York Times,* September 27, 2017 at https://www.nytimes.com/2017/09/27/technology/twitter-russia-election.html (accessed October 13, 2017).

It is obviously not an accident in light of the military origins of the Internet in DARPA and the fact that the military was intimately involved in its construction that military metaphors abound in the construction of the Internet and social media and that it would eventually become a terrain of cyberwar. Yet linking the Internet with the military can be exaggerated, and there is a body of literature that argues that the Internet and cyberspace were produced in university technoscience culture and then obviously business culture. In this survey, I've suggested a wide range of more domestic, natural, soft, corporate, travel, and adventure, metaphors used to denote computer culture and cyberspace, but I suppose that there are traces of the military construction of computers and cyberspace in the dominant metaphors, and today we can no longer ignore the fact that cyberwar is an increasingly salient feature of our era, one with potentially catastrophic consequences.

However, we need to turn to metaphors of technoculture to better capture the novelties and discontinuities of our computer technology and culture in relation to previous forms of technology and society.

Metaphors of Technoculture

After charting some anthropomorphic metaphors, ones drawn from home and work, from nature, from activities of everyday life, from travel, adventure, and the military, we come to a set of metaphors that are more technological and more novel, taking us from everyday life and common vernacular to new cultural spaces and discourses. Let's begin this discussion with the **Internet** itself which is a very interesting metaphor that connects the social (inter) with the constructed and natural net (which can be constructed, or natural, as can a web). Some technophobes claim that the very concept of the Internet carries the connotations of getting caught into the Net, being trapped in a net, that there is no outside, that one is "inter" a net that closes one off from nature, the social world, and other people. On the other side, technophiles claim that the net or web connects and liberates people and opens up new worlds and new adventures.

In any case, the "inter" of Internet connotes an inside with no coordinates, a being immersed in a space without poles or structures. Negatively, one might stress the impossibility of getting outside when one is in the Internet, a being trapped in a virtual world, while, positively, one is in a new world of adventures,

---

Of course, it should be noted that the U.S. had intervened in many elections over the years against Russian candidates, their surrogates in other countries, and allies during the Cold War; see Stockwell 1984.

pleasures, information, potential friends, and so on. For some, you are just there everyday in the net, but when you think about it there's no there, although for others you are there, and for many it's where it's at and where it's happening.

On one hand, the terms web and net connote a catching something, as being caught in a web, or a net, and thus being trapped. MUD originally connoted multi-user dungeon, which game players who enjoyed "dragons and dungeons" would consider positive, while for others dungeons are a prison. For some the Internet itself connotes being imprisoned in virtual space, while for others it's a site of adventure and discovery. Indeed, a net or web also suggests a maze, which has its exciting labyrinths, its mysteries and its wonders, but it is also a place where one can become lost and trapped.

While the "net" connotes an object that one may be tangled up in, it is also a complex set of connections and links. "Inter" connotes interactive and thus social relations, connecting the individual to other people and perhaps communities. Links connote being connected and linked to others, as well as to new sources of information and culture.

One suggestive earlier political metaphor for the Internet that has evolved into different forms has been "netroots," whereby the net metaphor is linked with grassroots communities and organizing whereby the Internet and social media are used as organizing tool by left and right. Some demonized the netroots as technofascism, using the Internet as a weapon to propagandize or slander and assail one's opponents, and this does happen on both sides. Yet netroots have also been effective in bringing new voices and groups into politics and may have decisive effects on the future of electoral politics in the US as well as constituting as alternative realm of the political (see Chap. 5 and 6 below).

Other metaphors deriving from webs include blogs which have evolved from web diaries or weblogs into a distinctive new form of political communication.[15] However, in divided societies, like the US in the twenty-first century, the blogosphere is one of conflict and political warfare with flaming, troll attacks, and the sort of vile nastiness associated with trolls, flamers, and aggressors, just as Donald Trump latter associated Twitter with cyberwar against his many enemies. Further, there is something of an implosion between the blogosphere and netroots as a wide variety of groups on both left and right and in many spaces between

---

[15] See Richard Kahn and Douglas Kellner, "Oppositional Politics and the Internet: A Critical/Reconstructive Approach," *Cultural Politics*, Vol. 1, Issue 1, 2005: 75–100 and "Technopolitics, Blogs, and Emergent Media Ecologies: A Critical/Reconstructive Approach," in *Small Tech. The Culture of Digital Tools*, edited by Byron Hawk, David M. Rider, and Ollie Oviedo. Minneapolis: University of Minnesota Press, 2008: 22–37.

use the blogosphere to mobilize an online community for specific political actions and agendas in a highly contested space of cyberculture.

Further in the Internet, there is a collapse of time/space coordinates, which provides instant communication anywhere and thus transforms time/space categories and experiences which may be confusing and overwhelming but also exhilarating. The Internet is indeed itself a new form of culture and space and one of the more central metaphors to take hold of the new information/communication/multimedia terrain of the new technologies is **cyberspace**. Both cyberspace and hyperspace in ordinary language connote power, enhancement, and novelty, with new selves traversing new cultural spaces. While hyper derives from the root more than, the root of cyber comes from the Greek word for control. Perhaps part of the entry of "cyber" into ordinary language originally from the concept of cybernetics in the 1940s which has the connotations of technical control, of a technological steering system, of seeing social systems as technical systems that could be controlled and managed. This is the sense of Norbert Wiener (2013; 2nd ed.) in his book on cybernetics and the concept of cyber has since entered the vernacular as well as giving rise to an intellectual discipline and field of inquiry of cybernetics.

Yet some of the connotation of cybernetics seems inappropriate for Internet cyberspace which is wild, anarchic, and uncontrolled—as the frontier metaphor suggests. Yet there are some deep connections here as cybernetics is originally controlled by **programming** and this became a dominant term for computer world. Computer programmers were once the techno-elite of the new cyber-world and early responses to computers included fear of being programmed. "Do not fold, spin, or mutilate" was a '60s Berkeley countercultural response to big computers and visions of a computerized society, with fears that it would be a programmed society. Films of the epoch such as *2001, Colossus: The Forbin Conspiracy* and George Lucas' early *THX 1138* all evoked fears of computers taking over, programming individuals and "programmed" was early on a negative metaphor that supported technophobia.[16]

Part of programming is **word-processing**, a term that is geeky, technical, and to sorts like myself, initially, off-putting. No serious writer wants to process words, or be a word-processor, although I don't know if this is still part of cyberdiscourse or that of everyday life, as terms have evolved and some disappeared since I first involved myself in cyberculture since its very beginnings in the 1980s.

While there are authoritarian connotations of cybernetics it also has connotation of feedback. Although originally the concept was somewhat authoritarian

---

[16] See Chap. 9 below on Kubrick's *2001* and technophobic culture and the fears of computers taking over.

and menacing with emphasis on control and programming, the new cyberculture makes it sound friendly and empowering, with emphasis on feedback, user-friendly, interaction, and interconnectivity with an emphasis on individual agency rather than cybernetic control.

Possibly cyberspace entered the vernacular through William Gibson's cyber-punk novel *Neuromancer* (1984) where cyberspace is described as a "consensual hallucination" which to me doesn't seem quite right since cyberspace is real and communal in many ways.[17] The technologically mediated sphere of cyberspace is indeed a real one, and there is nothing hallucinatory about computer databases, or web sites, or even our e-mail folders. Indeed, to some extent the spaces of the telegraph, the telephone, and the radio and television are cyberspaces in which technologically mediated space overcomes distances of time and space and allows shared communication and interaction.

Yet while Gibson and cyberpunk fiction popularized cyberspace, the root cyber has for the most part in current usage suggested techno or technologically mediated, as in the concepts of cyberspace, cybersociety, cyberdemocracy, cyborg or perhaps cyberpunk, a technologically mediated fictive world. My one-time colleague and friend at the University of Texas Michael Benedikt edited the first book on *Cyberspace* (1991) which in his reading has affinities with architecture, computer-programming, culture, and denotes a polysemic new technologically mediated space, a technospace, a space of new modes of communication, interaction, and adventure.

Since there are many connotations of cyberspace, as the Benedikt book indicates, it too is a **contested term** as are all the metaphors of cyberspace with different writers providing positive or negative valence of some of the key metaphors. Positive discourses and metaphors of cyberspace have been promoted by digerati like Sherry Turkle, or *Wired* magazine, and celebrate new technosubjects, empowered by connection with new technologies which generate expanded and multiplied selves. I recall in the early days of the technoculture, John Perry Barlow pronouncing the word "cyberspace" as a magical space where democracy, empowerment, and great new things would happen. Cyberpunk fiction engages the more disturbing and strange sides of cyberspace and, among other things, how corporations, the state, and crime can become involved in the new spaces which provide new forms of power and profitability as well as experience.

Cyberspace and techno-metaphors like "wired" assimilate the power and magic of electricity, harnessing the mystery of electricity to computer interaction—as

---

[17] On the novel *Neuromancer* see the chapter in Best and Kellner (2001); on cyberpunk see the first edition of my book *Media Culture* (Kellner 1995).

well as assimilating countercultural drug use connotations (i.e. "dude, I'm really wired"). There are several **key electrical metaphors**: being wired, plugged-in, connected, and interconnectivity, which also resonate with '60 s counterculture; i.e. being turned on, tuned in, and wired. Its interesting that '60 s countercultural types like Timothy Leary promoted computers as the turn-on of the contemporary era, pointing to countercultural roots of the current digerati who are helping produce these metaphors[18]–Stewart Brand, Kevin Kelly, and others associated with *Wired* magazine were all part of 60 s's countercultural movements and bring the communalism, but also individualism, romanticism, and liberationism, which has mutated into libertarianism in a more neoliberal age.[19]

The "wired" and "nethead" metaphors also appeal to a new empowered subject, and I suppose it's no accident that Bill Gates's dominant mantra is that computer technology is empowering; that you're connected and wired, and can do all sorts of great stuff and maybe become great yourself ("great" seems to be Gates' favorite word).

To criticize these metaphors, one could argue that computers are not just an extension of our self, are not always that friendly, are as hard as they are soft (in several senses), and that many metaphors ignore the economic forces and technical structures that organize the field of electronic culture and communication. Many of the metaphors I've been discussing are the hype of the computer and information industries, and are marketing devices trying to get you to buy into computer culture that therefore should be viewed with suspicion.

Now in discussing metaphors of cyberspace, our focus has shifted from personal, anthropomorphic, soft and friendly, and a wide range of metaphors drawn from everyday life to metaphors of technoculture that produce high tech metaphors like Internet, wired, cyberspace, and a whole repertoire of Geek speak that is possibly beyond my comprehension (I'm also curious to know how many of the metaphors I've discussed today are still in circulation).

It is curious today that people are deriving metaphors from cyberculture to describe everyday life. For example, more and more people speak of "interfacing," "networking," or "multitasking." Or, we say things like "click on that," or

---

[18] Edward Rothstein, "*A Crunchy-Granola Path From Macramé and LSD to Wikipedia and Google,*" September 25, 2006 at https://www.nytimes.com/2006/09/25/arts/a-crunch ygranola-path-from-macrame-and-lsd-to-wikipedia-and-google.html (accessed January 25, 2021).

[19] See again the Barbrook and Cameron article "The Californian Ideology" and on Stewart Brand see Turner (2008), and on how '60 s counterculture mutated into libertarianism, with Kevin Kelley as a key figure, see Turner (1998) and the critique of Kelley in Best and Kellner (2001).

"my cache is full," or "I can't seem to boot up this morning." This signals that cyberculture is itself becoming the dominant culture that is producing the next round of new metaphors for everyday life. So click on that.

One of the more problematic set of metaphors that has emerged to describe cyberspace is that of **liquid, fluids and flow,** nature metaphors that naturalize the technologically constructed sphere of cyberspace. These metaphors derive from poststructuralism, especially Deleuze and Guattari, and have been taken up by Zgymunt Bauman in his book *Liquid Modernity*, by John Urry, Scott Lash, Mark Poster, and others to describe the mobile, dissolving, implosive, and protean nature of the Internet and cyberspace.

These are somewhat misleading metaphors for cyberspace and I prefer the "network" metaphor that points to the infrastructure of the technologically-mediated society, to digitization that emphasizes its technological roots, and cyberspace to point to the architecture and organization of the new electronically-mediated spaces. Liquid and flows are just too loose and too ideological, caught up with the libertarian vision that the market provides a groovy free flow of capital that makes everyone rich, just like the Internet provides a free flow of information that makes everyone informed. Liquidity too is a capitalist metaphor that in this case makes it seem that everyone is rich, we all have liquidity and are part of liquid modernity. Quite frankly, I don't see the point of water metaphors like liquid and flow for the Internet that is very technologically mediated.

I might note that the "liquid modernity" metaphor also gives rise to technophobic visions and discourses where all is dissolving all the time, as with Baudrillard you have a dissolving of reality, meaning, subjectivity, and all the stabile points of modernity. Its curious, though, that the "all that is solid melts into air" metaphor comes from Marx and Engels' "Communist Manifesto" which characterized capitalist modernity (see the reading in Berman 1982), while the liquidity and flow discourse for the Internet comes from poststructuralism and is often, as in the case of Baudrillard, connected with postmodernity.

Parenthetically, I would say that poststructuralist and postmodern theory can be very valuable in illuminating cyberspace and digital culture, but you need to contextualize it in materialist discussions of technology, political economy, and history rather than dissolving everything in Deleuzian flows or Baudrillardian hyperreality or implosion. I have argued that the term "postmodern" is useful in signaling changes and discontinuities but needs to be given specific grounding to make sense and is often used to make gross exaggerations as with digital sublime discourse and those that make the Internet a vehicle of salvation or argue for a radical discontinuity. I have, by contrast, argued for a dialectic of continuity and discontinuity in discussing contemporary forms of economy, society, culture, and politics and have argued that we live in a space between the modern and the postmodern (see Kellner 1995, Best and Kellner 2001).

As I wind down, I should note that I'm not addressing the more spiritualistic digital sublime argument that claims cyberculture overcomes space, time, class, gender, race and nation and thus portends the end of history, politics, and modernity. Vincent Mosco has done a solid critique of this ideological discourse in *The Digital Sublime: Myth, Power, and Cyberspace* (2004) and I have nothing much to add to this critique. I am also not engaging the critique of the technofantasy that our data and/or selves can be digitized and put on-line guaranteeing us an eternal existence, although this is a fruitful subject for science fiction literature, films, fantasy, and metaphysical speculation.[20]

I should also point out that we are going beyond a computer-centered cyberspace today in the wireless era with digital devices like smart cell phones, I-Pods, Blackberries, light notebook computers and other devices we can carry around to be connected at all times with the technoculture, though I might note it is curious that many devices are named using nature metaphors evoking cells, pods, berries, or ThinkPads (which I am now using).

We are also moving into an era of convergence, where television, computer, phones, and other objects of everyday life will converge, and as this intensifies no double new metaphors will abound making the task of analyzing and interpreting the metaphors of cyberspace perhaps an unending one.

Some Concluding Comments

In this chapter, I have argued that our new worlds of culture and communication are worlds highly defined by metaphor: first the PC and screen of desktop icons, with homey, anthropomorphic and business metaphors; then the Internet, information superhighway, web, and cyberspace which connoted travel and adventure. Metaphors for interacting in this space include surfing, navigating, cruising, browsing, and other activities which in sum are quite different and all of which have specific connotations and meanings in everyday life, which have been sometimes useful and sometimes problematical in relation to technoculture.

---

[20] The moment of merging of the human and the technical creating a new superior technohuman has been called "the singularity"and has been promoted by Kurzweil 2005; the Wikipedia page on "Technological Singularity" has a full discussion of the conceptions antecedents, theorists, issues, and debates at https://en.wikipedia.org/wiki/Technological_singula rity (accessed on January 25, 2021).

I've indicated some of my preferences, but my main task here was more of a mapping, a charting, an interrogating of dominant metaphors than a full scale ideological critique, let alone a full-scale conceptual reconstruction—although I think both remain to be done. For now, here are some concluding comments:

1) We see that the field of electronic culture and communication absorbs all sorts of metaphors; anthropomorphic, natural, cultural, technological, militarist and so on, drawing on a wide range of domains and cultures.

   The multiplicity of metaphors from disparate sources show the complex roots of the Internet itself in military culture, university science-technology-engineering culture, counterculture hackers and digerati, and business culture. Obviously, these four cultures overlap as individuals from each enter into others but I want to resist reducing cyberculture to a military, business, university, or hacker culture as do some writers and digerati do, and believe that its multiple origins and heterogeneity of metaphors points to its multidimensionality and complexity.

   Hence, with its panorama of metaphors, the new technologies absorb a wide range of fields, are all-embracing, incorporating and transforming vast realms of experience and of everyday life into its universe. Further, arguably everything that comes into the orbit of the new pervasive and dominant cyberculture is absorbed and incorporated, pointing to the real technological revolution going on that is transforming every realm of society and experience—economy, polity, social life, everyday life, identity, etc.—and producing a new digitized and technologically-mediated society and culture.

   I would thus argue that the diversity of origins of metaphors shows the diversity of activities of which computers are colonizing like everyday life and the home, office and work, social life, travel and adventure, the military, and new forms of technoculture that evolve as we ponder what has already happened.

2) The ubiquity and scope of metaphors in the expanding and proliferating discourse of the information and communication technologies points to the centrality of metaphor in our life. To some extent, we live our metaphors and our intellectual and spiritual life is metaphorical: the life of poetry, literature, philosophy, all use and absorb metaphors, which we use to make sense of everyday experience.

   Seen from a semiotic perspective religion is metaphorical life. In Christianity, God the Father, Son, and Holy Ghost are arguably metaphorical although fundamentalists would argue against this and for literality. I might note that certain forms of Judaism are against images and metaphors while many Muslims and worshippers of other religions today are very sensitive to what metaphors and images are used and there is a big battle in Islam over what Jihad means. Indeed, almost all major world religions have their fundamentalisms, their literalities and would deny the metaphoricity of religion.

Politics too can be seen as a battle of metaphors. George Lakoff has discussed the importance of correct metaphors for politics, arguing that Republicans have been better at marshalling metaphors than Democrats who need to rethink their discourse and key metaphors (Lakoff 2016, 3rd Enlarged ed.), although the Trump era subverted much Republican discourse and left behind a pile of rancid metaphors and slogans like "America First," "Make America Great Again," "Build that Wall," and "Lock her up," which may be applied to Trump himself as he continues to be investigated by Congress, federal agencies, and state and urban agencies in cities like New York and Washington.

3) As I have argued, many cybertheorists and digerati have theorized with metaphors the interaction between the body and machines, and technoculture and everyday life, with many tending to naturalize and celebrate it. Given the ubiquity of metaphor in everyday life, therefore, it is important to choose appropriate metaphors and to be critical and conscious of the existing dominant metaphors. One should also see that similar metaphors may be marshaled for different valorizations in the discourse on new technologies, inflected in either a technophobic and technophilic direction. Many computer metaphors shift nature to culture, naturalize the new technologies and present it in positive, friendly and homey terms—that is the commercial strategy of Microsoft and Apple and the communication/information industry so one should be suspicious and critical of all such uses.

Negative metaphors by contrast are marshaled for technophobia, as we see most dramatically with Baudrillard and Virilio. These include either military metaphors or bad nature or animal ones like viruses (although digerati also use these), and technophobes turn upside down the positive valorization of the Internet and cyberspace to make it a space of entrapment, a prison.

4) The Metaphors of cyberspace are thus a contested terrain. How one talks about it, what metaphors one uses, helps construct how one sees it, uses it, feels about it, and reproduces it. Finally, I've not been arguing either that metaphors are themselves bad or misleading; metaphors are necessary and it is just a question of which ones we use. Nor have I argued that all of the dominant metaphors are necessarily bad and that we need completely new ones; I've presented a more descriptive than prescriptive analysis although a systematic reconstructive discourse of technoculture could be developed although it would be bound to be contested.

Further, some techno-metaphors can create positive insight into where digital culture comes from and what it does; some domesticate and teach us how to do things with computers; for instance, Donna Haraway's use of the cyborg Fig. (1990) point to ways that humans and technology are interacting and changing both; a theme Mark Poster (1990) has taken up with human–machine metaphors.

Finally, I've argued that metaphors are very important, so in conclusion as one confronts new technologies and new realms of culture and experience one needs new metaphors to cope with the novelties and challenges. Yet, ultimately, metaphor is not enough, we need critical social theory and cultural studies to map the changes in our present day world—that is related to the global restructuring of capitalism as well as technological revolution—and I would also suggest that science fiction is a useful genre in illuminating the weirdness and strangeness of the new technological culture and society, and thus will have a chapter on media culture and science fiction in this book (see Chap. 9).

Yet I've also insisted that we see the emerging technoculture through our metaphors, as our metaphors help constitute our experience and understanding, and that we should therefore **mind our metaphors**!

Technology and Alienation

3

> *"Human beings make their own history, but not under circumstances of their own choosing". Karl Marx*
>
> *"They who control the Microscopick, control the World". Thomas Pynchon*

Abstract

Chapter 3 "Technology and Alienation" explores the downsides and potential alienating effects of immersion in a high-tech world, as well as possibilities for dis-alienation and empowerment. The developing countries are currently undergoing a perhaps unprecedented technological revolution that has given new credence and life to the concept of alienation after a period of relative decline in which Marxian, existentialist, and other modern discourses were replaced with postmodern perspectives skeptical or critical of the concept of alienation. In this paper, I want to suggest that digital information and communication technologies and the restructuring of global capitalism require us to rethink the problematics of technology and alienation. If it is true that we are undergoing a Great Transformation, one of the epochal shifts within the

For discussions of an earlier version of this chapter and suggestions for revision I would like to thank Lauren Langman and Richard Kahn. This article was first published as "New Technologies and Alienation: Some Critical Reflections," *The Evolution of Alienation: Trauma, Promise, and the Millennium,* edited by Lauren Langman and Devorah Kalekin-Fishman. Lanham, Maryland: Rowman and Littlefield, 2006, p. 47–68. It has been revised and updated for this volume.

D. Kellner, *Technology and Democracy: Toward A Critical Theory of Digital Technologies, Technopolitics, and Technocapitalism,* Medienkulturen im digitalen Zeitalter, https://doi.org/10.1007/978-3-658-31790-4_3

history of capitalism, that digital technologies and social media are taking us into a novel field of cultural experience, and that the very nature of human identity and social relations are changing, then obviously we need to develop fresh theories to analyze these changes and politics to respond to them

Keywords

Technology and Alienation • Dis-alienation and empowerment • Marxian existentialist and other modern discourses • Postmodern perspectives • Great Transformation

The developing countries are currently undergoing a perhaps unprecedented technological revolution that has given new credence and life to the concept of alienation after a period of relative decline in which Marxian, existentialist, and other modern discourses were replaced with postmodern perspectives skeptical or critical of the concept of alienation. In this paper, I want to suggest that digital information and communication technologies and the restructuring of global capitalism require us to rethink the problematic of technology and alienation. If it is true that we are undergoing a Great Transformation, one of the epochal shifts within the history of capitalism, that digital technologies and social media are taking us into a novel field of cultural experience, and that the very nature of human identity and social relations are changing, then obviously we need to develop fresh theories to analyze these changes and politics to respond to them.[1]

For many, the changes underway on a global scale are as thorough-going and dramatic as the shift from the stage of market and competitive and laissez-faire capitalism theorized by Marx to the stage of state monopoly capitalism critically analyzed by the Frankfurt School in the 1930s.[2] Theorizing this ongoing and epic transformation requires critical social theory to engage anew the relations between the economy, state, culture industry, science and technology, social institutions and everyday life as radically as the Frankfurt School revised classical Marxism in the 1930s. In this context, talking about technology and alienation is not just an academic affair, the latest twist in the discourse of alienation or of technology, but

---

[1] Postmodern theories claim that we are undergoing dramatic changes and mutations in the transition from modernity to postmodernity (see Baudrillard 1993 [1976]; Jameson 1984, 1991; Harvey 1989 and the discussions in Best and Kellner 1991, 1997, 2001). Castells (1996, 1997, 1998) argues that information and communication technologies are creating a novel form of the global economy and networked society.

[2] On the various stages of development of the Frankfurt School, see Kellner 1989a and for reflections on the roles of emergent technologies in the current stage of capitalist development see Best and Kellner 2001; Kellner 2003a.

rather concerns the fate of the human being in the contemporary world and thus requires serious reflection and discussion whether the changes in society, culture, and human existence are or are not beneficial, and what we can do to promote a positive outcome and prevent a harmful one. Yet before we can talk intelligently about the emergent technologies and their impact on human and social life, we need to reject right from the beginning the two dominant ways of talking about contemporary technologies and need to develop a critical theory of technology to adequately address the issue of technology and alienation.

Technophobia vs. Technophilia

In studying the exploding array of discourses which characterize digital technologies and social media, I noted in the Introduction how they were characterized by either a technophilic discourse which presents digital technologies and social media as our salvation, as a force of progress that will empower individuals and produce a new high-tech society, or they embody a technophobic discourse that sees technology as our damnation, demonizing it as the major source of alienation and domination which empowers technology, megacorporations, and the state in exploiting, alienating, and dominating the individual. In this chapter, I will explore the ways that digital technologies and social media do and not produce alienation and social domination, as well as ways that digital technologies can empower and enhance individuals. Dialectics of digital technology: current forms of social media can help control, manipulate, alienate, and dominate individuals and social groups, while at the same time they can empower individuals and groups, be used to overcome certain forms of alienation and domination, and can be deployed as instruments of democratic transformation aimed at increasing democracy, freedom, and social justice.

In fact, we will see that digital technologies and social media are ambiguous and contradictory phenomena and can do positive and negative things at the same time. Against one-sided technophilic or technophobic approaches, I am arguing in this book for a critical theory of technology intended to sort out positive and negative features, the upsides and downsides, the benefits and the losses in the development and trajectory of digital technologies, devices, and social media as well as contradictions and ambiguities.[3] It is necessary to counter promises of technological utopia, that computers will solve current problems, produce jobs

---

[3] On the project of developing a critical theory of technology, see Feenberg 1991, 1995, 1999, 2017; Kellner 1997; Best and Kellner 2001.

for everyone, generate a wealth of information, entertainment, and education, connect everyone, and overcome boundaries of gender, race, class, and region. Yet a critical theory of technology also needs to counter technological dystopia which argues that computers are fundamentally vehicles of alienation, or mere tools of capital, the state, and domination.

Both one-sided approaches reveal the need for a dialectical critical theory a la Hegel and Marx that plays off extremes against each other to generate a more inclusive position, indicating how technology can be used both as instruments of domination and emancipation, as tools of both hegemonic societal powers and of individuals struggling for democratization, education, and empowerment, and thus both as forces of self-alienation and self-valorization and development. A critical theory of technology requires a substantive vision of what technology is, what it does and what it could do, as well as a normative perspective that delineates positive and negative uses, noting contradictions and ambiguities as well.

The critical theory of the Frankfurt School, which I am drawing upon here, criticizes existing institutions, social relations, and phenomena from a normative standpoint through which existing realities can be judged deficient and oppressive.[4] I suggest that those forms and uses of technology that enhance positive values such as democracy, community, freedom, self-development, and the like should be deemed life-enhancing and meritorious, while those forms and uses of technology which promote domination, oppression, alienation, and social injustice, while undermining positive values, should be criticized as blameworthy. Of course, often one cannot make such a clear distinction, there can be unintended consequences of introducing digital technologies and social media, and technologies are often highly ambiguous and contradictory, combining positive and negative functions and effects.

Moreover, societies and technologies evolve over time, so both normative standards and evaluative analyses will change as societies develop and new technologies appear and evolve. Yet we must oppose two forms of essentializing technology which deny its historical and social origins that a critical theory of technology should reject. An extremely common instrumentalist view understands technology as a neutral instrument that human beings use for a variety of purposes. Habermas, for instance, follows the German philosopher Arnold Gehlen in viewing technology and instrumental action as identical, as anthropological constants in which humans use technology to dominate nature (Habermas 1970, p. 87)

---

[4] On the various standpoints and strategies of critique of the Frankfurt School, see Kellner 1989a. Although the critical theorists are sometimes associated with a technophobic critique of technology as domination, in their best works they develop more dialectical perspectives; see Kellner 1989a.

by using in as a tool in which individuals carry out instrumental action to achieve a goal, like using a shovel to dig a hole, or a typewriter to produce a philosophical treatise. Yet there are different versions of this anthropological-essentialist position. In one extravagant and uncritical version of this position, technology is interpreted as an extension of the human being and technological environments are perceived as natural products of human evolution (McLuhan 1964). A less metaphysical version of the instrumentalist position simply posits technology as a neutral instrument used by humans for human purposes.

This latter position is held by social scientists who view technology as socially constructed, as dependent on specific social structures and cultural values, thus covering over the tremendous force and power of technology in the contemporary era.[5] Such social constructivist theories separate analysis of technology from theories of society and engage in empirical analysis of specific technologies, abandoning philosophy of technology which conceptualizes it as a key constituent of the contemporary world and attempts to articulate and critically engage its defining features and major effects. Likewise, dominant currents in social philosophy and the mainstream of academic philosophy also neglect philosophy of technology, displacing the problematic to a marginalized subdiscipline.[6]

There is, however, a more technophobic version of the essentialist view that perceives technology as intrinsically opposed to the human, which interprets instrumentality as a threat to human purposes, norms, and values. Some contemporary philosophical critiques of technology take this position and operate with highly dualistic and usually ontologized categorical distinctions between things such as technique and being (Heidegger), technical action and social interaction (Habermas), devices versus focal things and practices (Borgmann), and instrumentality and meaning (Simpson). In these theories, the former term is devalued as modes of technological domination and alienation, whereas the latter is valorized as the authentic sphere of human meaning and value. This mode of critique thus negatively ontologizes technology and excludes it a priori from the essential forms

---

[5] For a survey of social constructionist views of technology, see Philip Brey, "Social Constructivism for Philosophers of Technology: A Shopper's Guide" *Scholars Library*, Spring–Summer 1997 at https://scholar.lib.vt.edu/ejournals/SPT/v2n3n4/brey.html (accessed December 29, 2020). See also the page on "Social Construction of Technology" which notes that "SCOT developed like any normal scientific program: its agenda, central concepts, and even unit of analysis shifted in response to research findings and discussions among contributing scholars." From: *International Encyclopedia of the Social & Behavioral Sciences* (Second Edition), 2015 at https://www.sciencedirect.com/topics/social-sciences/social-constr uction-of-technology (accessed January 25, 2021).

[6] See the dual critique of academic philosophy of technology and social science positions in Feenberg 1991, 1995, 1999, and 2017.

of human being, as if technology were anti-human and opposed to human values and purposes. Such approaches separate technology from culture and society, and reify a notion of technical or instrumental action in which all action that involves technical imperatives follows a logic of things, of instrumentality, abstracted from human purposes and meaning. They therefore fail to note how technology itself is subject to human purposes, can be constructed or reconstructed in line with human projects and values, and can thus contribute to human development.[7]

Major currents in the philosophy of technology thus essentialize technology, decontextualize it, and abstract it from culture, social practices, and the construction of human projects, and thus fail to grasp its social and historical embeddness. Such essentialist and instrumentalist conceptions fail to perceive how technology itself changes, develops, and is socially constructed and reconstructed, viewing it as essentially instrumental, objectifying, and domineering. Moreover, instrumentalist views of technology as neutral are close in some ways to this essentialist view, although most philosophical essentialist discourse is negative, while some forms of instrumentalist discourse are positive, or merely descriptive. Such views, however, fail to articulate the extent to which specific societal biases, interests, and ideologies go into the very construction of technology and that therefore technology requires a historically specific mode of critique and reconstruction.

Both essentialist and instrumentalist conceptions of technology should thus be distinguished from a critical theory of technology that regards technology as socially constructed, embodying historically specific social biases and values, that criticizes distinctive technologies and their uses in concrete socio-historical contexts, that promotes the reconstruction and refunctioning of technology to serve positive values like democracy or human development, and that are ecologically sensitive. Technology can either be an instrument of domination and destruction, or creative and life-enhancing, depending on the technology in question, its specific uses in particular contexts, and the values and goals that are being pursued in particular situations. For example, broadcasting can be a tool of manipulative propaganda and narcotizing entertainment, or of education and genuine

---

[7] Marx, for example, in his conception of a humanized world, a world more fit more human beings, included industry and technology in such a schema. On Marx and technology, see Amy_Wendling, *Karl Marx on Technology and Alienation* Palgrave Macmillan; 2009 is available on-line at https://dl.uswr.ac.ir/bitstream/Hannan/130465/1/Amy_Wendling_Karl_Marx_ on_Technology_and_Alienation__2009.pdf {accessed on December 29, 2020). See also Mike Healy, "Alienation and information communications technology," *Semantic Scholar*, 2014, on-line at https://www.semanticscholar.org/paper/Alienation-and-information-communicatio nsHealy/20e6e4ea73d98e3e3e9e1958a4ddbf09bb78d601 (accessed December 29. 2020).

political debate. Critically analyzing reactionary television programs in the class-room is very different from viewing them home alone. And computers and digital technologies can be used either for programming nuclear weapons and government surveillance of citizens, or as vehicles of lively political discussion and educational research.

Yet it should also be noted that technologies are often highly ambiguous, that their positive and negative aspects are often interconnected, and that it is thus often extremely difficult to appraise and evaluate specific technologies, let alone technology in general. The ambiguity in part derives from the centrality of technology in human life, its deep embeddedness in every integral dimension of human life ranging from the economy, to the polity, to social and everyday life, and culture and human subjectivity itself. Indeed, as Andrew Feenberg argues (1999), the social constructivist view often fails to note the extent to which technology is deeply involved with what human beings are, and that humans are products of their technologies just as technologies are products of human beings in specific social situations. From this perspective, after centuries of using technologies, human beings are technical beings, technologies are extensions of human faculties which in turn come to shape human thought, behavior, and interaction. Technology is pivotally embedded in the human adventure from the start, and is thus bound up with the nature of the very beings that we are. For this reason, social constructivist conceptions of technology miss the depth and pathos of technology, its centrality in human experience, and the extent to which it influences the organization of human society and culture in all known historical periods.

Social constructivist views thus tend to have too narrow and instrumentalist a conception of technology and downplay its central importance in the construction of modernity and, for some, the transition into postmodernity (Baudrillard 1993; Jameson 1991; Best and Kellner 2001). A critical theory of technology, by contrast, develops what Feenberg (1999) calls a "substantive" theory of technology, that theorizes its centrality in contemporary society, without, however, falling into either technophobia or technophilia, as do most instrumentalist and essentialist theories of technology.

Yet in one sense, technology is socially constructed, specific societal biases and interests are encoded in technology, and the social relations in which technologies are produced and used will help determine their nature and uses. Hence, a critical theory of technology is concerned to articulate the potentials of specific technologies, to develop a substantive vision of the role of technologies in human life, and to project ways that technologies can serve human self-development, democratic values, and the creation of a more cooperative and ecologically viable social organization. A critical theory of technology in the workplace, for example, should articulate dialectical perspectives that can distinguish between technologies that further life-enhancing and fulfilling work and social relations, opposed to technologies that create a less creative, democratic, and more authoritarian social order, or produces products that are destructive of human beings and nature.

A critical theory of technology will critique the oppressive and authoritarian forms and uses of technology and sketch ways in which the restructuring and refunctioning of technology can promote progressive social change and the creation of the good life and the good society. Thus, a critical theory of technology is driven by a philosophical vision of normative conceptions of ethics, aesthetics, and politics, judging technology according to normative criteria, and regarding the construction and reconstruction of technology as fundamental to human experience. Overcoming one-sided conceptions of technology, a critical theory of technology recognizes in the mode of historicism the social constructedness of technology, but interprets it as fundamental to human life and history, and thus develops a substantive philosophy of technology adequate to its importance and centrality in human life.

Calling for dialectical normative appraisal of its positive and negative aspects is not to reject radical critiques of technology, or of specific technologies, out of hand, for often the critiques are valid and important. All technology has its biases, its built-in interests, and its predispositions to certain uses. Some technologies are inherently harmful and destructive such as nuclear weapons or nuclear energy which contain the potential for catastrophe devastation. Other technologies can be used for good or evil, depending on who is using them, how, and to what purposes. Television and film can be great instruments of education and enlightenment, or of manipulation and debasement. Computers and digital technologies can be used to promote progressive or regressive ideas, and emancipatory or oppressive social forces and interests.

It is a mistake, however, to dismiss technology per se as merely a mode of domination and oppression, though it may be so in many cases and threaten positive values. Technologies, like the computer, were initially used and developed by big government, corporations, and the military as centralized instruments of social control and power and were, with much justice, criticized in the 1960s for contributing to state and corporate institutional domination, the dehumanizing and disempowerment of humans, and the proliferation of destructive and life-threatening bureaucratic systems and weapons of mass destruction. Yet in the 1980s and 1990s, computers were recreated, made "personal," and are significantly different in their constitution and effects than their earlier incarnations (Turkle 1995).

A critical theory of technology thus develops a historically specific and normative critique of technology. It not only attacks life-negating, oppressive, and destructive aspects of technology, but valorizes empowering, democratizing, and ecologically positive forms and uses. Crucially, it attempts to discover and invent ways that technology can serve the interests of human emancipation and well-being, while aspiring to delineate ways that technology can be used to create a better world. A critical theory of technology may deploy strategies of immanent critique, taking existing norms and values as the standpoint of critique, but may wish to develop stronger normative conceptions of democracy, freedom, and the good society than notions currently in play and should carry out critiques of restricted and ideological notions of democracy, empowerment, and freedom being promoted by the avatars of new computer and multimedia technologies.

In *Dialectic of Enlightenment*, however, Horkheimer and Adorno (1972) argued that Enlightenment values had turned into their opposite. For Adorno and Horkheimer, rationality, democracy, culture, and other bourgeois ideals had shifted from serving as a form of emancipation and progress to that of oppression and domination. In their view, science, technology, industry, and instrumental rationality had created a machinery of war, death camps, and nuclear annihilation; bourgeois democracy voted in fascist regimes; and culture, supposed to be emancipatory, was built-into totalitarian systems of social control and oppression. Henceforth, Adorno and Horkheimer attempted to develop innovative strategies of critique and opposition to the emergent forms of technological domination and power. The first generation Frankfurt School, however, never was able to create adequate theories of democracy, a task taken up by Habermas and his followers (1970, 1991), and other of us in the third generation Frankfurt School (see Kellner 1989).

A critical theory of technology may also deploy strategies of immanent critique, taking existing norms and values as the standpoint of critique. Yet emancipatory theory may wish to develop stronger conceptions of democracy, freedom, and the good society than notions currently in play and carry out critiques of restricted and ideological notions of democracy, empowerment, and freedom being promoted by the avatars of computer and multimedia technologies. This, of course, is an immense task and my present reflections in this chapter can only contribute to making a few observations on developing some criteria to indicate ways that ICTs can be said either to produce forms of alienation or contribute to disalienation and overcoming social forms and activities often labelled as "alienation."

Alienation and Technology

As my discussion of the technophobic discourses indicates, many critics have argued that ICTs have distinctive alienating effects and are creating novel forms of alienation. Others maintain that the very notion of alienation is bound up with modernity and with essentialist forms of theory and that the very discourse of alienation should be discarded. The argument is that theories of alienation assume non-alienated forms of possibilities of human being and thus assume a human essence from which one is alienated. I should confess here that I never bought the fashionable poststructuralist/postmodern argument that alienation is connected with essentialism or mystification per se, although it can be in some discourses. Marx's analysis of alienated labor in the *Economic-Philosophical Manuscripts of 1844* (2007) always seemed perfectly cogently for me in his argument that labor under capitalism was fragmented, repetitive, controlled by supervisors and a labor apparatus, and coerced, while human beings were many-sided beings with multiple potentials, needs, and possibilities who needed to engage in a wealth of free activity to attain fulfillment and self-development. Working two summers in factories in the mid-1960s to earn money for college convinced me that Marx was right and I determined I would neither work in a factory or a corporate office like my father where one was forced to go to work every day for much of the waking day and was subject to a coercive labor apparatus and system of domination.

Yet, for the concept of alienation to make sense one must specify—as with Marx—what one is being alienated from, how this is happening, what, if anything, is wrong with this, and how one might overcome what is described as alienation. While idealist conceptions of alienation assume something like an invariable human essence from which one is alienated, Marx arguably develops a more concrete concept of alienation in his account of the alienation of labor that is much less metaphysical than idealist-humanist concepts. Marx presents in his early writings a normative concept of the human being as many-sided, creative, and at once individual and social (1975). The young Marx began seriously studying economics in Paris in 1843–1844, and after an encounter with Engels in Paris in 1844, he intensified his economic studies. Convinced that the rise of capitalism was the key to modern society and history, Marx sketched out his analysis of capitalism in his *Economic and Philosophical Manuscripts of 1844*, which present his initial perspectives on modern societies in terms of a vision of the alienation of labor under capitalism and its projected emancipation under socialism (1975, p. 231 ff).[8]

For Marx, labor under capitalism was alienated because one was estranged from one's potentials as a many-sided being, as people were forced to engage in specialized and one-sided labor. Moreover, individuals were alienated from their human potentialities of creativity, self-realization through labor, and the development of one's full range of human potentials since labor under capitalism was external, directed from the outside, coercive, and necessary, as one was forced to work to survive. Further, work was fragmented and routinized and not free and inventive. Finally, individuals under capitalism were alienated from other people in that labor was competitive and not cooperative, society was divided into masters and slaves, and thus a system of domination and exploitation prevented human, social and self-realizing labor.

It is important to note that for Marx alienated labor was largely a function of capitalist social relations and not technology and could be overcome with the transition from capitalism to socialism where workers would own and control the means of production; organize labor cooperatively and democratically; and engage in many-sided activity rather than the one-sided activity of capitalism in a division of labor, celebrated by Adam Smith, where one engaged in the same repetitive task or function day-after-day.[9] In his most radical vision of an emancipated society, Marx envisaged a realm of freedom made possible by the developments of modern technology and industry. In the *Grundrisse*, he sketched a theory of a possible rupture between capitalist and post-capitalist societies that would be as radical as those between pre-capitalist and capitalist ones. On Marx's account, capital generates factories, machine production, and eventually an automated system of machinery (1978, p. 278 ff.).

In his famous analysis of automation, Marx sketches out an audacious vision of the development of a fully automated system of production under capitalism that brings it to an end and produces the basis for an entirely different social system. In Marx's vision, the "accumulation of knowledge and of skill, of the

---

[8] These notebooks were never published during Marx's life and their printing in 1932 caused a sensation, presenting a vigorous philosophical and humanist Marx quite different from the economic theorist and "scientific socialist" championed by the official Marxian working class movements. On the importance of the Paris *Economic and Philosophical Manuscripts of 1844* for the interpretation of Marxism, see Marcuse 1972 [1932], p. 3–48.

[9] On Smith, Marx, and the division of labor and alienation under capitalism, see Douglas Kellner, "Capitalism and Human Nature in Adam Smith and Karl Marx," in Jesse Schwartz, ed, *The Subtle Anatomy of Capitalism* (Santa Monica, Cal.: Goodyear Publishing Company, 1977), p. 66–86.

general productive forces of the social brain" are absorbed into capital and produce machinery which "develops with the accumulation of society's science, of
the productive force generally" (1978, p. 280). As machinery and automation
develops, the worker becomes more and more superfluous, in contrast to the growing power of machines and big industry. On the other hand, machines free the
worker from arduous and backbreaking labor. In this situation: "Labor no longer
appears so much to be included within the production process; rather, the human
being comes to relate more as watchman and regulator to the production process
itself.... He steps to the side of the production process instead of being its chief
actor" (1978, p. 284).

Marx saw that technology both provided capital with powerful forces of domination, exploitation, and the realization of profit, and workers with instruments
that could be used against capital and to organize labor in accord with interests
of democracy, justice, and human self-development. Technology for Marx was
thus complex and his analysis of alienation always focused on those features that
could be overcome and eliminated in a non-alienated society. While Marx offers
a precise and concrete account of the alienation of labor and its overcoming, in
the discussions of emergent technologies and alienation, such careful distinctions
or concreteness are rarely attained in most discussions of alienation and technology. In the claims, for example, that in cyberspace, we are alienated from other
people, our bodies, nature, and "real life" (RL), or that we are lost in hyper- or
virtual reality (VR) and disconnected from the real, there is rarely any detailed
analysis or specification of what one is alienated from, why this is bad, and how
such alienation should be overcome. Or, the accounts given are not particularly
persuasive.

Take the claim that we are alienated from "real life," other people, or the
body in cyberspace. To begin, I would resist the extreme binary distinctions between reality and hyperreality, as if they were two distinct zones, as if, as Sherry
Turkle seems to imply (1995), there is a great divide between RL and VL. For
Turkle and other digerati, virtual life is positive, enabling individuals to experiment with new identities, to enter into novel types of social relations, and to form
virtual communities,[10] while for technophobes like Borgmann or Stoll, cyberlife
is inferior to real life and cyberspace is depicted as a realm of alienation. Both

---

[10] Turkle (2011 and 2014) later had second thoughts about her celebration of the technoculture
and published critical reflections on decades of observations of her students uses of technology
and the evolving technoculture, which I discuss below.

extremes, however create an illicit distinction between everyday life and cyber-life, with technophiles celebrating the virtual life as liberating and exciting while technophobes claim it is derivative, secondary, banal, harmful, and alienating.

I was once visiting a family in Silicon Valley whose teen-age son was immersed in a game that was more real to him than his family and their guests. When I tried to get him to explain the game, he immersed himself deeper in the virtual world, muttering comments that indicated he wanted to focus on his game, so I let him continue his game. Such immersion in virtual worlds could be a pleasurable diversion, or escape from boring guests or home-work, or it could become an obsession that destroys the players' life as in famous stories, of various Korean, Japanese, and American gamers playing without break day after day and finally dying,[11] or kids failing school and losing friends because of their obsession with games or VL.

Yet I would argue against either technophobia or technophilia and would see virtual life as a dimension of, as part of real life, that may or may not be empowering or alienating depending on its nature, effects, and contexts. Indeed, for the concept of alienation to have force, the discourse should be elaborated and specified. The term "alienation" initially derived from the Roman alienatio which signified to "transfer," a sense included in the early economic sense of "to alienate" as to sell or transfer ownership. Analogically, this conception could be operative in discussions of technology, alienation, and labor, or other forms of activity, in which technology performs activities once performed by humans. Yet, as in Marx's analysis in the *Grundrisse*, it might be a positive development to let technology execute socially necessary labor that machines could accomplish, thus freeing creative life-activity for individuals and social groups.

However, Marxian, existentialist, and other philosophically-grounded conceptions of alienation also connotated separation, as when in alienated activity one is separated from control over the means of activity as in alienated labor under capitalism, or for existentialists one was subject to control by the Other in alienated labor—which could be applied to our obsessive gamer in discussions of computer, video, or VR games. The conception of alienated labor also often signified surrender and relinquishment, as when one gives up control or power to alienating forces; and thus the strong concept of alienation often involves a loss of something fundamental and important to human life. Hence, with Marx, one's fundamental life-activity, one's very labor power, was surrendered to the capitalist. On Marx's view, in labor under capitalism, workers lost their potential for free and creative

---

[11] For documentation, see "10 Gamers Who Tragically Died Playing Videogames—YouTube" at https://www.youtube.com/watch/El7kwIpjrtQ (accessed January 14, 2021).

activity, and labor was thus subjugated, exploited, and alienated from the potential for free self-valorizing and creative activity. Moreover, for Marx, individuals under capitalism were trapped in a mode of life that was organized around labor and required most of an individual's life to be centered on labor instead of many-sided creative activities that were in Marx's view necessary for full self-realization and development.

I would argue that such a strong sense of alienation is not necessarily operative in activities with technologies in cyberspace, that such activity may be empowering, as one finds valuable research material, engages in relatively non-constrained dialogue and discussion, or enjoys harmless play, banter, and surfing for information. Furthermore, it might be positive to be "alienated" in the weaker sociological sense from other people and one's environment.[12] For instance, a teenage kid spending 10 hours a day in cyberspace might be better off than participating in banal or destructive social activities in one's "real life," such as gangs, drugs, or just wasting time. Someone who has difficulty in connecting socially with those around oneself may be able to make contact with people in cyberspace who share their interests or situation, which might help with an individual's self-esteem or lead to better contacts and social relations in the real world. For example, there are accounts of gay teens who found support on-line that helped them accept their homosexuality,[13] and many other people who had self-esteem problems because

---

[12] When the concept of alienation began circulating in the 1950s and 1960s in Marxist, existentialist, religious, and sociological discourses, the implication was always that alienation itself is bad, that it constitutes a danger to human beings that should be overcome in the transformation to a non-alienating form of life. At the time, against this view, I believed that it was good to be alienated in some senses from the dominant society of the period, so that in the '50s and '60s when the discourse of alienation began circulating in sociological, philosophical, and even public circles, I always thought it was positive to be alienated from an other-directed, conformist, and repressive society. Consequently, when I was doing my doctoral dissertation in philosophy in the 1960s and early 1970s I argued that in some senses alienation and authenticity were equivalent, that you couldn't be an authentic self, in contemporary U.S. and advanced capitalist societies, without being alienated from it in some ways, and thus alienation from the dominant society was a necessary step in creating a new life and society—an argument I made in my 1973 Ph.D. dissertation *Heidegger's Concept of Authenticity* which I successfully defended in the Columbia University Philosophy department and got my Ph.D. and an academic career. I agree with Marx and Heidegger, however, that in other senses, however, alienation signifies a harmful condition that should be overcome, thus a discourse of alienation must specify whether the condition described is positive or negative and if the latter how it can be superseded.

[13] For documentation, see "*A Community for LGBTQ + Teens*" at https://www.qchatspace. org/?gclid=Cj0KCQiA9P__BRC0ARIsAEZ6iriqd666bvt6UBfFflQ3xiNe9r89lNObETim vOreUcXZGjYNpmlNaZ8aAvo6EALw_wcB (accessed January 14, 2021).

of a perceived difference or oddity found support in communities of like-minded people.

I do not, however, want to romanticize the net and computer culture. Computerized labor can be highly alienating both for those forced to produce the technologies in the system of production and for those laboring in some workplaces that use technologies as modes of surveillance and domination, or who are tied to their desks and digitized work-stations for hour upon end engaged in repetitive activity. Obviously, in certain relations of production activity in cyberspace one is not free or creative, but is controlled by the workplace and corporations or the state. Moreover, no doubt, many youths are seriously alienated from school, peer social groups, and the world around them and became lost in cyberspace playing games hour after hour. Obviously, significant amounts of concrete bodily interaction with other people is necessary to creative self-development and fulfillment, and it might be harmful to individual development to spend too much time in cyberspace and not relating to other people. It is, however, still debated whether there is a correlation between immoderate time in cyberspace and poor social relations within one's immediate environment, or negative psychological consequences for the individual. Although a study was released in August 1998 that claimed to make a correlation between increased time spent in cyberspace on the Internet and increased feelings of depression and loneliness, this study was soon subjected to sharp critique and its findings were put into question.[14] Yet later findings did document many studies in which there was a correlation between excessive on-line activity and depressive, and suicidal behavior.[15]

Likewise, claims that we are alienated from our body in cyberspace, strike me as unconvincing, as my body includes my hands, my eyes, and other senses that are fully active in computer-mediated activity. The assertion that the body does not count in cyberspace, that it is devalued and that we are alienated from it, perhaps derives from William Gibson's distinction between "meatlife" and cyberlife

---

[14] See Kraut, et al. 1998, and the discussion of criticisms of the study in Denise Caruso, "Technology; Critics are picking apart a professor's study that linked Internet use to loneliness and depression." *The New York Times*, September 14, 1998 at https://www.nytimes.com/1998/09/14/business/technology-critics-are-picking-apart-professor-s-study-that-linked-internet-use.html (accessed January 25, 2021).

[15] Jean M. Twenge, Thomas E. Joiner, and Megan L. Rogers,[2]
"Increases in Depressive Symptoms, Suicide-Related Outcomes, and Suicide Rates Among U.S. Adolescents After 2010 and Links to Increased New Media Screen Time." *Sage Journals*, First Published November 14, 2017 at https://doi.org/10.1177/216770261772 3376 (accessed December 29, 2020); this study contains a large bibliography of articles documenting various studies of obsessive media time and on-line behaviors, as well as depression and suicides in youth.

and his cybercowboys' view that the body is just "meat" and that real action is out there in cyberspace conceived as a "consensual hallucination" or mindspace (1984). Yet while many of Gibson's insights are prescient, his denigration of the body as "meat" is problematic. As I write these reflections, or participate in cyberspace discussions, e-mail, or surfing the web, my fingers are rapidly pounding the keys, my eyes, and my body are active and participatory in the experience; Websites increasingly contain images, graphics, and sounds and thus there is now an aesthetic dimension to the Internet experience.

Yet there are studies and analyses that suggest that there are growing forces of alienation among some users of digital technology and social media. While earlier Sherry Turkle promoted computers and digital technologies in books like *The Second Self* (1984) which I note above, in later books she became more critical. In *Alone Together* (2011), Turkle explores some of the ways in which previous social forms of communication and social interaction are being degraded through constant exposure to computer screens and digital culture, leading to increased alienation between people. In her following book *Reclaiming Conversation: The Power of Talk in a Digital Age* (2015), Turkle advocates social relations fostering face-to-face interaction and engaged conversations.

I'm going to pass over cybersex in this discussion since I'm inadequately experienced in this domain, although I have skimmed some articles and books on the topic, but from what one reads people have bodily sexual experiences in computer-mediated interaction, as they do in phone sex—although people have lost their jobs because their computer sex activity on apps like Zoom were recorded! Of course, one could argue that on-line sexuality is an alienation of sexuality, a form of alienated sex, and if one is against masturbation and has a certain normative and probably heterosexual conception of what sex should be, the argument for the alienation of the body in cybersex probably has some force, as does Frank Sinatra's bon mot "whatever gets you through the night." Yet without questionable assumptions about "normal" sexuality, the argument linking cybersex and alienation is none too compelling. On the other hand, clearly predatory pedophilia is highly objectionable and there are by now many cases and much discussion of the dangers of sexual predators in cyberspace. There is a general consensus that such activity is wrong and laws and law enforcement agencies are prosecuting such behavior.

In any case, and this is the philosophical point, to make strong claims about alienation and technology, one must specify exactly how one is being alienated, what is wrong with this, and what should be done about it, rather than condemning cyberspace or ICTs per se as a domain of alienation. On the whole, therefore, in regard to assertions concerning alienation in cyberspace, I would hold that arguments that in cyberspace one is alienated from other people, one's body, nature, and the real world all have problematic metaphysical assumptions, or are not adequately specified, and thus are not convincing.

Moreover, although it seems plausible that alienation of sorts can occur if one is too heavily engaged in cyberspace, this is also true of individuals excessively engaged in watching television, reading books, playing sports, or any obsessive activity. And while it is true that computers and digital technologies are major fetish objects of our era, and highly addictive, they are also potentially democratizing tools that can be used to empower individuals and groups traditionally subordinate and oppressed. Further, they can promote positive values such as democracy, justice, and equality. For example, while digital technologies might be forces of further alienation and inequalities in the political sphere, they can also be empowering, democratizing and thus disalienating, as I will argue in following chapters and have argued before (see Kellner 1995, 1997, 1998; Best and Kellner 2001). Indeed, computers are a potentially democratic and democratizing technology. While broadcast communication tends to be one-way and unidirectional,[16] computer communication is bi- or omni-directional. Where TV-watching is often passive, computer involvement can be interactive and participatory. Individuals can use computers to send e-mail to communicate with other individuals, or can directly communicate via sites like Skype or Zoom which visually link individuals with each other in interactive networks.

In an earlier era at the beginning of computer culture, modems could tap into community bulletin boards, Web-sites, computer conference sites, or chat rooms, that made possible alternative forms of interactive public information and communication, while today social media sites can provide these functions, enabling individuals to engage in public discussions, post articles or art works, engage in live on-line debates, or produce forums for discussion, or sites for organizing social movements.

---

[16] As technologies and media institutions and practices evolve, possibilities for feedback and critique of media programming is possible. For example, call-in and talk radio and television, as well as electronic town meetings, can involve two-way communication and participatory democratic discussions. In the social media age, fans watched shows live and commented on and critiqued the programming. Hence, theorists like Baudrillard who argue against television and the media on the grounds that they promote only one-way, top-down communication essentialize the media and freeze the then-current forms of the media into fixed configurations, covering over the fact that media evolve and can be reconstructed, refunctioned, and transformed into more democratic modes of communication.

Thus, the Internet makes possible becoming a producer as well as consumer, an active participant in the production of culture as well as a passive receiver of media and digital images and messages. Once one gains a minimal degree of computer literacy, one can post messages to Web-sites, participate in chat-rooms or list-serve discussions, and create your own Web-sites and blogs and become active in social media and the netroots.[17] This technology makes possible materialization on a vast scale of Walter Benjamin's dream of "The Artist as Producer,"[18] where artists, activists, and others can participate in bringing new cultural forms and intervening in the reconstruction of a mode of cultural production, as well as sending out alternative messages and engaging in political debate, organization, and struggle. It also realizes Bertolt Brecht's vision in his essay on radio theory that anticipated the Internet, which has helped realize his call for reconstructing the apparatus of broadcasting from one-way transmission to a more interactive form of two-way, or multiple communication, a form first realized in CB radio and then electronically-mediated computer interaction.[19]

Digital technologies and social media also make possible the revitalization of democracies that have been dangerously atrophying in an era of spectacle politics dominated by television and the politics of image and spin, which I shall discuss in the next chapter. Democracy involves democratic participation and debate as well as voting. In the Big Media Age, most people were kept out of political discussion and were rendered by broadcast technologies passive consumers of infotainment. Access to mainstream broadcasting media has been controlled by big corporations and only a limited range of voices and views have been allowed to circulate. In the Internet Age, by contrast, everyone with access to a computer, a service provider, and social media sites and can thus participate in discussion and debate, facilitating access to evolving public spheres to large numbers of individuals and groups kept out of the democratic dialogue during the Big Media Age. Consequently, a technopolitics can unfold in the public spheres of cyberspace and provide a supplement, though not a replacement, for intervening in face-to-face public debate and discussion. For instance, many social media and Web-sites have a political debate conference where individuals can type in their opinions and other individuals can read them and if they wish respond. Other sites have live real-time chat rooms where people can meet and interact. These

---

[17] I will discuss netroots, digital activism, and using social media for progressive social change in more detail in this and in following chapters.

[18] For a recent biography of Walter Benjamin that gives an excellent detailed account of his life and works, see Eiland and Jennings 2014. On Benjamin's media theory, see Kang 2016.

[19] See Brecht's radio theory collected in Silberman 2000, p. 41 ff.

forms of cyberdemocracy constitute a innovative modes of public dialogue and interaction, and take place in new public spheres, thus expanding our conception of democracy (see Chap. 5 and 6 below).

Social media have expanded rapidly in recent years and encompass interactive digital technologies that facilitate the creation or sharing of information, ideas, career interests, and other forms of expression via virtual communities and networks. They involve interactive Web 2.0 Internet-based applications, and encompass text posts or comments, digital photos or videos, and are structured by ever evolving apps and social media sites like Facebook, TikTok, WeChat, Instagram, Weibo, Twitter, Tumblr, Baidu Tieba and LinkedIn. Other popular social media services include YouTube, Quora, Telegram, WhatsApp, LINE, Snapchat, Pinterest, Viber, Reddit, Discord, VK, Microsoft Teams, and more.

Since social media encompass all data generated through one's online interactions, critics worry that they are becoming forms of surveillance in which giant corporations or governments can collect data used to track and manipulate consumers and citizens. While some social media sites are built up by users like Wikipedia and a variety of Wikis, others are maintained by corporations like Google, Amazon, and Facebook which have amassed over the years tremendous amounts of data and user profiles (see Galloway 2017). Social media can become effective marketing tools for megacorporations and advertising, but are also used by entrepreneurs, non-profit organizations, advocacy groups, and netroots advocacy groups, as well as individuals.

Yet social media can also be used by political parties, politicians, and governments, and thus have become a major force of political struggle and activism in recent years, as well as major forces of manipulation and disinformation by politicians and government. State and corporate surveillance of citizens has been one of the most negative features of the Internet and it intensified in the 2000s as states and corporations created powerful instruments to track and record information from its citizens (Lyon 1988, 2018; Zuboff 2020).

Widespread government surveillance was dramatically revealed by a highly proficient member of the U.S. national security apparatus Edward Snowden in 2013 with his release of NSA documents that showed that the U.S. government was capable of tracking all of its citizens' telephone, computer, digital device use, and other forms of electronic communications, and that major corporations like Google, Yahoo, Facebook, and others were complicit—or were surreptitiously breached by government programs and spies (Snowden 2019).

Snowden, a 29-year old employee of the National Security Agency (NSA), leaked in 2013 a vast archive of data when he was a Central Intelligence Agency

(CIA) employee and subcontractor for a U.S. security firm. Like Daniel Ellsberg with the Pentagon Papers, which revealed secrets of U.S. involvement in the Vietnam war, Snowden collected and leaked highly classified information from the most secret government surveillance programs. He revealed numerous global surveillance programs, many run by the NSA and with the cooperation of telecommunication companies and European governments, and prompted an intense debate about national security and individual privacy, still raging today.

Snowden says he gradually became disillusioned with the government surveillance programs with which he was involved, and that he tried to raise his ethical concerns through internal channels but was ignored. On May 20, 2013, Snowden flew to Hong Kong after leaving his job at an NSA facility in Hawaii, and in early June he released thousands of classified NSA documents to journalists Glenn Greenwald, Laura Poitras, and Ewen MacAskill. Filmmaker Laura Poitras taped interviews with Snowden which explained how he became grieved that the Internet was being destroyed, in his view, by government surveillance of all communications passing over it and how telephone companies, Internet companies like Google, Facebook, Yahoo, Apple, and other major companies were involved in this surveillance—while these companies denied complicity, it was later revealed that lax security protections allowed the U.S. government and others to mine and collect individual's data from these High-Tech Megacorporations (see Greenwald 2014; Harding 2014).

Snowden's revelations in 2013 became a sensation after stories based on his files revealing jaw-dropping programs of surveillance appeared in *The Guardian*, *The Washington Post*, and other publications. Soon after, Poitras' documentary film *Citizen Four* (2014) showed how she, and journalists Glenn Greenwood and Ewen MacAskill from the *Guardian* newspaper, met secretly with Snowden in his Hong Kong hotel room, where he shared his files of data with them and told them his story. Subsequently, *The Guardian* and other papers published the most sensational revelations concerning the nature and scope of government surveillance, information also gleaned from other sources by *Washington Post* reporters. Reporters Greenwald (2014) and Luke Harding (2014) of the *Guardian* published books on the spy-thriller nature of Snowden's collection and dissemination of the data, taking it via computers to Hong Kong, and arranging with reporters and filmmaker Poitras to disseminate the information, with Snowden eventually publishing his own memoir of the story (2019).

The revelations generated intense discussion within the Obama administration quickly passing laws protecting communications privacy for citizens, but also charged Snowden of two counts of violating the Espionage Act of 1917 and theft of government property, while revoking his passport. Following these actions, a vigorous debate raged over whether Snowden was a hero or villain —— debates continuing to this day. Snowden himself remains in exile, stuck for months in a Russian airport en route to hopeful exile in Latin America, but then released to live freely in Russia, where he remains today, after he was granted permanent residency.

As noted, Google, Amazon, Microsoft, and various social media companies were used by the government to mine data on citizens and while the companies, sometimes vehemently, denied being willingly complicit, there is no doubt that they mined data from their users to sell to corporations who might wish to market products to them based on their interests, and that this was their very business model (Zuboff 2020). Indeed, Google and other "free" social media sites had decided that selling the eyeballs and clicks that accessed their sites to advertisers would be a good method to accumulate massive profits from their product, produced by their users!, and thus social media are part of a major government-corporate surveillance apparatus that has built the most massive surveillance machine in history rendering privacy a quaint eccentricity of an earlier era.[20]

Appropriate and Sustainable Technology

Critics of digital technologies, technopolitics, and social media repeat incessantly that it is by and large young, white, middle or upper class males who are the dominant players in the cyberspaces of the present, and while this was true at one time, statistics and surveys indicate that many more women, people of color,

---

[20] Social media companies also sold their data to political interests as I discuss in Chap. 6 where I document how Facebook and other social media companies sold their data to the Trump campaign, helped them with the Russians to win the election for Trump, and, in general, how various political interests use data mined from social media to direct political ads to individuals based on their user profiles. Indeed, while the Russians used this data to help elect the nefarious Trump, he has currently undergone his second impeachment as I write this in the dark days of early 2021 just after Trump's thug army of Storm Troopers invaded and desecrated the Halls of Congress while terrorizing its members who were forced into lockdown as the marauders trashed offices, broke windows, and beat up guards in a melee in which at least five people were killed, helping to bring the Trump era to its not unexpected implosion, and leading Trump to be permanently banned from Twitter, Facebook, and other social media companies which had fed his insatiable narcissism, mobilized his tribe of cultists, and unleashed the forces of darkness with a twitch of his Twitter finger and the widespread retweeting by his Legions of Darkness, many of which are now fugitives from the law and who face decades in prison, while Trump was impeached by Congress on January 13, 2021.

seniors, and other minority categories are becoming increasingly active, and that the digital divide, while still an important issue has become quite different by 2021.[21] Moreover, it appears that computers are becoming part of the standard household consumer package in the overdeveloped world and will perhaps be as common as television sets in the near future, and certainly more important for work, social life, and education than the TV set. In addition, there are plans afoot to wire the entire world with satellites that would make the Internet and communication revolution accessible to people who do not now even have telephones, televisions, or even electricity.[22]

---

[21] The "digital divide" emerged early on in the 1990s as the buzzword for perceived divisions between information technology have and have nots in the U.S. economy and society. A U.S. Department of Commerce report released in July 1999 claimed that the digital divide in relation to race was dramatically escalating and the Clinton administration and media picked up on this theme (See the report "Americans in the Information Age: Falling Through the Net" at https://www.ntia.doc.gov/ntiahome/digitaldivide/ (accessed January 7, 1999). A critique of the data involved in the report emerged, claiming that it was outdated; more recent studies by Stanford University, Cheskin Research, ACNielson, and the Forester Institute claim that education and class are more significant factors than race in constructing the divide (see http:cyberatlas.internet.com/big-picture/demographics [accessed January 2, 2021] for a collection of reports and statistics on the digital divide). As I revise this chapter in Winter 2020–2021, after many schools and Universities have been put on lockdown, including mine, because of the COVID-19 pandemic, and classes are being offered on-line, many students are deprived of access because they do not own computers or have digital access at home, and thus the digital divide remains an issue as important and burning than ever. Indeed, it continues to be clear that there is a gaping division between information technology haves and have nots, that this is a major challenge to developing an egalitarian and democratic society, and that something needs to be done about the problem. My contribution involves the argument that empowering the have nots requires the dissemination of new critical digital and media literacies (see Kellner-Share 2019 and Chap. 10 below) and that access alone is not enough to empower groups and individuals previously excluded from economic opportunities and socio-political participation. Wikipedia has an increasing comprehensive page on the digital divide and another on the digital divide between various countries; see https://en.wikipedia. org/wiki/Digital_divide (accessed on November 30, 2020).

[22] It was announced in April 1997 that Boeing Aircraft joined Bill Gates in investing in a satellite communications company, Teledesic, which plans to send up 288 small low-orbit satellites to cover most of the Americas and then the world in 2002 that could give up to 20 million people satellite Internet access at a given moment. See *USA Today*, April 30, 1997; in May 1998, Motorola joined the "Internet in the Sky" Project, scrapping its own $12.9 billion plan to build a satellite network capable of delivering highspeed data communications anywhere on the planet and instead joined the Teledesic project, pushing aside Boeing to become Teledesic's prime contractor (*New York Times*, May 22, 1998). An "Internet-in-the-Sky" would make possible access to digital technologies for groups and regions that did not even have telephones, thus expanding the potential for democratic and progressive uses

Theorists ranging from Lewis Mumford, Herbert Marcuse and Murray Book-chin to Ivan Illich and E.M. Schumacher have called for appropriate and sustainable technology. The UN and World Economic Forum have used the criteria of "sustainability" to evaluate whether certain policies or technologies serve the interests of developing countries.[23] Moreover, theorists of an alternative globalization, or globalization from below, have also called for technologies that serve people's basic needs, protect the environment, empower individuals and groups to participate more fully in labor and social activity, and that are self-valorizing rather than just enhancing capital and dominant such groups.

Moreover, there are dangers for developing countries in adopting new technologies too fast or without adequate preparation and protection. For instance, introducing industrial technology can notoriously damage the environment and using technology to replace or deskill workers can harm the working class, just as replacing traditional cultures with globalized and "modern" ones can destroy traditional practices and cultures. Thus in some cases a "precautionary principle" should be adopted that carefully evaluates the effects and consequences of introducing new technologies before they are implemented.[24]

---

of information and communications technologies; as of the present, however, such plans have failed to materialize and some are skeptical that they will, while others see wireless and satellite networks as the next stage of development. For current discussion, see the proceedings of "10th EAI International Conference, WiSATS 2019, Harbin, China, January 12–13, 2019, Proceedings, Part I," Springer Links at https://link.springer.com/book/10.1007/978-3-030-19153-5 (accessed December 12, 2020).

[23] While the term "sustainability" goes back to post-World War Two attempts to develop technologies appropriate to human beings and the quality of the environment, the term "sustainable development" was adopted in the Agenda 21 program of the United Nations unveiled at the 1992 Earth Summit. Critics, however, saw this concept as promoting excessive development and sought to define sustainability as "a means of configuring civilization and human activity so that society and its members are able to meet their needs and express their greatest potential in the present while preserving biodiversity and natural ecosystems, and planning and acting for the ability to maintain these ideals indefinitely." See https://www.wikipedia.org/wiki/Sustainability (accessed December 13, 2020).

[24] The precautionary principle was introduced by environmentalists in the 1980s, "and is reflected in the 1992 Rio Declaration on Environment and Development (signed at the United Nations Framework Convention on Climate Change)." The principle indicates that "even if there is scientific uncertainty regarding a risk and its consequences, preventative measures may be justified. This principle is often invoked when the consequences are considered great enough to require expensive amelioration, even when the risks are considered low." In practice, it involves sorting out costs and benefits of introducing new technologies and deploying preventive measures to control harmful effects and consequences. See https://www.wikipedia.org/wiki/Precautionary_Principle (accessed November 30, 2020).

Bill Joy (2000) has made a similar argument for developed countries that are considering certain forms of cloning and genetic engineering, biotechnology, or nanotechnology before these technologies are fully understood or their effects can be understood and charted. Just as scientists have called for voluntary relinquishment and restricting certain technologies like human cloning, so too should developing countries be careful in their promotion and widespread adoption of information and computer technologies (ICTs).

Yet many developing countries and progressive groups and individuals within them have used ICTs in emancipatory and positive ways. There are by now copious examples of how the Internet and cyberdemocracy have been used in oppositional political movements.[25] A large number of insurgent intellectuals are already making use of these technologies and public spheres in their political projects. The peasants and guerilla armies struggling in Chiapas, Mexico from the beginning used computer databases, guerrilla radio, and other forms of media to circulate their struggles and ideas. Every manifesto, text, and bulletin produced by the Zapatista Army of National Liberation who occupied land in the southern Mexican state of Chiapas in 1994 was immediately circulated through the world via computer networks.[26] In January 1995, the Mexican government moved against the movement and computer networks were used to inform and mobilize individuals and groups throughout the world to support the Zapatistas struggles against repressive Mexican government action. There were many demonstrations in support of the rebels throughout the world, prominent journalists, human rights observers, and delegations traveled to Chiapas in solidarity and to report on the uprising, and the Mexican and U.S. governments were bombarded with messages arguing for negotiations rather than repression. The Mexican government accordingly backed off their repression of the insurgents and over succeeding years, they have continued to negotiate with them, and there not been the type of repression usually seen in regard to oppositional movements.

ICTs have also famously been used in anti-corporate struggles against global megacorporations such as Nike and McDonald's, have been used by oppositional social movements ranging from gays and lesbians to environmentalists, and have

---

[25] See Kellner 1997; Best and Kellner 2001; Downing 2001; Couldry and Curran 2003; Elias G. Carayannis and Campbell, editors, 2014, and, below, Chap. 4–7 and 10.

[26] See Cleaver 1994, the documents collected in Zapatistas 1994, and Castells 1997. The Zapatista's continue to exist and thrive in the present, see Hilary Klein, "A Spark of Hope: The Ongoing Lessons of the Zapatista Revolution 25 Years On. What are the lessons of the EZLN's revolutionary struggle for Indigenous autonomy, a quarter-century after declaring war on Mexico and global capitalism?," *NACLA*, January 18, 2019 at https://nacla.org/news/2019/01/18/spark-hope-ongoing-lessons-zapatista-revolution-25-years

helped generate and sustain the anti-corporate globalization movements and peace movements (see note 20). And in the Trump era, groups like Black Lives Matter, the Dreamers, the #MeToo movement, and the anti-Trump resistance used social media to organizes movements against racism, sexism, homophobia, fascism, and to protect the environment in an era of intense climate crisis (see Kellner and Satchel 2020 and Chap. 4 below).

Yet, obviously, reactionary groups can and have used the Internet to promote their political agendas as well. Rightwing extremist groups and authoritarian leaders throughout the world are using Twitter and social media to disseminate fake news and disinformation, that is emerging as a major problem of our times. Social media promote these authoritarian views and the ultraright is extremely active on many computer forums, as well as their radio programs and stations, public access television programs, fax campaigns, video and even rock music productions have been manipulated by demagogic politicians or leaders, as we have seen with the resurgence of global authoritarian populism and rightwing extremism during the Trump era.

The Internet is thus a contested terrain, used by Left, Right, and Center to promote their own agendas and interests. The political battles of the future may well be fought in the streets, factories, parliaments, and other sites of past struggle, but all political struggle is already mediated by media, computer, and information technologies and will increasingly be so in the future. Those interested in the politics and culture of the future should therefore be clear on the important role of the new public spheres and intervene accordingly.

Such netroots technopolitics and activism, however, should not replace political struggle in the real world and the danger exists that virtual democracy will become a closed in space and world in itself in which individuals delude themselves that they are active politically merely through exchanging messages or circulating information. Further, there are dangers of social media increasing political inequalities, producing new elites, and disfranchising even further the disenfranchised. Obviously, much of the world does not even have telephone service, much less computers, and there are vast discrepancies in terms of who has access to computers and social media, and who participates in the technological revolution and cyberdemocracy today.

Concluding Comments

In conclusion, I would suggest that there is an objective ambiguity inhering in the connection between emergent technologies and alienation. While ICTs and social media are in some cases empowering of individuals and groups in opposition to the dominant social order, they also increase the power of ruling social forces and can be used as instruments of domination which produce the alienation of the citizen and can divide the population as happened during the Trump era. While information and communication technologies may give a chance for members of subordinate class, race, gender, and regional formations to gain more power and equality vis-a-vis hegemonic forces, they may also increase inequality, domination and social alienation. Thus, in both individual and group use, digital technologies and social media may arguably produce alienation in some forms, but may also contribute to disalienation.

In this complex situation, to make serious claims concerning alienation and digital technologies and social media, one must specify in more detail than in dominant technophobic discourses on ICTs and social media exactly what sort of alienation is being produced, how this is happening, what is bad about it, and how it can be overcome. I have suggested that many claims concerning technology and alienation in the mushrooming literature on ICTs are not as convincing as the classic Marxian discourse of the alienation of labor under capitalism, and thus while the emergent technologies provide new life and substance for the discourse of alienation, one needs more substantive analysis and critique to redeem claims being made concerning technology and alienation.

Finally, it may be that alienation is the human condition and that it can never be fully overcome as the complexity and conflicts of human life make it impossible to reconcile all separations and differences between human beings, nature, and cultures and technologies. Yet we should be aware that technologies ranging from the machine, to assembly lines, to broadcasting media, to ICTs, to social media and to biotechnology profoundly transform human beings. They all arguably produce specific forms of alienation that can be delineated, attacked, and in some cases overcome. Hence, we should always be aware that new technologies may produce novel forms of alienation and thus challenges for critical theory to articulate and radical politics to confront. In the following chapter, I will discuss how the politics of spectacle and social media provide threats to democracy and how these threats can be confronted and combatted to produce a stronger, more participatory democracy, engaging the theme of "the demands of democracy" which will run through this book.

The Media, Democracy, and Spectacle 4

> *A popular government without popular information, or the means of acquiring it, is but a prologue to a farce or a tragedy, or perhaps both.*
>
> —*James Madison*

Abstract

Chapter 4 provides analysis of "The Media, Democracy, and Spectacle," which will begin exploring the notion of technopolitics and how we can use digital technologies and social media for social transformation and to meet the demands of democracy, a theme that will run throughout the book. I document shifts In the 2000s with blogs, wikis, Facebook, MySpace, and other new media and social networking sites, such as YouTube and Twitter, further proliferating within the ubiquitous and omnipresent media matrix. Whereas during the previous decades there were few broadcasting media sources in the U.S. and much of the world, where state controlled media often reigned, suddenly there was an explosion of information sources from broadcasting, the Internet, social media, and multiplying apps. While, on one hand, this situation increased democracy as a conversation between a wide range of voices representing major groups, interests, and politics, on the other hand, the proliferating news sources became a source of disinformation and propaganda for various groups and political forces. In this situation, the demands of democracy require critical media and digital literacies to utilize reliable sources and avoid disinformation and lies.

D. Kellner, *Technology and Democracy: Toward A Critical Theory of Digital Technologies, Technopolitics, and Technocapitalism*, Medienkulturen im digitalen Zeitalter, https://doi.org/10.1007/978-3-658-31790-4_4

Keywords

Media, Democracy, Spectacle • Blogs, wikis, Facebook, MySpace • New media and social networking sites • YouTube • Twitter • Critical media and digital literacies

A democratic society requires a separation of powers in which the media can serve as a check and balance against excessive power or corruption of the state and other major institutions, as well as to help to create informed citizens who can intelligently participate in public affairs. Sovereignty, in this framework, thus rests both in the constitutional order and with the will of the people.[1] A free press was deemed vitally necessary to maintain a democratic society, and it is often claimed by champions of democracy that freedom of the press is one of the features that defines the superiority of democratic societies over competing social systems.

This concept of a free press was extended in the twentieth century to the broadcast media, which were assigned a series of democratic responsibilities. In countries like Britain, which developed a public service model of broadcasting, radio and then television were considered part of the public sector, with important duties to reproduce the national culture and provide forums of information and debate for citizens (Tracey 1998). Even in the United States, where a private industry model of broadcasting came to dominate, in the Federal Communications Act of 1934 and subsequent legislation and court decisions, broadcasting was to serve the "public interest, convenience, and necessity, ascribing certain democratic functions to the media," until the setting aside of these strictures in the 1980s and 1990s.[2]

In the era of intensifying globalization in the 1990s and into the new millennium, market models of broadcasting generally emerged as dominant in the neo-liberal world, and a series of global mergers took place that consolidated media ownership into ever fewer hands. The result has been that a shrinking

---

[1] The conception of democracy upon which I am drawing here has been developed in Kellner (1990, 2001). In this chapter, I largely focus on the configuration of the media and democracy in the past several decades in the United States and the ways that broadcasting technology and ICTs have impact on democracy and political and social life.

[2] See the discussion of the media and democracy in Kellner (1990), Chap. 2 and 3; on the Federal Communications Act of 1934 and on the battle for democratic media in the 1930s, see McChesney (1993). For history and overview of corporate media and politics in the United States, see Halberstam (1979) and McChesney (1993, 2013).

number of giant corporations have controlled a widening range of media in corporate conglomerates that control the press, broadcasting, film, music, and other forms of popular entertainment, as well as the most accessed Internet and social networking sites.[3]

Especially broadcasting media have been increasingly organized on a corporate business model, and competition between proliferating commercialized media have provided an impetus to replace news with entertainment, to generate a tabloidization of news, and to pursue profits and sensationalism rather than public enlightenment and democracy.[4] Since the 1960s in the U.S., corporate control of broadcasting and the decline of television documentaries and public affairs programming has arguably helped to produce a less informed electorate, more susceptible to political manipulation.

Democracy requires vigorous public debate of key issues of importance and an informed electorate, able to make intelligent decisions and to participate in politics. Corporate control of the media meant that corporations could use the media to aggressively promote their own interests and to cut back on the criticism of corporate abuses that were expanding from the 1970s to the present. The tabloidization of news and intense competition between various media meant that the corporate media ignored social problems and focused on scandal and tabloid entertainment rather than issues of serious public concern.

During the Clinton era (1992–2000), for instance, the media focused intensely on the O. J. Simpson scandals in the mid-1990s and then turned toward the Clinton sex scandals (Kellner 2003). Although previously, corporate media tended to support presidents in office, and had been especially uncritical of the ruling administration in the Reagan and Bush Senior years, during the Clinton era the media became fierce watch dogs, pouncing on every potential scandal involving the Clintons and feasting on the sex scandals, which eventually became dominant in the mainstream corporate media in the 1990s. This was an era in the U.S. of right-wing talk radio, the rise of conservative television networks like Fox, and the proliferation of the Internet, which had many anti-Clinton activists and gossips like Matt Drudge, whose website first broke the Clinton sex scandals.[5]

---

[3] On global media consolidation and its impact, see McChesney (2000); Compaineand Gomery (2000); Baker (2007).

[4] On media consolidation and its impact in the U.S. over the past decades, see Herman and Chomsky (1988); Schiller (1990); Kellner (1990); McChesney (1993, 1997, 2000, 2004); Bagdikian (2004); and McChesney and Nichols (2012).

[5] To this day, Drudge is continuing to trudge along, daily attacking the Democrats and taking the Republican Party attack line of the day; see https://www.drudgereport.com/ (accessed on August 4, 2019).

In this era, **media spectacle** emerged as a dominant form in which news and information, politics, war, entertainment, sports, and scandals were presented to the public in the United States and then globally, which circulated through the matrix of old and new media and technologies.[6] By "media spectacles" I am referring to media constructs that present events which disrupt ordinary and habitual flows of information, and that become popular stories which capture the attention of the media and the public. These media spectacles circulate through broadcasting networks, the Internet, social networking, cell phones, and other new media and communication technologies centering public attention on certain events. In a global networked society, media spectacles proliferate instantaneously, become virtual and viral, and in some cases become tools of socio-political control, while in other cases they can become instruments of opposition and political transformation, as well as mere moments of media hype and tabloidized sensationalism.

Dramatic news and events are presented as media spectacles and dominate certain news cycles. In the first decade of the 2000s, stories like the September 11, 2001 terror attacks, Hurricane Katrina, and Barack Obama and the 2008 U.S. presidential election were produced and distributed throughout the media and technoscape as media spectacles which were central events of their era in the United States. In 2011, the Arab Uprisings, the Libyan Revolution, the UK riots, the Occupy movements and the other major media spectacles, engaged in my book *Media Spectacle and Insurrection, 2011: From the Arab Uprisings to Occupy Everywhere*, cascaded through broadcasting, print, and digital media, seizing people's attention and emotions, and generating complex and multiple effects that may make 2011 as memorable a year in the history of social upheaval as 1968 and perhaps one as significant (Kellner 2012).

The infrastructure of media spectacle that generates its proliferation was, initially, global cable and satellite television which emerged in the 1980s era of neoliberalism and deregulation, and increased media monopoly and competition between different media corporations and new media technologies. The period marks the rise of cable news networks that broadcast news 24/7 and used media spectacle to capture viewers. In the 1990s in the United States, new media and politicized forms of media spectacle proliferated, including Talk Radio, *Fox News*, and highly partisan Internet sites. Increasingly politicized mainstream media continue to heat up and expand from the 1990s through the present in the U.S., illustrated by the battles between *Fox News* on the Right and *MSNBC TV* cable

---

[6] On the rise and escalation of media spectacle as a dominant form of news, information, and culture, see Kellner (2003, 2012).

news channels on the Left, as well as within the Internet which has become a contested terrain used by left, right, and everyone in-between (Best and Kellner 2001; Kahn and Kellner 2005).

The 1990s in the U.S. thus exhibited the explosive rise of the Internet as well as contentious news cable channels and Talk Radio, providing new forms of political media spectacle that captured the attention of the general public. Hence, news and information were technologically driven with cable and satellite broadcasting producing a dramatic expansion of new sources of news and information, although corporate mergers limited the ownership and control mechanisms and initially did not promote much diversity in mainstreaming broadcasting of news and information. Initially, the Internet created a whole new ecology of news and information, significantly democratizing news and proliferating sources of news, information and opinion, although the net too was colonized and corporatized by mega-broadcasting corporations and a diverse number of groups on the right, left, and alternative sources creating a diversity of news and information with Big Corporate Media and High-Tech companies ultimately coming to dominate the Internet as well as broadcasting.

The 1990s was also an era in which media spectacle accelerated in the fields of sports, entertainment, fashion, and consumer culture, which were always a domain of the spectacle. In addition, the 1990s witnessed the spectacle of globalization and anti-globalization movements, the global commodity spectacle such as the McDonald's and Nike spectacle, NBA basketball, the World Cup, and other global sports spectacles (see Kellner 2003). This was also a period in which spectacle came to play an even greater role in Hollywood film during the blockbuster era, an aesthetic form appropriate for the neoliberal capital of the era which use spectacle to promote its goods, services, and the consumer society as a whole.

Although the mainstream media in the United States tended to be largely uncritical of Reagan and Bush in the 1980s, they were *attack dogs* against Clinton and his administration in the 1990s (Alterman 2003; Kellner 1990, 1992, 2001, 2003a; Miller 2004). Thus, it was not surprising that during the 2000 election key sectors of the media would be highly critical of Democratic Party candidate Al Gore and give George W. Bush, son of the former president, an easy time (see Kellner 2001, 2005, Chap. 1). During the Bush-Cheney administration, the corporate media tended to be *lap dogs,* failing to investigate in any depth the scandals of Bush and Cheney, their bogus claims about weapons of mass destruction in Iraq, and the destructive consequences of their domestic and foreign policies. Thus, the corporate media in the United States had largely abandoned their role as a "fourth estate" or *watch dogs,* investigating economic and political scandal and corruption in the public interest.

In the following sections, I will first document shifts in the 2000s with blogs, wikis, Facebook, MySpace, and other new media and social networking sites, such as YouTube and Twitter, further proliferating within the ubiquitous and omnipresent media matrix. Whereas during the previous decades there were few broadcasting media sources in the U.S. and much of the world, where state controlled media often reigned, suddenly there was an explosion of many news and information sources from broadcasting, the Internet, social media, and multiplying apps. While, on one hand, this situation increased democracy as a conversation between a wide range of voices representing major groups, interests, and politics, on the other hand, the proliferating news sources became a source of disinformation and propaganda for various groups and political forces. In this situation, the demands of democracy require critical media and digital literacies to utilize reliable sources and avoid disinformation and lies (Kellner and Share 2019)—a theme I shall expand upon in Chap. 10 below.

Hence, the political economy and communications technology infrastructure of media spectacle have generated a proliferation of cable and satellite television, followed by the dramatic eruption of digital technologies like the Internet, social media, virtual gaming, and multiple forms of tech-app culture. This information overload, on one hand, made it possible for everyone to voice opinions and to circulate news and information through ever-expanding new media and social media sites, in which Facebook, Twitter, YouTube, I-Phones and I-Pads, and other digital technologies, devices, and social media enable everyone to become part of the spectacle (if you can afford and know how to use the technology).

Hence, today, everyone, from Hollywood and political celebrities to Internet activists in Egypt and Tunisia, or terrorists like al-Qaeda or deranged killers, or aggrieved Trumpie White Dudes, can create their own media spectacles, or participate in the media spectacle of the day—as the North African Arab Uprisings, European movements against global capital, and the Occupy movements demonstrated on a global scale in 2011, as well as the Trump mob occupation of the U.S. Capital on January 6, 2021, which led to a call for Trump's second impeachment, and portended major Trump media spectacles in the Post-Trump era.

The epoch of neoliberalism in which media spectacle triumphed exhibited the rise of **infotainment**, with the implosion of news and entertainment (i.e. the O. J. Simpson trial, Clinton sex scandals, celebrity scandals and the like; see Kellner 2003). Fierce competition for ratings and advertising led information and news to become more visual and engaging, bringing codes of entertainment into journalism. News accordingly became more narrative and tabloid, with scandals and ever-multiplying segments on fashion, health, entertainment, and items of personal interest. In this media environment, hard politics and international news are now declining on the major U.S. television networks like ABC, CBS, and NBC, while the cable news networks are dominated by media spectacle and often partisan political talk shows.

The 1990s in the U.S. were an era of escalating social problems caused by globalization and the abuses of corporate capitalism, ecological crisis, decline in public health, growing inequality between rich and poor, and dangerous corporate practices that would eventually explode in 2002 in the Enron, WorldCom, and other corporate scandals, and in the 2008 global financial crisis. It was an era of neoliberalism in which not only were the broadcasting media deregulated, but so too were corporate practices, financial markets, and the global economy, creating new forms of technocapitalism.

The media tended to celebrate the "new economy" and the period of economic boom and growing affluence, and ever more-powerful Silicon Valley corporate interests helped produce new media and technologies, creating an expansion of technocapitalism in which technology continued to be an important element of the infrastructure of a global networked society. Further, new technologies and social media also became part of the forces of production that were driving the new forms of technocapitalism and a new libertarian neo-liberalism became the new dominant ideology, driving economic development and creating new sources of wealth and power (see Kelly 1995 and the critique in Best and Kellner 1999, as well as Turner 2006).

The exuberant celebration of the "new economy" overlooked the dangers of an overinflated stock market, an unregulated economy, growing divisions between haves and have nots, and rising inequalities between rich and poor. During this era, the corporate media thus neglected social problems and social reform in favor of celebrating the capitalist economy and technological revolution. The media also overlooked the growth of terrorism, dangerous consequences of the growing division between the haves and the have-nots, growing poverty throughout the world, and escalating ecological problems.

In the political mediascape of the United States in the past decades with the rise of new digital and social media which generated partisan television networks and radio shows of the left and the right, along with new alternative media and social media that provided a wide diversity of opinion and critique of the dominant political and corporate order. I then argue that this shift in media technology and rise of social media in the U.S. mediascape helped elect Barack Obama, which in turn intensified contestation within media politics, which helped produce the candidacy and victory of Donald Trump in the 2016 U.S. presidential election—and then contributed to his defeat in 2020.

My writings on U.S. media and politics from the 1980s through the first years of the 2000s stressed how mainstream corporate media largely served the interests of the Republican Party within U.S. politics from the 1980s up to the Obama era, but now I would present the current U.S. mediascape as a highly contested terrain with increasingly important roles taken on by **technopolitics**, as digital technologies and social media became increasingly important for campaigning and governing.[7] Next, I indicate how Barack Obama used the media and media spectacle to successfully win the 2008 election, while Donald Trump mastered new media and broadcasting television to help him win the 2016 U.S. presidential election, concluding with a discussion of how digital and alternative media can help to promote genuine democratic debate and help disseminate the full range of information and ideas necessary to have a robustly democratic social order.

The Rise of Partisan Corporate Broadcasting Programs and Networks and the Obama Era

What is the role of a free and independent press in a democratic society? Is it to be a passive conduit responsible only for the delivery of information between a government and its people? Is it to aggressively print allegation and rumor independent of accuracy or fairness? Is it to show boobies? No. The role of a free press is to be the people's eyes and ears, providing not just information but access, insight and, most importantly, context.

—The Daily Show with Jon Stewart

A democratic social order can best be maintained if the media in a country discuss issues of public concern and social problems from a variety of viewpoints and foster spirited public debate, accompanied by the development of vigorous and competent investigative and alternative media. The democratic imperative in the

---

[7] I show how the mainstream corporate media benefited the Republican Party and promoted their politics, candidates, and ideologies during the 1980s in Kellner (1990, 1992). I document how the corporate media benefited Republicans and the Bush-Cheney administration in the 2000 U.S. presidential election in Kellner (2001), and more generally document how the corporate media in the U.S. served Republican Party interests throughout the Bush-Cheney era in Kellner (2005). My analysis in this chapter will focus on the role of technopolitics and digital technologies and social media during the Obama and Trump eras, marked by a proliferation of new digital and social media, intensified political struggle, and a growing division into political tribes. This analysis suggests an especially intense political division and conflict fueled in part by more complex and contested media constellation in the U.S. in the contemporary moment.

United States that the mainstream corporate press and broadcasting provide a variety of views on issues of public interest and controversy was increasingly sacrificed from the Reagan era of media deregulation through the Bush-Cheney era, as has their responsibility to serve as a check against excessive government or corporate power and corruption. Mainstream corporate media tended to promote the interests of the corporations that own them, which tend to be pro-market and anti-regulation and have largely advanced the interests of corporate media and other institutions and conservative politics. As I have documented (Kellner 1990, 2001, 2003, 2005), there has been a crisis of democracy in the United States in part because the mainstream corporate media have been biased toward Republicans and conservatives from the 1980s into the first decade of the 2000s when due to failures in the Bush-Cheney administration in the 2003 Gulf war and what was seen as a failed administration helping produce economic crisis on 2007, the national television networks ABC, CBS, and NBC appeared to be less biased toward Republicans and more open to liberals and Democrats while cable television, the Internet, and emergent social media became increasingly contested terrains (Kellner 2005, 2008).

I suggest that four convergent trends have seriously undermined U.S. democracy from the 1970s into the 2000s: the corporate control of mainstream media, which biases dominant media toward conservativism and profit; an implosion of information and entertainment and rise of a culture of media spectacle, which makes politics a form of entertainment and spectacle; the rise of a right-wing Republican media propaganda and attack apparatus, which systematically deploys lies and deception to advance the agenda of conservative groups and interests; this has produced a tribalized culture where conservatives get their information from rightwing media sources and liberal and progressives find news sources and social media that articulate their views and politics.

In a situation of information overload, we suffer an infopandemic in which lies and disinformation comes to dominate in certain sectors of the **news and social media corporate post-industrial infotainment complex**. An ever-growing right-wing Republican media machine, ranging from the *Wall Street Journal* and the conservative press to the Rupert Murdoch—owned *Fox News*, talk radio, and the extreme right sector on the Internet, all disseminate propaganda of a scope and virulence never before seen in U.S. history.[8] Expanding significantly since

---

[8] The rise and growing influence of a right-wing Republican media propaganda and attack apparatus has been well documented in Alterman (2000, 2003); Brock (2004); Conason (2003); Miller (2005); and Stelter (2020). In Kellner (2005), I update and expand my critique of right-wing and corporate media and show how they generally promoted the agenda of the Bush/Cheney administration and discuss how the media promoted the Trump candidacy

the 1980s, the Republican propaganda machine has cultivated a group of ideological storm troopers who loudly support Republican Party and conservative policies, and attack those who criticize them. These extremists are impervious to argument, ignore facts and analysis, and demonize as unpatriotic anyone who challenges Bush-Cheney, or later Trump, policies. Groomed on *Fox TV* and right-wing talk radio, they verbally assault anyone who did not march in lockstep with the Bush/Cheney and Trump administrations and waged ideological war against the socialists, liberals, people of color, feminists, gays and lesbians, and other dissenters. These ideological warriors allowed no disparagement of the Bush and Cheney or Trump administration policies and refused civil dialogue, preferring denunciation and invective (see Brock 2004; Brock et al. 2012; Stelter 2019).

Although the mainstream corporate media are vilified as "liberal" by the right-wing attack machine, in fact, mainstream journalists have been easily intimidated when the right-wing army e-mails, calls, writes, and harasses any corporate media source that goes too far in criticizing the Bush-Cheney or Trump regime. The mainstream corporate media have been largely subservient to corporate interests, follow the sensation of the moment, and rarely engage in the sort of investigative journalism that was once the ideal and that now takes place largely in the alternative sphere. Corporate media increasingly promote entertainment over news and information, like the tabloids framed by codes of media spectacle (Kellner 2003, 2005).

One interesting development within the corporate media in the U.S. during the 1990s, however, was the rise of competing partisan media broadcasting networks with the politicization of the liberal *MSNBC* TV news network as a counterbalance to the rightwing Fox News Network. In October 1996, *Fox News* joined *CNN* as a 24/7 news channel, but unlike *CNN*'s centrist approach, *Fox* was aggressively rightwing from the beginning. Owned by Rupert Murdoch and with former Nixon administration political operative and rightwing activist Roger Ailes as its President, *Fox News* provided a consistently pro-Republican and conservative perspective on the news, although it claimed it was "fair and balanced." Against this claim, Robert Greenwald's 2004 documentary film *Outfoxed: Rupert Murdoch's War on Journalism* convincingly shows that Fox News has a strong conservative and pro-Republican Party bias, a critique supported by many books and scholarly articles.[9]

---

and presidency in Kellner (2016, 2017), although I also discuss in this book how the media contributed to taking Trump down.

[9] On Fox News conservative bias, see Amann (2007); Brock et al. (2012); Muto (2013); Stelter (2020). For my take on Murdoch and his global media empire, see Kellner (2012, 2013).

MSNBC cable news TV network was founded as a spinoff from *NBC News* in a partnership with Microsoft in 1996, but during its first decade was not appreciably different from the centrist *CNN*, attempting to provide a balance of slightly left and rightwing hosts, commentary, and punditry. Beginning in 2005, however, *MSNBC* added *Countdown With Keith Olbermann* who heavily satirized *Fox News*, and in particular his nemesis Bill O'Reilly, and put a left-liberal spin on his news presentations and discussion of contemporary politics. By 2007, Olbermann's nightly attacks on the Bush-Cheney Gang were so savage, brilliantly satirical, and comprehensive that I abandoned my *blog-left* which had provided daily attacks on the Bush-Cheney regime from 2002–2007, but was rendered irrelevent by Olbermann's fierce attacks.

Olbermann was the "last angry man," raging with nightly commentaries on the horrors of the day perpetrated by the Bush-Cheney Gang, and systematically criticizing their policies, while ridiculing media pundits and others who supported them. Almost all his guests shared his left-liberal views, and perhaps for the first time ever there was a nightly TV news show seriously critical of the Republican Party and the rightwing in the U.S. *MSNBC*'s nightly show *Hardball with Chris Matthews* also began accelerating attacks on the Bush-Cheney administration and provided strong liberal commentary and critique.

Rachel Maddow, a political analyst, *Air America Radio* host, and Olbermann protégé, was given in 2008 *The Rachel Maddow Show*, and provided left-liberal, feminist, and gay perspectives on the news of the day, while *MSNBC* added *The Ed Show* in 2009, in which Ed Schultz provided a left-liberal populist and strongly pro-labor take on news and politics. Hence, by 2009, *MSNBC* provided a liberal counterbalance to *Fox News*. Since that time, *Fox News* shows have attacked the Democrats and boosted the Republican party line of the day, while *MSNBC* attacked Republicans and usually took left-liberal pro-Democratic Party positions (as long as the Dems took properly liberal positions).

A new cable TV channel, The Comedy Channel's *The Daily Show with Jon Stewart* (1999–2015) broke new ground in providing critique of conservative politicians and media. Comedian Jon Stewart focused his satirical news commentary on politics and the national media, providing nightly satire of recent news stories, political figures and commentators, and the media and entertainment industries. It has won 24 Primetime Emmy Awards and became the longest running program on Comedy Central, and prospered as a popular source of news for younger audiences.[10]

---

[10] For Stewart's take on the media and contemporary U.S. politics, see Stewart and The Writers of *The Daily Show* (2004). On Stewart and *The Daily Show*, see Steve Gennaro, "*The Daily*

One of Stewart's regular "correspondent" satirists, Stephen Colbert, launched a Comedy Central spin-off show *The Colbert Report* in 2005, brilliantly satirizing rightwing commentators by playing a pompous rightwing pundit himself, thus expanding comedic critique of Republican Party and the rightwing attack machine.[11] *Politically Incorrect with Bill Maher*, whose late-night political talk show ran on *Comedy Central* from 1993 to 1997 and on ABC from 1997 to 2002, was followed by *Real Time with Bill Maher* which ran on cable channel HBO from 2003 to the present. Maher initially marketed himself as a libertarian who mocked all "politically correct" dogma of the Left and Right, but increasingly labeled himself a "progressive" and aggressively supported the Democratic Party and attacked conservatives and Republicans as the years have gone by.

Many youth today get their news and opinions from late night comedy shows which are becoming increasingly subversive. A study by the Pew Research Center stated that "61 percent of young people reported regularly or sometimes learning campaign information from comedy television."[12] Likewise, studies have shown that young people who regularly watch late night comedy that sharply satirizes politics are more likely to participate in politics.[13]

Show: The Face of American News in 2005," *Kritikos*, Volume 2, April 2005 at https://intert heory.org/gennaro.htm (accessed December 22, 2020).

[11] In 2013, Colbert won the Emmy for Outstanding Variety Series "The Daily Show" which Jon Stewart had won in this category for many years. In 2015, Colbert went over to CBS to do *The Late Show with Stephen Colbert* playing himself rather than satirizing rightwing pompous media personalities, and during the 2020 COVID-19 quarantine has been doing his show at home, under the rubric *A Late Show*, sometimes with his wife, children and dog making appearances. For Colbert's current show see https://www.cbs.com/shows/the-late-show-with-stephen-colbert/ (accessed December 22, 2020). See also the excellent analysis of Colbert during COVID-19 and Trump, Joe Hagan, "'Look At What We Love. It's on Fire': Stephen Colbert on Trump Trauma, Leadership, and Loss. The late-night host is ready for a little less excitement: 'If Joe Biden is a pair of khaki pants inside a manila envelope, that would be great.'" *Vanity Fair*, December 1, 2020 at https://www.vanityfair.com/hollywood/2020/11/stephen-colbert-on-trump-trauma-leadership-and-loss?itm_content=footer-recirc (accessed December 22, 2020).

[12] See Baumgartner and Morris (2007).

[13] Amy Bree Becker, "Political satire makes young people more likely to participate in politics. Trevor Noah's *The Daily Show* is likely to continue that trend." LSE, US Centre, August 26, 2016 at https://blogs.lse.ac.uk/usappblog/2016/08/26/social-media-and-online-communities-expose-youth-to-political-conversation-but-also-to-incivility-and-conflict/ (accessed December 22, 2020).

With the expansion of left-liberal broadcasting programs and even channels during the past decade, in the 2008 U.S. Presidential election and into the Obama and Trump years and beyond, there have been an increasing amount of sites within the corporate media that have criticized the Republican Party and supported the Democratic Party and liberal policies and politicians. As noted, there are now two competing corporate cable TV news networks that take strongly opposed political views, providing alternative universes of U.S. politics, with *Fox News* a reliable supporter of Republicans and conservative views, and *MSNBC* taking more left and liberal positions. Consequently, during the 2008 and 2012 elections, *Fox* fiercely attacked Barack Obama and repeated Republican Party lines-of-the-day, while *MSNBC* promoted Obama and attacked the Republicans[14]—trends continuing into the present.

During the 2008 presidential election, it was widely perceived that the mainstream media favored Barack Obama over Republican candidates.[15] Obama mastered the art of media spectacle and his message of hope and change resonated with large sectors of the mainstream media as well the public. There was also an impressive Internet spectacle in support of Obama's presidency. Obama raised an unprecedented amount of money on the Internet, generated more than two million friends on Facebook and 866,887 friends on MySpace, and reportedly had a campaign listserv of over 10 million e-mail addresses, enabling his campaign to mobilize youth and others through text-messaging and e-mails.[16] Videos compiled on Obama's official campaign YouTube site were accessed over 11.5 million times (Gulati 2010, p. 195), while the YouTube (UT) music video "Obama Girl," featuring a young woman singing about why she supports Obama interspersed with images of his speeches, received well over 5 million hits and was one of the most popular in the site's history.[17]

Indeed, grassroots campaigns for Obama illustrate the impact of YouTube and Internet spectacle for participatory democracy. Among the enormous numbers of Internet-distributed artifacts for the Obama campaign, Will.i.am's "Yes We

---

[14] For my take on Obama and the 2008 presidential election, see Kellner (2009).

[15] See, for example, the Pew Research Center's Project for Excellence in Journalism report *Winning the Media Campaign: How the Press Reported the 2008 Presidential General Election*, October 22, 2008 at https://www.journalism.org/analysis_report/winning_media_camp aign (accessed on August 4, 2014).

[16] On Obama's mobilization of Facebook and the Internet, see Gulati (2010), Cornfield (2010), and Kellner (2012).

[17] See the video at https://www.youtube.com/watch?v=wKsoXHYICqU (accessed on August 4, 2014). For my analysis of the role of new media in the 2008 presidential election, see Kellner (2009a).

Can" music video manifests how grassroots-initiated media artifacts can inspire and mobilize individuals to support Obama. In addition to this video made by professional musicians, there emerged grassroots-based videos made by ordinary people who produced their own videos and narratives to support Obama, collected on a YouTube Web site.[18] Traditionally underrepresented youth and people of color enthusiastically created YouTube-style self-made videos, containing their personal narratives and reasons why they support Obama for President, and used these videos as an innovative platform for grassroots political mobilization with which to inspire and consolidate potential Obama supporters online and off-line.

Throughout major cities like Los Angeles, hundreds of Obama art posters and stickers appeared on stop signs, underpasses, buildings and billboards, with Obama's face and the word "HOPE" emblazoned across them. Even street artists began creating Obama graffiti and urban art in public places with Obama's image competing with those of Hollywood stars, sports figures, and other celebrities as icons of the time.

So in terms of stagecraft and spectacle, Obama's daily stump speeches on the campaign trial, his post-victory and even post-defeat speeches in the Democratic primaries, and his grassroots Internet and cultural support demonstrated that Obama was a master of the spectacle. In the 2008 campaign against Republican candidates Sarah Palin and John McCain, Obama used media spectacle as a major instrument in his campaign, and again effectively used new media and social networking to raise money, organize, and circulate his message. During September, 2008, Obama raised an unprecedented $150 million, much of it from small Internet and personal donations, and was soaring in the polls, which showed him pulling ahead of McCain nationally and in the significant battleground states. As he entered the last weeks of the campaign in November, Obama presented the spectacle of a young, energetic, articulate candidate who had run what many considered an almost flawless campaign and attempted during the election's final days to project images of hope, change and bringing the country together to address its growing problems and divisions—exactly the message that Obama started off his campaign with.

On election night, in Grant Park in Chicago—the site of the spectacle "The Whole World is Watching" during the Democratic convention in 1968, when the police tear-gassed antiwar spectators, and the site a year later of the Weather Underground abortive "Days of Rage" spectacle—in 2008 Chicago hosted a peaceful assembly of a couple of hundred thousand spectators, mostly young and

---

[18] See https://www.dipdive.com/dip-politics/ywc/ (videos 2 to 30, accessed January 30, 2009). For detailed analysis of the YouTube videos assembled here, see Kellner and Kim (2009).

of many colors, that had assembled to celebrate Obama's historical victory. In the crowd, television networks showed close-ups of celebrities like Jessie Jackson, tears streaming down his face, a jubilant Spike Lee, a solemn and smiling Oprah Winfrey, and others who joined the young crowd to hear Obama's victory speech. The park hushed into silence as McCain gave his concession speech and the audience nodded and applauded respectfully.

When Obama, his wife, Michelle, and his two beautiful girls took stage, the crowd went wild and the eyes of the world were watching the spectacle of Barack Obama becoming president of the United States. Television networks showed the spectacle of people celebrating throughout the United States, from Times Square to Atlanta, and even throughout the world. There were special celebrations in countries like Kenya and Indonesia where Obama had relatives or had lived and his connections to these countries were producing national shrines that would be tourist destinations. Obama had become a global spectacle and his stunning victory would make him a world celebrity superstar of global media and politics.

In the 2008 U.S. Presidential Election, Barack Obama's mastery of media spectacle thus helped generate a coalition based on hope and change that produced a decisive victory over his Republican opponents.[19] Obama was seen by his followers as the One, a candidate of youth and charisma who inspired a generation to believe that meaningful change could be created in the United States.

In 2012, media spectacle in the U. S. Presidential election focused, by contrast, on the struggle between the two major candidates and their parties. Hence, while the personality and charisma of Barack Obama was arguably at the center of the 2008 presidential election, in 2012 I suggest that the spectacle resided in the campaign between Obama and his Republican challenger Mitt Romney. In much of the presidential race, the Republican Romney became at times a negative media spectacle, while the Democrat "no drama Obama" was often presented in the media as calm, cool, professional and very much the President. Consequently, while Obama's personality and charisma were perhaps decisive in winning the 2008 campaign, in 2012, it was arguably his campaign team and strategy that was central to his victory, using Obama's presidential record and agenda to push the campaign, with Obama largely serving as a prop, as he himself acknowledged during the course of the election (Kellner 2012).

It is too soon to assess the Obama era and the ways that increasingly fragmented mainstream media and new media and social networking affected his presidency, the limits of governing through media spectacle, and Obama's actual accomplishments and failures. It is clear, however, that Obama benefitted from

---

[19] See Chap. 1 of Kellner (2012).

new media and social media as did Trump who emerged as a master of Twitter. In my discussion of digital technology and social media in the Trump era, I will argue that while U.S. society was more divided and contested than ever before in my lifetime, I also want to advance some reflections on how alternative and social media played an important role in keeping democracy alive during the Trump era and beyond, just as they helped promote Trump's candidacy and cult of personality, making broadcasting and social media a highly contested terrain in the twenty-first century.

The Trump Era

Explaining the Donald Trump phenomenon is a challenge that will occupy critical theorists of U.S. politics for years to come. My first take on the Trump phenomenon is that Donald Trump won the Republican primary contest and then the 2016 U.S. Presidential Election because he is the **master of media spectacle**, a concept that I've been developing and applying to U.S. politics and media since the mid-1990s.[20] In this study, I will first discuss Trump's use of media spectacle in his business career, in his effort to become a celebrity and reality-TV superstar, his successful 2016 political campaign, and his presidency. Then I shall examine how Trump embodies Authoritarian Populism and has used racism, nationalism, xenophobia, and the disturbing underside of American politics to mobilize his supporters in his successful Republican primary campaign and in the hotly contested win in the 2016 general election. Finally, in line with the conception of the dialectic of digital culture I am exploring in this book, I will stress how Trump has used digital culture to win and advance his presidency, but that his use of digital culture and media spectacle contributed to his undoing, suggesting that digital culture in the contemporary era can empower and destroy, has both an emancipatory and a regressive and destructive side, and thus creates a highly unstable and unpredictable social and political situation in the contemporary era.

In the 2016 U.S. presidential election, obviously Donald Trump emerged as a major creator of media spectacle. He has long been a celebrity and master of the spectacle with promotion of his buildings and casinos from the 1980s to the present, his reality-TV shows, self-promoting events, his presidential campaign, and then his presidency. Hence, Trump was empowered and enabled to run for the presidency in part because celebrity and media spectacle has become a major force in US politics, helping to determine elections, government, and more broadly the

---

[20] On my concept of media spectacle, see Kellner (2001, 2003a, 2003b, 2005, 2008).

ethos and nature of our culture and political sphere, and Trump is a successful creator and manipulator of the spectacle.

Further, Donald Trump may be the first major Twitter presidential candidate and then President, and certainly he is the one using it most aggressively and frequently with arguably contradictory results. Twitter was launched in 2006, but I don't recall it being used in a major way in the 2008 election—Obama used Facebook and his campaign bragged that he had over a million "Friends" and used Facebook as part of his daily campaign apparatus (Kellner 2015). I don't recall, however, previous Presidential candidates using Twitter in a big way like Donald Trump, although many politicos now have accounts.

Twitter is a perfect vehicle for Trump as he can use its 140 (later 280) character framework platform for attack, bragging, and getting out simple messages or posts that engage receivers who feel they are in the know and involved in TrumpWorld when they get pinged and receive his tweets. When asked at an August 26, 2015, Iowa event as to why he uses Twitter so much, he replied that it was easy, it only took a couple of seconds, and that he could attack his media critics when he "wasn't treated fairly." Trump has also used Instagram—an online mobile photo-sharing, video-sharing and social networking service that enables its users to take pictures and videos, and share them on a variety of digital social networking platforms, such as Facebook, Twitter, Tumblr and Flickr.

Twitter is perfect for General Trump who can blast out his opinions and order his followers what to think, and in some cases what to do. It enables Business-man and Politician Trump to define his brand and mobilize those who wish to consume or support it. Trump Twitter gratifies the need of Narcissist Trump to be noticed and recognized as a Master of Communication who can bind his warriors into an on-line community. Twitter enables the Pundit-in-Chief to opine, rant, attack, and proclaim on all and sundry subjects, and to subject TrumpWorld to the indoctrination of their Fearless Leader.

Hence, it is no surprise that political campaigns are being run as media spectacles, and that Knights of the Spectacle, like Donald Trump, are playing the spectacle to win and then perform the presidency. However, it is arguable that Trump fell victim to the spectacle, after years of relentless criticism of him by the mainstream media, a resistance movement that began the day after his 2016 election (Kellner 2017), the Mueller probe, and copious media and Congressional investigations, Trump lost his ability to overcome scandal through spectacle and, as we shall see later in the book, Trump lost the presidency to Democratic Party candidate Joe Biden in 2020 (see Chap. 5), and then suffered shame and the humiliation of a second impeachment after his encouragement of an ultra-right mélange of his supporters stormed and occupied the Capital on January 6, 2021,

encouraged by Trump and his League of Darkness, and rampaging through the
Capital, while threatening to kill politicians, including Trump's Vice-President
Mike Pence.[21]

Hence, Trump has mastered new digital media as well as dominating television and old media through his orchestration of media events as spectacles and
daily Twitter Feed. Yet the Trump resistance movements have also mastered social
media and used digital technology effectively against Trump in an era of unparalleled protest and struggle in the U.S., organized, mobilized, and enabled through
digital media.

The Trump Resistance

With the announcement of Trump's election in 2016, vigils and protests flared
up across the country the day after the election, as opponents of President-elect
Trump displayed their anger and rage over the election results, highlighting continued division in the country and that the election was not over and the country
was far from united. The Trumps got to view the protests up front and close,
as thousands of protesters marched up Fifth Avenue toward the Trump Tower in
midtown Manhattan the day after the 2026 election. Trump Tower was surrounded
by giant garbage trucks filled with sand, armed police, and security guards, and a
crowd of thousands gathered in front of the President-elect's building with angry
demonstrators chanting: "Fuck your tower! Fuck your wall!" Several blocks of
Fifth Avenue were shut down for traffic, making New York appear a city under
siege.

Elsewhere in the country, protestors held marches and sit-ins from sea to shining sea on election night and in some cases for days thereafter. College students
gathered in spontaneous marches and asked university leaders to schedule meetings to assure students of color, Muslims, women, and others denigrated and
threatened by Trump and his followers that they would be protected. Following
Trump's victory speech, more than 2,000 students at the University of California,
Los Angeles, gathered on campus and marched through the streets of Westwood.
There were similar protests at the University of Southern California in Los Angeles, with rivals UCLA and USC united in their horror of Trump. Other campuses

---

[21] The January 6, 2021 White Riot occurred while I was completing this book and I will follow
it in several contexts in other chapters as news of this stunning assault on U.S. democracy
and Trump's Second Impeachment after his inciting the riot and his defeat by Joe Biden in
the 2020 election continue to unfold as I complete this book.

in the University of California system in Berkeley, San Diego and Santa Barbara held protests, as did other Universities throughout the country.

High school students also stormed out of class and held raucous anti-Trump demonstrations. Students walked out of classes in Arizona on election day to protest Trump and notorious Sheriff Joe Arpaio, infamous for his aggressive anti-immigrant policies and draconian treatment of prisoners (Arpaio lost his bid for re-election and was under criminal investigation for his policies, but was later pardoned by Trump in 2020 in a Trump Pardon Dump during the last weeks of his presidency). On Facebook, a page titled "Not My President" called for protesters to gather on Inauguration Day, January 20, 2017 in the nation's capital. "We refuse to recognize Donald Trump as the president of the United States, and refuse to take orders from a government that puts bigots into power. We have to make it clear to the public that we did not choose this man for office and that we won't stand for his ideologies."

Thousands of anti-Trump protesters took to the streets all over the country to protest on election night, and the day after the election, there were major demonstrations with protestors marching and chanting "not my president," while shutting down roadways, freeways, and downtown areas in major cities like Los Angeles, New York, Washington DC, and Philadelphia. Other demonstrations, fueled by social media, took place in Seattle, Portland, Oakland, Denver, Minneapolis, Milwaukee, Portland, Oakland, and dozens more US cities. While the demonstrators were mostly peaceful, there were effigies of Trump burned, a piñata of Trump beaten to shreds close to Trump Tower, small fires in the street and broken windows in some cities that featured clashes between demonstrators and police.

As the day of Trump's inauguration and the coronation spectacle of King Donald and his Demons of Darkness approached, it appeared that Trump was coming in as the least popular U.S. president in history, as his inauguration was one third less the size of Obama's first inauguration, the parade was the shortest and most sparsely attended in recent history, with many bleachers almost empty, and the concerts, the balls, and other events were the most paltry and pathetic in memory, as major celebrities, artists, and performers were all boycotting the ceremonies.

Furthermore, there were copious protests and disruptions in Washington, as throughout the city, groups of people protested the inauguration with signs, chants, and tweets against the Trump presidency. In Washington, protest groups gathered at each of the 20 security checkpoints where attendees entered the Mall and some got in to protest. Anarchists and angry young people smashed Starbucks windows and attacked a limo, burned trash, and fought with police leading to many arrests.

On the whole, however, protest in Washington and elsewhere was peaceful but defiant with RESIST emerging as the keyword for opposition to Trump.

Throughout the country, there were large marches and demonstrations in every major city and part of the country with thousands marching through driving rain in Los Angeles, angry masses gathered in John Lewis' Atlanta outraged at Trump's racist remarks about their city, and defiant groups gathered in state capitals from Austin to Madison, representing a diversity of issues and groups, as part of a week of anti-Trump protests of diverse kinds.

The big demonstration, however, was scheduled for Saturday the day after the inauguration, and already tens of thousands of women were descending on Washington for the Women's March. Charter buses began arriving from around the country, and joined by husbands, boyfriends, children, and male protestors, women demonstrators occupied terrain around the National Museum of the American Indian. March organizers, who originally sought a permit for a gathering of 200,000, said that they now expect as many as a half million participants—dwarfing Trump's inaugural crowd. Indeed, as Saturday arrived and protestor's swarmed in for the Women's March by mid-morning Washington authorities were predicting that over 500,000 were assembled, making it one of the largest DC protests in history and a spectacle in favor of U.S. democracy and against Trump.

In addition, millions of people gathered in cities around the world as part of an international day of action in solidarity with the Women's March on Washington and other protests in the U.S. Estimates indicated that over 400,000 demonstrated in New York, with a crush of demonstrators surrounding Trump Tower. An equal amount of marchers was estimated showing up in a boisterous rally in Los Angeles, and March organizers estimated that there were many other Women's March events taking place in the U.S. with more than 670 related demonstrations taking place throughout the world.

The Trump resistance continued to the end of his presidency with new social movements resisting Trump to the end with such groups as Latino Dreamers, Black Lives Matters, Muslim resistance groups, the #Me Too Movement, and others that I will discuss in later chapters dedicated to technopolitics,[22] global revolt, youth movements, and the democratic transformation of politics, society, and culture in the contemporary era. First, pursuing the theme of media spectacle and global health pandemic with the eruption and devastating proliferation of the deadly COVID-19 virus, I will indicate how Trump's denial, attacks at distraction, and dangerous inaction helped made the U.S. the global epicenter of the virus and Donald Trump a one-term president.

---

[22] On the Trump Resistance, see Kellner and Satchel 2020.

The COVID-19 Spectacle and the Downfall of Trump

As deaths and panic from the COVID-19 virus expanded in the U.S. by March 2020, producing a media spectacle that would help take Trump down, he renamed the virus "the China virus," and used the crisis to deflect blame on China, the World Health Organization, and other global entities, as he tried to deny the intensity of the crisis. Indeed, the virus was global in scope, illustrating the dark side of globalization that could transmit globally deadly viruses as well as technologies, goods, democracy, and interpersonal communication (see Kellner 2002, 2014). Scientific experts believed that the COVID-19 virus arose in Wuhan animal markets which trafficked in illegal animals, like bats, which have previously conveyed deadly viruses to humans, as well as exhibiting dangerous interactions between humans and animals in what are called "wet markets."[23] This called attention to the dangers of production of mass animal harvesting in animal breeding/feeding operations in factory farms in China which contributes to global environmental crisis, as well as the slaughter of species of animals and dangers of viruses being transmitted from animals to humans.

The COVID-19 crisis thus illustrates what the Frankfurt School called "the revenge of nature," as the destruction of animals, plant life, and the earth itself through the project of the domination of nature in which nature is subjected to exploitation and ravages as human being colonize animals, plant life, and the earth for human use and profit.[24] Since the mass production of animals takes place throughout the globe, it intensifies species extinction, global eco-crisis, and the spread of diseases from one country throughout the world in an era of global commerce, trade, and population movement. The COVID-19 virus quickly spread through Asia, Europe and the United States. After the outbreak in December 2019 in Wuhan, China, The World Health Organization (WHO) declared the outbreak a "Public Health Emergency of International Concern" on January 30, 2020 and a "pandemic" on March 11, as the COVID-19 virus spread through China, Taiwan, Italy, Iran, South Korea, Japan, and other countries from Asia to Europe.

The first cases in the United States were reported in January 2020 and continued to spread, but Trump refused to acknowledge any dangers, and assured

---

[23] "Wildlife Markets and COVID-19," Humane Society International, April 19, 2020 at https://www.hsi.org/wp-content/uploads/2020/04/Wildlife-Markets-and-COVID-19-White-Paper.pdf (accessed on August 11, 2020). For background on this issue, see Quammen 2013.

[24] See the 1985 book from my University of Texas student C. Fred Alford, *Science and the Revenge of Nature:Marcuse and Habermas*. University Press of Florida.

Americans that he had the pandemic under control and that it would soon disappear—a line he continued to take up to the November 2020 election. By mid-March 2020, the Trump administration was forced to acknowledge the enormity of the crisis, created a Pandemic Response Team, and started conducting daily press briefings at the White House. Critical media voices pointed out that Trump had shut down the pandemic response group that Obama had established and ignored a pandemic presidential response plan that the Obama administration had produced in 2015, and thus the U.S. government had no coherent crisis response to the pandemic, a situation that has continued through summer and fall 2020 as COVID-19 cases continued to multiply and deaths continued to rise.[25]

Moreover, Trump has repeatedly uttered falsehoods regarding the pandemic, contributing to the more than thirty thousand confirmed lies he had told as president as of January 2021.[26] One theme of Trump's falsehoods is exaggeration of the constructive measures allegedly taken by his administration to control the virus and the great achievements of the private sector to produce a vaccine, under his leadership. Trump has also understated the projected time to produce a vaccine and promoted unapproved treatments such as hydroxychloroquine and chloroquine, even to the point of claiming that he has been taking hydroxychloroquine to protect himself against COVID-19, despite claims by Dr. Anthony Fauci and other experts that it doesn't work.

Trump's false medical advice feeds into an "infodemic" that describes an overload of information from public officials, media, the Internet, and social media. False information about the virus leads people to attempt dangerous medical solutions, often with fatal results (Bellware 2020; United Nations 2020). Facebook, Twitter, and responsible social media sites and medical authorities are forced to

---

[25] Abigail Tracy, "How Trump Gutted Obama's Pandemic-Preparedness Systems." *Vanity Fair,* May 1, 2020 at https://www.vanityfair.com/news/2020/05/trump-obama-coronavirus-pandemic-response (accessed August 10, 2020). Bill Chappell and Alana Wise, "More Than 150,000 People Have Died From Coronavirus in the U.S." *National Public Radio,* July 29, 2020 at https://www.npr.org/sections/coronavirus-live-updates/2020/07/29/896491 060/more-than-150-000-people-have-died-from-coronavirus-in-the-u-s (accessed August 11, 2020).

[26] According to *The Washington Post* "Fact Checker" team "In four years, President Trump made 30,573 false or misleading claims," Updated Jan. 20, 2021 at https://www.washin gtonpost.com/graphics/politics/trump-claims-database/ (accessed on January 22, 2021). The Wikipedia entry on "Veracity of statements by Donald Trump" cites other data bases collecting his lies and offer well-documented examples of Trump's stunning amount of lying throughout his career at https://en.wikipedia.org/wiki/Veracity_of_statements_by_Donald_ Trump (accessed on January 22, 2021).

fight and respond to the dangerous misinformation, but in an infodemic it is difficult to get false information under control—and indeed one could argue that social
media have been dangerously negligent in stressing false COVID-19 information,
including anti-vaccination misinformation.[27]

Trump repeatedly refused to admit mistakes as reporters confronted him with
false statements or erroneous claims about the COVID-19 virus and crisis, instead
blaming many others. The *Washington Post* estimated that around 15 percent of
Trump's April 6–24, 2020 speeches were spent attacking others, with the most
frequent targets being Joe Biden and Democrats, followed by the media, state
governors, and China (Bump and Parker 2020).

Accompanying the anxiety and deaths in the COVID-19 crisis has been a global economic crisis with massive unemployment, jobs disappearing, and sectors
of the economy brought to a halt with economic futures impaired.[28] In the United
States, the economy was largely shut down for many months in large parts of the
country, but because of the lack of a national plan, different states and even cities
had wildly different shut-down and opening-back-up processes with uneven economic impact. Even though the Congress and the Trump Administration produced
a relief package for families in the lowest economic categories, and loans to some
small businesses through July 2020, many families and regions suffered economi-
cally. Further, Congress and the Trump Administration could not agree on a relief
plan to move forward and the U.S. economy crashed into its worst situation since
the Great Depression and many families face bankruptcy, losing their homes, and
worse, as the new Biden administration struggles to get Republican support for
their COVID-19 relief package in January 2021.

Moreover, as schools across the United States began to open in August 2020,
parents, teachers, and citizens were forced to make difficult decisions whether
to open schools and send their kids. Trump continued to urge in daily tirades to

---

[27] Rebecca Heilweil, "Social media companies are already losing the vaccine misinformation
fight. Posts that discourage and make fun of Covid-19 vaccination are racking up engagement." *Vox.com*, December 19, 2020, at https://www.vox.com/recode/22188680/facebook-
twitter-youtube-misinformation-anti-vaccine-covid-19 (accessed December 30, 2020). See
also Brandy Zadrozny, "Anti-vaccination groups target local media after social media crack-
downs," *NBC News*, December 17, 2020 at https://www.nbcnews.com/tech/tech-news/anti-
vaccination-groups-target-local-media-after-social-media-crackdowns-n1251485 (accessed
December 30, 2020).

[28] David Evans and Amina Mendez Acosta, "The Economic Impact of COVID-19: After
Record Unemployment, Countries around the World Begin to Reopen Industries," Center for
Global Development, June 5, 2020 at https://www.cgdev.org/blog/economic-impact-covid-
19-after-record-unemployment-countries-around-world-begin-reopen (accessed on August
11, 2020).

"open up the schools," just as he as for months urged opening up businesses and the economy, with dire effect. As schools began to open, there were predictable reports of outbreak of COVID-19 in the schools, followed by quarantines and widespread panic and anxiety as individuals and regions debated how to protect their children as the President ranted about opening the schools, leading to the possible slaughter of innocents.[29]

Despite the death, destruction, and chaos of the COVID-19 crisis, Trump was publicly optimistic throughout the pandemic, although his optimistic messaging often contradicts that of his administration's public health officials and medical science experts. From January to mid-March, Trump downplayed the threat posed by the coronavirus to the United States, as well as the severity of the outbreak. He presented himself as a "cheerleader for the country," claimed repeatedly that he had the virus under control, and from February into the present, Trump continually asserted that the coronavirus would "go away," and simply disappear.[30]

It didn't and Trump decisively lost the 2020 U.S. Presidential Election to Joe Biden who got 306 votes in the Electoral College and beat Trump by more than seven million votes in the popular vote count. Trump had declared early on that he was a "war-time President" and waging a war against the COVID-19 virus, and it is now clear that Trump lost the war, since as of January 26, 2021, there are now over 400,000 U.S. deaths because of COVID-19, more than 17 million people have contracted the virus in the U.S., and the United States on Trump's "watch" has more documented cases than anywhere in the world.[31] Surely, Trump's shamefully weak response to the pandemic contributed to his loss of the U.S.

---

[29] Robin Foster and E.J. Mundell, "As Schools Reopen, Report Shows 97,000 U.S. Kids Infected With COVID in Late July." *U.S. News and World Report,* August 10, 2020 at https://www.usnews.com/news/health-news/articles/2020-08-10/as-schools-reopen-report-shows-97-000-us-kids-infected-with-covid-in-late-july (accessed on August 13, 2020) and Adam K. Raymond, "As Schools Open, Coronavirus Outbreaks Follow," *New York Times magazine,* Aug. 7, 2020 at https://nymag.com/intelligencer/2020/08/as-schools-open-corona virus-outbreaks-follow.html (accessed on August 13, 2020).

[30] Jessica McDonald, "Trump Baselessly Claims Coronavirus Will 'Go Away' Without Vaccine," *Fact Check,* May 19, 2020 at https://www.factcheck.org/2020/05/trump-baselessly-cla ims-coronavirus-will-go-away-without-vaccine/ (accessed on August 11, 2020).

[31] See the Google COVID-19 page at https://www.google.com/search?source=hp&ei=O4Q MYLnCEbDE0PEP19-u0AE&q=covid-19&oq=CoV&gs (accessed January 23, 2021), and World Health Organization daily statistics are found at "WHO Coronavirus Disease (COVID-19) Dashboard" at https://covid19.who.int/?gclid=EAIaIQobChMIqPWSqb-T6wIV8R-tBh 3xJQHvEAAYASAAEgKVyfD_BwE (accessed January 26, 2021). See also worldometer coronavirus at https://www.worldometers.info/coronavirus/country/us/ (accessed December 16, 2020).

election and a significant part of his legacy will be his abject failure to deal with the COVID-19 pandemic.

Indeed, from March 2020 through the election and into 2021, the COVID-19 Pandemic was a daily media spectacle that has dominated the news and people's lives despite Trump's daily antics of distraction, diversion, and denial. Many of us that have been on lockdown through the pandemic are angry at Trump and his enablers for not properly dealing with the pandemic and millions of families have suffered medical crises and deaths from the pandemic and lost jobs, businesses, and lives because of its deadliness. While the mainstream broadcasting media and social media broadcast Trump's lies and evasions, they also broadcast the spectacle of daily mass death at the hands of the COVID-19 virus.

As I write in early 2021, there are more deaths from COVID-19 every day in the U.S. than died in the terror attacks of 9/11/2001, or the Pearl Harbor attack that propelled the U.S. into World War II.[32] This spectacle of horror will be forever associated with Donald Trump's failed presidency, and while his mastery of media spectacle and social media propelled his rise to power the same technologies that won and sustained his presidency helped bring about its downfall, confirming again that those who rise and triumph via media technology and spectacle can also be destroyed by it.

And yet as globalization spread the dread virus and the production of meat in a Chinese meat market was allegedly the origin of the virus, we now are standing in 2021 in a moment where we hope that vaccines have been created and that vaccinations will be significantly carried out which will cure the virus. Already people are getting the vaccine and the trials have been encouraging. Yet many of us are waiting impatiently to have access to the vaccine, in Trump's America and elsewhere in the world, the vaccination process has been chaotic, and yet there is now hope that science and a vaccine technology will provide a cure to a horrific global pandemic that has characterized 2020 as the Annus Horribilis. Hence, while technology can create horrors it can also create wonders, and it is our fate to play out the dance and drama of technology in our lives.

---

[32] Summer Lin, "Daily COVID deaths in US likely to exceed 9/11 death toll for months, CDC director says." *Miami Herald,* December 11, 2020 at https://www.miamiherald.com/news/cor onavirus/article247775935.html (accessed December 15, 2020).

Intellectuals, Citizens, and Digital Technologies in a New Era of Struggle

5

Abstract

In Chap. 5, I discuss "Intellectuals and Digital Technologies in a New Era of Struggle," focusing on the political challenges of a high-tech world and how individuals can be active participants in the great adventures of the digital revolution. I discuss some challenges to the classical conceptions of the critical intellectual and some of the ways that digital technologies and new public spheres offer new possibilities for democratic discussion and intervention, which call for a redefinition of the intellectual in the ever-proliferating technological matrix. Consequently, I will discuss some changes in the concepts of intellectuals and the public sphere, and how digital technologies and new spheres of public debate and conflict suggest some new possibilities for redefining intellectuals in the present era and the ways that digital technologies and social media can be used as instruments of progressive social change. I argue that the demands of democracy require that intellectuals today master and use digital technology and social media to fight for a more just, egalitarian, and democratic society in an era in which democracy is under attack on a global scale.

Keywords

Digital revolution • Critical intellectual • Digital technologies and new public spheres • Democracy

Critical intellectuals were traditionally those who utilized their skills of speaking and writing to denounce injustices and abuses of power, and to fight for truth, justice, progress, and other positive values. In the words of Jean-Paul Sartre (1974,

© The Author(s), under exclusive license to Springer Fachmedien Wiesbaden GmbH, part of Springer Nature 2021
D. Kellner, *Technology and Democracy: Toward A Critical Theory of Digital Technologies, Technopolitics, and Technocapitalism*, Medienkulturen im digitalen Zeitalter, https://doi.org/10.1007/978-3-658-31790-4_5

p. 285), "the duty of the intellectual is to denounce injustice wherever it occurs." The modern critical intellectual's field of action was what Habermas (1989) called the public sphere of democratic debate, political dialogue, and the writing and discussion of issues of public concern in newspapers, journals, pamphlets, and books. Of course, not all intellectuals were critical or by any means progressive. With the rise of modern societies, there was a division between physical and mental labor, and intellectuals became those who specialized in mental labor, producing and distributing ideas and culture, with some opposing and some legitimating the established forms of society.

Thus, intellectuals were split into those critical and oppositional individuals who opposed injustice and oppression, as contrasted to those producers of ideology and hegemony who legitimated the forms of state, class, race, and gender domination and inequality in modern societies. Both ideologues who produced discourses of individualism, the market, and capitalism legitimating the dominant social order, as well as critical intellectuals who fought for a more just, egalitarian, and democratic society, used the technology of print media for centuries, so intellectuals in the modern era are associated with writing, print, and a culture of books in which reading and writing became important social skills for the intellectual class and eventually all social classes.

In the following reflections, I want to discuss some challenges to the classical conceptions of the critical intellectual and some of the ways that digital technologies and new public spheres offer new possibilities for democratic discussion and intervention, which call for a redefinition of the intellectual in the ever-proliferating technological matrix. Consequently, I will discuss some changes in the concepts of intellectuals and the public sphere, and how digital technologies and new spheres of public debate and conflict suggest some new possibilities for redefining intellectuals in the present era and the ways that digital technologies and social media can be used as instruments of progressive social change. I argue that the demands of democracy require that intellectuals today master and use digital technology and social media to fight for a more just, egalitarian, and democratic society in an era in which democracy is under attack on a global scale.

The Public Sphere and the Intellectual

Jurgen Habermas' concept of the public sphere described a space of institutions and practices between the private interests of those in civil society and the realm of state power. The public sphere thus mediates between the domains of the family and the workplace—where private interests prevail—and the state which often exerts arbitrary forms of power and domination. What Habermas (1989) called the "bourgeois public sphere" consisted of the realm of public assemblies, pubs and coffee houses, literary salons, and meeting halls where individuals gathered to discuss their common public affairs and to organize against arbitrary and oppressive forms of social and public power. The public sphere was nurtured by newspapers, journals, pamphlets, and books which were read and discussed in social sites like pubs, coffee houses, libraries and other public spaces. The bourgeois public sphere was thus the locale—often alongside universities—in which intellectuals were produced and functioned. Emerging forms of democracy required forms of public discussion and debate of the issues of the day, and intellectuals came to specialize in writing, speaking about, and debating those issues of common concern and importance.

Bourgeois societies split, of course, across class lines and different class factions produced different political parties, organizations, and ideologies with each party attracting specialists in words and writing known as intellectuals. Some insurgent intellectuals also produced, or wrote for, newspapers, making the technology of the daily newspaper one of the key elements in the process of eighteenth century revolution and nineteenth and twentieth century politics and culture in all dimensions. Hegel epitomized the newspaper as the key guide to worldly existence, writing: "Reading the morning newspaper is the realist's morning prayer. One orients one's attitude toward the world either by God or by what the world is. The former gives as much security as the latter, in that one knows how one stands."[1]

Many oppressed groups also developed their own insurgent and what Gramsci called organic intellectuals, ranging from representatives of working class organizations like Karl Marx, who largely supported himself writing for newspapers, to women like Mary Wollstonecraft fighting for women's rights, to leaders of oppressed groups of color, ethnicity, sexual preference, and so on who emerged as leaders of new oppositional movements in the 19th and 20th centuries. These insurgent intellectuals often produced their own newspapers and journals, as well as books, contributing to a progressive political public sphere promoting democracy and justice.

Insurgent intellectuals attacked oppression and promoted action that would address the causes of oppression and domination, linking thought to action, theory to practice. Whereas the intellectuals who defended and legitimated the existing

---

[1] Georg Wilhelm Friedrich Hegel, *Miscellaneous Writings, Quotes, at* https://www.goo dreads.com/quotes/1012429-reading-the-morning-newspaper-is-the-realist-s-morning-pra yer-one (accessed December 31, 2020).

society produced affirmative discourses, celebrating modern societies, critical and insurgent intellectuals were specialists in critique and negation, who produced critical and oppositional discourses and in some cases attempted to link discourse to political action. Yet many critical intellectuals were independent of all political organization and limited their range of activity to perpetual criticism, to putting into question, focusing their activity on writing and publication, debate and discussion. These specialists in ideas and debate, these critical intellectuals, were nonetheless often denounced by those in power as merely negative, as ineffectual dreamers,—or as subversive challengers of the existing order.

This pejorative sense of the intellectual was earlier anticipated in Napoleon's denunciation of "ideologues" who are in his view mainly ineffectual although dangerous theoreticians who repose in the realm of ideas and are not grounded in practical affairs or of use to the general public (Barth 1976). Another negative concept of intellectuals as ideologues for the ruling classes shaped a pejorative concept of ideology which Marx himself took up to develop with Engels (1975; 1845) in *The German Ideology*. Marx and Engels denounced intellectuals who served the ruling powers, specializing in legitimation and defense of the existing bourgeois order. And yet Marx, Engels, and most of the spokespeople for the emergent socialist movement were themselves intellectuals, specialists in ideas and criticism who studied, wrote, spoke, and organized against the dominant capitalist society and state.

Indeed, Karl Marx himself was a newspaper editor in the Rhineland before Prussian authorities shut down his paper and he was forced to go into exile in France where for some years he wrote for immigrant newspapers, before moving permanently to London and writing regularly for the *New York Herald* Tribune. Intellectuals in modern societies were thus conflicted beings with contradictory social functions. The classical critical intellectual—represented by figures like the French Enlightenment ideologues, Thomas Paine, Mary Wollstonecraft, and later figures like Heine, Marx, Hugo, Dreyfus, Du Bois, Sartre, and Marcuse—was to speak out against injustice and oppression and to fight for justice, equality, and the other values of the Enlightenment. Their medium was the book, the pamphlet, using print media as a weapon against oppressive figures, institutions, and systems. Indeed, the Enlightenment itself represents one of the most successful discourses of the critical intellectual and a triumph of print culture, with its multifarious books, publications, and political pamphlets and movements which assign intellectuals and print media key social functions. And yet conservative intellectuals attacked the Enlightenment and its prodigy the French Revolution, and produced discourses and texts that legitimated every conceivable form of oppression from class to race, gender, and ethnic domination, and from legitimation of institutions like the monarchy, military, church, and other centers of power and domination, making print culture and the public sphere a contested terrain.

Sartre, the Public Intellectual, and the Postmodern Challenge

Perhaps it is Jean-Paul Sartre who provides the most consequent and probing conceptualizing of the tensions of the classical modern critical intellectual and writer. For Sartre, the domain of the critical intellectual is to write and speak within the public sphere, denouncing oppression and fighting for justice, human rights, and other values. On this model, a critical intellectual's task is to bear witness, analyze, expose, and criticize a wide range of social evils. The sphere and arena of this intellectual is the word and his or her function is critical and negative: to describe and denounce injustice wherever it may occur. For instance, Sartre himself denounced French torture in Algeria in the 1960s, leading to the bombing of his house; he attacked US policy in Vietnam, joining the Bertrand Russell Peace tribunal and other national and international bodies against the US intervention; indeed, he canceled a lucrative lecture tour to the US in the 1960s and rejected a Nobel Peace prize to dramatize his opposition to the Vietnam war.

Moreover, Sartre was a model of a public intellectual who wrote, spoke, and intervened in the public sphere. He founded and edited a journal, *Les temps modernes*, which involved itself in all the political and social issues of the day, and in the 1960s and 1970s Sartre helped found and support a series of radical newspapers like *La cause du peuple* and *J'Accuse*, which he sold in the streets, and *Liberation*, the first independent leftist daily newspaper in France. Sartre also frequently gave interviews to newspapers, journals, and the radio, and late in his life participated in television programs and even movies to publicize his political positions. For the most part, however, for Sartre his pen was his sword and his privileged activity was writing—essays, novels, plays, film scripts, and philosophical treatises to promote his ideas and his politics.

For Sartre, "the true labor of the committed writer… is to show, demonstrate, demystify, and dissolve myths and fetishes in a little bath of critical acid" (1974, p. 375), and his own critical acid became more corrosive and burning than ever during his last decade of activity in the 1970s. For Sartre, the committed intellectual engages on the side of freedom and fights for expanding the realm of freedom to all.

Against Sartre and this notion of the committed intellectual, Foucault complained that Sartre represented an ideal of the universal intellectual who fought

for universal values such as truth and freedom, and assumed the task of speaking for humanity (1977). Against such an exalted and in his view exaggerated
conception, Foucault militated for a conception of the specific intellectual who
intervened on the side of the oppressed in particular issues in specific contexts,
not claiming to speak for the oppressed, but to intervene as an intellectual in
specific issues and debates. Out of this conception of the specific intellectual—
and a turn toward new social movements as the domain of contemporary politics
(Laclau and Mouffe 1985), replacing the state and the national realm of party
politics,—there has emerged a new conception of postmodern politics (see Best
and Kellner 2001).

For a postmodern politics, power is diffuse and local and not merely to be
found in macroinstitutions like the workplace, the state, or patriarchy. Macropolitics that goes after big institutions like the state is to be replaced by micropolitics,
with specific intellectuals intervening in spheres like the university, the prison,
the hospital, or for the rights of specific oppressed groups like sexual or ethnic
minorities. Global and national politics and theories are rejected in favor of more
local, micro theories and politics, and the discourse and function of the intellectuals is seen as more specific, provisional, and modest than in modern theory
and politics, subordinate to local struggles rather than more ambitious projects of
emancipation and social transformation.

In my view, such a binary distinction between macro and micro theory and
politics is problematical, as are absolutist commitments to either modern or postmodern theory tout court (Best and Kellner 1991; Kellner 1995). Using the
example of the events of 1989 that saw the collapse of communism, for instance, it
is clear that the popular offensives against oppressive communist power combined
micro and macropolitics, moving from local and specific struggles rooted in working class pubs, universities, churches, and small groups to mass demonstrations
forcing democratic reforms and even classically revolutionary change and mass
insurrection, as in Romania, which overthrew an oppressive regime. In these struggles, intellectuals played a variety of roles and deployed a diversity of discourses,
ranging from the local and specific to the national and general.

Thus, whereas I would argue that postmodern theory contains important criticism of some of the illusions and ideologies of the traditional modern intellectual,
it goes too far in rejecting the classical role of the critical intellectual, and thus
I argue that the modern conception of the critical and oppositional intellectual
remains useful—although in some contexts the specific intellectual can be called upon to meet the demands of democracy and challenges to struggle against
oppression in a local context. I would, in fact, reject the particular/universal
intellectual dichotomy in favor of developing a normative concept of the **public**

intellectual. The public intellectual—on this conception—intervenes in the public sphere, fights against lies, oppression, and injustice and fights for rights, freedom, and democracy—as with Sartre's committed intellectual. The terrain of struggle for the committed intellectual might be the global sphere, in which one fights for human rights, preserving the environment, or feeding the poor. The oppositional intellectual might also fight in the national realm, fighting oppressive ultranationalists who are racist, xenophobic, and who oppose the basic norms of democracy. Or, as some activists advise, often one should "think globally and fight locally," capturing the multiple dimensions of intellectual and political struggle.

I might note that my first experience of an exemplary public intellectual was hearing Noam Chomsky speak against the Vietnam war when I was a graduate student in philosophy at Columbia University in the mid-1960s. Chomsky was known to be as a philosopher who had a rather complex theory of mind and language, but his presentation of the history of Vietnam anti-colonial wars, first against France, and then the U.S., and his critical dissection of the war provided a lucid analysis based both on fact and critical theory and Chomsky emerged for me as a model of the critical public intellectual. Indeed, he kept writing regularly against the Vietnam war in *The New York Review of Books* and throughout his career wrote works speaking out against imperialism and U.S. foreign policy while continuing his philosophical studies as well.

Shortly thereafter, I met Herbert Marcuse who had also come to Columbia to lecture on the New Left and Vietnam war and was privileged to talk to him at a reception in the Columbia Philosophy lounge and then to accompany him to the West End bar for political and philosophical discussion. Marcuse's subsequent work with the New Left and philosophical-political writings continued his development of a critical theory of society and he continued his activism until his death in 1979. Thereafter, I wrote a book on his life and work *Herbert Marcuse and the Crisis of Marxism* (1984) and was chosen as editor by his family to edit a six volume collection of his writings published by Routledge from 1998 to 2014 and have continued to be active to this day in the International Herbert Marcuse Society.

With these models of a public intellectual before me I decided to become a public intellectual myself and in addition to writing books and articles, giving lectures, and going to conferences, I participated in a public access TV show *Alternative Views* from 1978 until I left the University of Texas in 1998, and the series was shown on public access systems throughout the country, while I participated in public access groups, as well as supporting many specific struggles and movements. Then when I came to UCLA in the late 1990s, I attended a conference on blogs and decided to research and use the new digital public spheres for political analysis and intervention—projects I discuss in the next section.

Yet a democratic public intellectual does not speak for others, does not abrogate or monopolize the function of speaking universal truths, but simply participates in discussion and debate, defending specific ideas, values, or norms or principles that may be particular or universal. If the idea or position advocated is universal, like human rights, they are contextual, provisional, normatively shared or consensual, and not valid for all time. Indeed, rights are products of social struggles and are thus social constructs and are not innate or natural entities—as the classical natural rights theorists would have it. Yet rights can be generalized, extended, and can take universal forms—as with, for instance, a UN charter of human rights that holds that certain rights are valid for all individuals—at least in this world at this point in time.

Consequently, one does not need all of the baggage of the universal intellectual to maintain a conception of a public or democratic intellectual in the present era. Intellectuals may well seek to occupy a higher ground than particularistic interests, a common ground seeking public interests and goods. However, intellectuals should not abrogate the right to speak for all and should be aware that they are speaking from a determinate position with its own biases and limitations. Moreover, intellectuals should learn to get out of their particular frame of reference for more general ones, to take the position of the other, to empathize with more marginal and oppressed groups, to learn from them, and to support their struggles. To perpetually criticize oneself, to develop the capacity for self-reflection and critique—as well as self-expression—is thus part of the duty of the democratic intellectual.

In concluding this section, I suggest that it is Jean-Paul Sartre who provides one of the most impressive articulations of the role of the intellectual in contemporary societies, and who best criticizes the limitations of traditional intellectuals. Then in the final sections, I argue, building on Sartre's conception, for an expansion of the role of the intellectual in our time. In the next section, I argue that digital technologies and social media producing new public spheres for intellectual intervention. Yet I would also argue that in text-based digital technologies reading and writing remain more important than ever in intellectual activity and socio-political struggle, and that digital technologies expand the field of activity of the engaged intellectual from book and print media to digital technologies and new public spheres.

New Technologies, New Public Spheres, and New Intellectuals

In Sartre's view (1970), the vocation of the intellectual was criticism and negation. The critical project required concern with values and ends and capacity for vision, seeing oppression and injustice and ways to fight it. In the following discussion, I will argue that although the public intellectual should assume new functions and activities today, the critical capacities and vision of the classical critical intellectual are still relevant, thus I suggest building on models of the past, rather than simply throwing them over, as in some types of postmodern theory.

In a certain sense, there were always important connections between classical intellectuals and technology. Intellectuals—especially scientific scholars like Leonardo de Vinci, Galileo, or Darwin—deployed technologies from the quill and pen to the printing press, and entire groups like the British Royal Society were concerned with technologies and were indeed often inventors and writers themselves. Some intellectuals used printing presses and were themselves printers and many, though not all, of the major intellectuals of the twentieth century probably used a typewriter and then computers, though I personally know of no major studies of the relationship between the typewriter or computer and intellectuals. Yet a classical intellectual did not have to intrinsically deploy any specific technology and there was thus no intimate connection between intellectuals and specific technologies.[2]

I now want to argue that in the contemporary high-tech societies there is emerging a significant expansion and redefinition of the public sphere and that these developments, connected primarily with media, computer, and digital technologies, require a reformulation and expansion of the concept of critical or committed intellectual to seize these new forces of production, to "refunction" them, and to turn them into instruments to democratize and revolutionize society. Sartre too worked on radio and television series and insisted that "committed writers must get into these relay station arts of the movies and radio" (1974, p. 177; for discussion of his *Les temps modernes* radio series, see p. 177–180).

Previously, radio, television, and the other electronic media of communication tended to be closed to critical and oppositional voices both in systems controlled by the state and by private corporations, as in the U.S. Public access and

---

[2] One could argue that Nietzsche's aphorisms, striking images, and what he called lightning flashes of thought were a result of his enthusiastic use of the typewriter; see the interesting Open Culture entry "Behold Friedrich Nietzsche's Curious Typewriter, the 'Malling-Hansen Writing Ball,'" December 11th, 2013 at https://www.openculture.com/2013/12/friedrich-nie tzsches-curious-typewriter.html (accessed December 3, 2020). The site indicates that Mark Twain was the first major writer to write a book with a typewriter.

low power television and community and guerilla radio, however, opened these technologies to intervention and use by critical intellectuals during the 1970s and 1980s. For some years, I urged progressives to make use of public access and alternative broadcast media, and then extolled Internet activism (Kellner 1979; 1985; 1990; Best and Kellner 2003; Kahn and Kellner 2007). As noted earlier, I was involved in a public access television program *Alternative Views* in Austin, Texas from 1978 to 1998 when I moved from the University of Texas to UCLA, and then for some years at UCLA worked on BlogLeft.[3] Since radio, television, and other electronic modes of communication were creating new public spheres of debate, discussion, and information, I believed that progressives who wanted to be where the people were at, who wanted to communicate with the general public, and who thus wanted to intervene in the public affairs of their society should make use of these technologies and develop new communication politics and new media projects.

In fact, one can argue that the victory of Reagan and the Right in the United States in the 1980s was related to the Right's effective use of television, radio, fax and computer communication, direct mailings, telephones, and other sophisticated political uses of new technologies. Furthermore, one could argue that Clinton's victory over Bush in 1992, and the surprising success of the Perot campaign, were related to effective uses of communication technologies. And in the 1990s in the US, the Republican and rightwing success in the 2000 and 2004 elections can be related to their use of talk radio, computer bulletin boards, and other technologies, while Barack Obama's surprising victory in 2008 can be attributed in part to his adept use of Facebook (over a million friends!) and other social media, while Donald Trump's use of Twitter helped him enormously in the 2016 U.S. presidential election (Kellner 2010, 2015).

---

[3] See the Alternative Views entry in Wikipedia at https://en.wikipedia.org/wiki/Alternative_ Views (accessed on December 3, 2020). It opens: "*Alternative Views* was one of the longest running Public-access television cable TV programs in the United States. Produced in Austin, Texas in 1978, it produced 563 hour long programs featuring news, interviews and opinion pieces from a progressive political perspective. Show founders and on-air hosts, Douglas Kellner and Frank Morrow, produced the show on virtually no budget using facilities at Austin Community Television (ACTV) and The University of Texas at Austin. They also pioneered an innovative syndication system that placed the program in almost 80 television markets around the country." Videos from the show can be found in *The Internet Archive* and "*AlternativeViewTV (YouTube Channel)*". Unfortunately, *BlogLeft* seems to have disappeared the Internet, but see Richard Kahn and Douglas Kellner, "New Media and Internet Activism: From the 'Battle of Seattle' to Blogging," *New Media and Society* 2004 at https://nms.sag epub.com (accessed December 3, 2020).

Consequently, I would argue that effective use of technology is essential in contemporary politics and that intellectuals who wish to intervene in the new public spheres need to deploy new communications media to participate in democratic debate and to shape the future of contemporary societies and culture. My argument is that first broadcast media like radio and television, and then computers, the Internet, and social media have produced new public spheres and spaces for information, debate, and participation that contain both the potential to invigorate democracy and to increase the dissemination of critical and progressive ideas—as well as new possibilities for manipulation and social control. Yet participation in these new public spheres—computer bulletin boards and discussion groups, talk radio and television, social media, and the always-evolving sphere of cyberspace democracy require intellectuals to gain evolving technical skills and to master digital technologies and social media.

I would suggest that there is a more intimate relationship between intellectuals and technology than in previous social configurations in the contemporary digital matrix. To be an intellectual today involves use of the most advanced forces of production to develop and circulate ideas, to do research and involve oneself in political debate and discussion, and to intervene in the digital public spheres produced by broadcasting and digital technologies. In a sense today almost everyone is an intellectual and the majority of people are increasingly using electronic and digital media to communicate their views, discuss issues of personal or public interests, or establish a presence on social media.

New public intellectuals should thus attempt to develop strategies that will use these technologies to attack domination and to promote education, democracy, and political struggle—or whatever goals are normatively posited as desirable to attain. There is thus an intrinsic connection in this argument between the fate of intellectuals and the forces of production which, as always, can be used for conservative or progressive ends.

For some decades now, critical intellectuals have been involved in the use of broadcast and digital technologies to develop alternative forms of culture and information, and today in particular younger intellectuals are involved in the design and use of digital devices and social media as new sources of information, expression, and discussion. In a sense, digital technologies are at least potentially more democratic and empowering than previous communication technologies that were more centralized, often inaccessible to public intervention, and involved in more one-way and top-down communication. Digital technology, to the extent that it is spread throughout the public, is more decentralized, accessible to participation, and thus both empowering and potentially capable of promoting democratic debate and discussion.

Thus, the critical public intellectual who wants to intervene in the new public spheres of the evolving high-tech society has to master the use of digital technologies and social media to be an effective intellectual who participates in some of the key debates and discussions currently going on. The public intellectual of the present must thus learn how to use computers, digital devices, video equipment, and other technologies in order to communicate with a broad public and to assume the role of critical intellectual in promoting democracy and progressive social change. This requires modifying our conception of what an intellectual is and seeing closer connections between intellectuals and the mastery of information and communication technology that was previously the case.

Of course, traditional humanist intellectuals who are intrinsically hostile to technology will not accept such arguments and many of those who identify themselves as progressives will dismiss such an argument as elitist or utopian. Yet I would maintain that the earlier argument that most individuals do not have computers or participate in computer-mediated discussion is now outdated as virtual digital-technology based learning is becoming more central to schooling,[4] and digital technologies are part of the "standard package" of consumer items in the home along with televisions, radios, refrigerators, telephones, and the like. More houses in the US, for instance, have television sets than indoor toilets and to the 99% + penetration by television, I have seen statistics that digital video-recorders are available in over 80% of US homes and will be universal by the twenty-first century, and Uncle Google has just informed me that: "In 2016, 89.3 percent of all households in the United States had a computer at home" (May 12, 2020 at https://www.google.com/search?ei=8x7JX7TwGIbl-gS637HgAQ&q= homes+with+computers+in+the+United+States (accessed on December 3, 2020).[5]

Initially, there was resistance by teachers to implementing computers in the schools. I gave workshops in teaching media literacy around the United States to grade school and high school teachers during the 1980s and 1990s and observed

---

[4] As I write this in December 2020, many schools and Universities and businesses have been closed for months and on-line education is becoming necessary in some parts of the country, and the world, as the COVID-19 pandemic continues to accelerate out-of-control as some governments, like the now-outgoing Trump administration, failed to take responsibility and effective action. This event is providing significant momentum for on-line education and may significantly affect the future of schooling, for better or worse….

[5] "Research: 80% US homes own connected TV device," *Advanced-Television,* June 8, 2020 at https://advanced-television.com/2020/06/08/research-80-us-homes-own-connected-tv-dev ice/ (accessed December 16, 2020) and Stephanie Prange, "Study: Nearly 80% of U.S. Households Subscribe to Netflix, Amazon Prime and/or Hulu," *Media Play News,* August 28, 2020 at https://www.mediaplaynews.com/study-nearly-80-of-u-s-households-subscribe-to-netflix-amazon-prime-and-or-hulu/ (accessed December 17, 2020).

the rapid penetration of computer technologies in the school on all levels, but also observed that, initially, some older teachers were resisting computer technologies and modes of learning. I noted that US corporations were donating their older computer models to poorer school districts for tax write-offs so that computers began to become a standard tool of education in many regional areas of the United States. Even the Republican party in the U.S. suggested in the 1990s that each family in the country should be given a laptop computer for access to the information "super highway," and studies of job prospects over the past decades suggest that without computer skills young people will not be able to enter the labor force and that therefore computer education is an essential part of future schooling.

Computer technologies thus have become more and more central to the home, workplace, social life, politics, and schooling, and thus intellectuals are increasingly forced to master these technologies. Moreover, video and multimedia technologies are becoming more wide-spread in public schools in the United States and elsewhere, suggesting that digital technologies may massively supplement books and print media, requiring that intellectuals who want to intervene in their country's politics and culture will have to learn to use critical media and digital literacies to communicate with the public and their students (Kellner and Share 2019), as they have for decades used computer technologies in their research, writing, and its dissemination. Indeed, with the proliferation of electronic books and journals through major publishing companies, such things as traditional paperbound books and journals may become obsolete, replaced by E-books, on-line electronic journals, and information data bases such as the one Google is amassing.

In fact, intellectuals, like myself, did major research for their books from the 1990s on from computer data bases,[6] and over the past decades I find that I am spending far more time in data bases or internet sources than in libraries for a variety of research projects, such as this one. Computer-mediated discussion forums, sites, and archives often involve the most up-to-date and lively material and discussion of issues concerning digital technologies, communications theory, public policy, and many other topics I am interested in, and I have also participated in enlightening discussions, and found much interesting material, on literature, popular culture, film, rock music, and other topics on discussion forums and digital archives. Likewise, philosophy and theory is archived and discussed on a wave of thinkers encompassing Marx, the Frankfurt School, Marcuse, postmodern

---

[6] In Kellner 1992 and 1995 I tell how I used emerging computer data bases and the Internet for research and writing books in a relatively early era of computer-based research.

theory, Baudrillard, and other topics—to name a few of my own interests. Such electronic archives and discussion forums allow instant communication of one's ideas without the mediation of gate keepers or watchdogs, as well as breaking down the hierarchies between professors and students, tenured and untenured, and so on that structure standard academic communication. And growing archives allow one to peruse in-depth texts and secondary material and to download on their own computer texts and material in areas that interest them—including Uncle Google who is making more and more books available at https://books.google.com/ (accessed January 15, 2021).

Of course, on-line discussion forums also have their limitations as those who have observed flame wars, trolls, and outright stupidity can attest. Indeed, in the past decades of a highly contested U.S. politics it is almost impossible to engage in rational political debate in the highly divided and toxic atmosphere of the Trump era, toxicity which continues into the Biden era. Hate speech was normalized by the President of the United States in Donald Trump's daily twitters which attacked groups, individuals, or nations that criticized him, or questioned his authoritarian proclivities and politics.

Yet one can also explore the higher ground of electronic technoculture. Indeed, as noted, there are more specialized archives in Shakespeare, Marx, cultural studies, media theory, and other topics all the time, as well as newspapers and journals from all over the world that often run intelligent and moderated discussion forums. Thus today's Net citizen is allowed instant access to tremendous amounts of material, so far often free of charge, for those with University, business, or government affiliations. One can also visit museum exhibits, specialized libraries, government archives, and many other sites via the Internet and other computer sites and routes. Indeed, as I suggest below, fighting for continued and expanded free access to computer data bases, archives, and information and communication services is an important political issue for the present and future.

Being an intellectual in the evolving high-tech societies thus requires the mastery of digital technologies and intervention in new public spheres. This process requires expansion of the roles and functions of the intellectual, new terrains and sites of interventions, and new challenges and possibilities—as well no doubt of new traps and illusions. I want to suggest some of these new possibilities in the succeeding sections. Yet first I want to affirm allegiance to the traditional functions of reading, writing, speaking, teaching, and interacting in face to face communication. The evolving digital technologies and public sphere will probably not replace but rather supplement the previous sites and modes of intellectual intervention, and I am personally still committed to traditional roles of reading and writing books, articles, and the like, as well as to face to face teaching and lecturing—which I have been doing more than 50 years in University settings. Yet even our roles as teachers and writers and activists in the traditional sense can be enhanced by digital technologies and social media which should be seen as essential supplements to our activities which involve new challenges and possibilities.

New Tasks for the Public Intellectual

Building on Sartre, I am thus proposing expansion and redefinition of the role of the critical, oppositional, and public intellectual who intervenes in the crucial issues and debates of the day. Indeed, an engaged intellectual, an intellectual who wishes to promote democracy and social justice, should participate in the potentially empowering and participatory spheres of public access television and talk radio, computer discussion forums, social media, and the other spaces of an always-expanding cyberspace democracy and culture. I am aware that these technologies and skills are not yet accessible to all and could be the basis for a new social division between information haves and have nots.

Yet I would argue that we begin thinking of the consequences of the always more powerful digital technologies and social media, and how they can be used to empower people, to promote democracy, and progressive social change—as well as producing new social problems and divisions which we will increasingly have to deal with, such as Trump and reactionary politicians' use of Twitter, Facebook and other social media; the monopolies on information and communication by Google and other Big Tech sites, and future issues that will continue to emerge in an era of always expanding technoculture.

Recognizing problems with the technoculture at large, I am arguing here that reflection on the roles of media and computer technologies in contemporary politics calls attention to the urgency of impending tasks for critical intellectuals that have been often been neglected or overlooked in the tumult and confusion of the past decades. On the positive side, we are living in exciting times in which evolving media and computer technologies are producing expanded possibilities for communication, cultural expression, and ways of living everyday life—at least for privileged individuals and those with access to digital technologies and social media. We should not forget, however, the misery of the vast majority of people around the world and should struggle so that they can attain the same opportunities as those more fortunate.

Moreover, we need to consciously come to terms with our digital technolo-gies and technoculture, and devise ways to use them to enhance our lives and to make them available to all. This requires reflection on media and technology and the demands of democracy, and the challenges and problems of living in an always-evolving digital media/technological society. With these concerns in mind, I would suggest that media and cultural studies need to address several topics that require reflection on expanded activity for critical intellectuals in the present era that I will expand on in this and following chapters.

I want to stress next that a variety of insurgent intellectuals have been making use of these digital technologies and public spheres in their political struggles. Earlier forms of technopolitics involved global phenomena like the peasants and guerilla armies struggling in Chiapis, Mexico who from the beginning of their movement in the 1990s used computer data bases, guerrilla radio, and other forms of media to circulate their struggles and ideas. Every manifesto, text, and bulletin written in Chiapis was immediately circulated through the world via computer networks. When the Mexican government moved against the insurgent movement, computer networks were used to inform and mobilize individuals and groups throughout the world to support their struggles against repressive Mexican government action, putting pressure on the Mexican government and preventing wholesale slaughter of the insurgents.

Earlier, in the late 1970s audiotapes were used to promote the revolution in Iran and to promote alternative information by political movements throughout the world. The Tiananmen Square democracy movement in China and various groups struggling against the remnants of Stalinism in the former communist bloc and Soviet Union used computer bulletin boards, as well as a variety of forms of communications in the 1980s and 1990s, to circulate their struggles. Thus, using new technologies to link theory and practice, to circulate struggles, is neither extraneous to political struggle nor merely utopian.

Indeed, a series of struggles around gender and race have long been mediated by new communications technologies. After the 1991 Clarence Thomas Hearings in the United States on his fitness to be Supreme Court Justice, Thomas's assault on claims of sexual harassment by Anita Hill and others (1998), and the failure of the almost all male US Senate to disqualify the obviously unqualified Thomas, prompted women to use computer and other technologies to attack male privi-lege in the political system in the United States and to rally women and men to support women candidates. The result in the 1992 election was the election of more women candidates than in any previous election and a general rejection of conservative rule.

Likewise, African-American insurgent intellectuals have made use of broadcast and computer technologies to circulate their struggles. John Fiske (1994) has described some African-American radio projects in the "techostruggles" of the 1990s and the central role of the media in recent struggles around race and gender. African-American "knowledge warriors" are using radio, computer bulletin-boards, and other media to circulate their ideas and counter-knowledge on a variety of issues, contesting the mainstream and offering alternative views and politics. Likewise, activists in communities of color—like Oakland, Harlem, and Los Angeles—are setting up community computer and media centers to teach the skills necessary to survive the onslaught of the mediazation of culture and computerization of society to people in their communities. And Black Lives Matter have effectively used media and technoculture to get out their message and organize their struggles against police violence and oppression of their communities, while black women have been especially effective in the group, and a variety of women and women's groups have been active in the #MeToo Movement against sexual harassment and violence against women (see the extensive #MeToo page on Wikipedia at https://en.wikipedia.org/wiki/Me_Too_movement); (accessed on December 3, 2020).

Consequently, a variety of insurgent intellectuals in the present are using the digital technologies and social media to circulate their struggles and information. The technologies of communication are becoming more and more accessible to young people and average citizens, and they should be used to promote democratic self-expression and social progress. Thus, technologies that have traditionally blocked the expansion of participatory democracy, by transforming politics into media spectacles and the battle of images (see Kellner 2003), could also be used to help invigorate democratic debate and participation.

Alternative and Oppositional Media

Over the past decade or more, the oppositional and investigative function of traditional journalism in the United States has been supplemented by alternative media and the Internet and social media with both positive and negative results.[7] I have argued in this study that the only way that a democratic social order can be maintained is for the mainstream media to assume their democratic function of critically discussing all issues of public concern and social problems from a

---

[7] For a variety of perspectives on alternative media, see Couldry and Curran (2003) and Lievrouw (2011).

variety of viewpoints and fostering spirited public debate, accompanied by the development of vigorous and competent investigative and alternative media. The democratic imperative that the mainstream corporate press and broadcasting provide a variety of views on issues of public interest and controversy has often been sacrificed, as has their responsibility to serve as a check against excessive government or corporate power and corruption. As I have documented (Kellner 1990, 2001, 2003a, 2005), there was a **crisis of democracy** in the United States in part because the mainstream corporate media have been biased toward Republicans and conservatives from the 1960s into the Obama era when the mainstream media, as alternative media, became bifurcated into warring camps, a story I am telling in this book. Mainstream corporate media tended for decades to promote the interests of the corporations that own them, which tend to be pro-market and anti-regulation and have largely advanced the interests of corporate institutions and conservative politics.

To remedy this situation, first of all there must be a strengthening of the media reform movement and recognition of the importance of media politics in the struggle for democratization and the creation of a just society, and support and development of alternative media.[8] Democratizing the media system will require development of a dynamic reform movement and recognition for all progressive social movements of the importance of invigorating the media system for forward-looking social change and addressing urgent social problems and issues. This process will involve sustained critique of the corporate media; calls for reregulation; and the revitalization of public television, cultivation of community and public radio, improved public access television, an expansion of investigative and public service journalism, investigations of and call to break-up Big-Tech monopolies, and full democratic utilization of the Internet and social media.

Since corporations control the mainstream press, broadcasting, and other major institutions of culture and communication, there is little hope that the corporate media will be democratized without major pressure or increased government regulation of a sort that is not on the horizon in the present moment in most parts of the world. The Internet and social media, by contrast, provide potential for a democratic revitalization of the public sphere. The Internet and social media makes more information accessible to a greater number of people, more easily, and from a wider array of sources than any instrument of information and communication in history. It is constantly astonishing to discover the extensive array of material available, articulating every conceivable point of view and providing news, opinion, and sources of a striking variety and diversity. Moreover, the Internet

---

[8] On the media reform movement, see Nichols and McChesney 2010.

and social media allows two-way communication and democratic participation in public dialogue, activity that is essential to producing a vital democracy.

One of the major contradictions of the current era is that for the wired world at least, and increasingly the public at large, a rich and diverse information environment is expanding, consisting of a broad spectrum of radio and television broadcasting networks; print media and publications; and the global village of the Internet and social networking sites, which contain the most varied and extensive sources of information and entertainment ever assembled in a single medium. The Internet can send disparate types and sources of information and images instantly throughout the world and has been used by a variety of progressive and oppositional groups (see Best and Kellner 2001; Kellner 1999; Kahn and Kellner 2003; and Lievrouw 2011).

Still, the majority of people in the USA today get their news and information from a highly ideological and limited corporate media—or highly biased and in some cases dangerous hate and conspiracy sites–, creating a major division between the informed and uninformed in the contemporary era. Of course, right-wing and reactionary forces can and have used the Internet to promote their political agendas as well. In a short time, one can easily access an exotic witch's brew of websites maintained by the Ku Klux Klan and myriad neo-Nazi assemblages, including the Aryan Nation, various militia groups, the right-wing Republican attack apparatus—and worse. Indeed, one of the most dispiriting features of the Trump era was how rightwing Republican media and social media from the hardright to the lunatic right spread disinformation of the most appalling sort, culminating in the Big Lie after Trump decisively lost the 2020 election to Joe Biden that the election was "stolen" and that "Trump won by a landslide," culminating in the January 6, 2021 assault on Congress that is still being processed as I finish this book, with hundreds arrested and being investigated and instigators like Trump and his Dark Alliance being investigated.

Hence, the Internet and social media are a contested terrain with progressive, reactionary, and corporate forces using the technology for their conflicting agendas. To be sure, much of the world is not yet wired, many people do not even read, and different inhabitants in various parts of the globe receive their information and culture in very dissimilar ways through varying sources, media, and forms. Thus, the type and quality of information vary tremendously, depending on an individual's access and ability to properly interpret and contextualize it.

Democracy, however, requires informed citizens and access to information and thus the viability of democracy is dependent on citizens seeking out crucial information, having the ability to access and appraise it, and to engage in public conversations about issues of importance. Democratic media reform and alternative media are thus crucial to revitalizing and even preserving the democratic project in the face of powerful corporate and political forces. How media can be democratized and what alternative media can be developed will of course be different in various parts of the world, but without a democratic media politics and alternative media, democracy itself cannot survive in a vigorous form, nor will a wide range of social problems be engaged or even addressed.

Alternative media need to be connected with progressive movements to revitalize democracy and bring an end to the current conservative hegemony. After the defeat of Barry Goldwater in 1964 when conservatives were routed and appeared to be down for the count, they built up a movement of alternative media and political organizations; liberals and progressives now face the same challenge. In the current situation, we cannot expect much help from the corporate media and need to develop ever more vigorous alternative media. The past decades have seen many important steps in the fields of documentary film, digital video and photography, community radio, public access television, an always expanding progressive print media, and an ever-growing liberal and progressive Internet, blogosphere, and social media. While the right has more resources to dedicate to these projects, the growth of progressive democratic public spheres has been impressive.

One result of both the 2008 and 2012 U.S. Presidential elections has been the decentering and marginalizing of the importance of the corporate media punditocracy by Internet, blogosphere, and social media sources. A number of websites and blogs have been dedicated to deconstructing mainstream corporate journalism, taking apart everyone from the right-wing spinners on *Fox News* to reporters for the *New York Times*. An ever-proliferating number of websites have been attacking mainstream pundits, media institutions, and misreporting. Further, during this period, bloggers like Bob Somerby on www.dailyhowler.com (now at https://dailyhowler.blogspot.com/; accessed August 3, 2014) and others savaged mainstream media figures, disclosing their ignorance, bias, and incompetence, while also criticizing relentlessly the Bush-Cheney Gang, Republican Party policies, and increasingly rightwing conservatives, and of course Donald Trump's four year autocratic assault on democracy (Kellner 2017, Stelter 2020).

As a response there were fierce critiques of the blogosphere and social media by mainstream media pundits and sources, although many in the corporate mainstream have developed blogs and social media, appropriating the genre for themselves, and at present blogs and social media are becoming mainstream themselves, used by commentators, politicians, celebrities, and ordinary citizens alike. Yet mainstream corporate broadcasting media, and especially television, continue to exert major political influence in the U.S. and other contemporary societies, and constant critique of corporate and state media should be linked with efforts at reform and developing alternatives, as activists continue to create ever more developed and distributed critical and oppositional media linked to ever-expanding social movements. New media and social networking thus continue to have democratic potential, for without adequate information, intelligent debate, criticism of the established institutions and parties, and meaningful alternatives, democracy is but an ideological phantom, without life or substance.

I have argued in this Chapter that the terrain of contemporary media is highly contradictory and contested. I would not argue that established state and corporate media are reactionary and undemocratic per se, while digital and social media are progressive, for the opposite could be the case in many contexts. Rather, within the current media spectrum, I am arguing that alternative media, social media, and netroots have democratic and progressive potential that should be recognized and utilized by citizens and activist intellectuals.

Throughout the contemporary global world, the media and citizens need to promote democracy by articulating a full range of opinions on matters of public importance, making sure that the government and dominant institutions are responsible and do not abuse power, and by seeing that powerful institutions are playing according to rule of law and a democratic social order. The political struggles of the future will be fought on the terrain of the media which are increasingly important forces within social and political life and which are a key component of democracy. Consequently, the use of digital technologies and alternative media, despite limitations noted above, can be a powerful force in promoting more democratic societies. Indeed, as we shall see in the next section, there was a global upsurge of democratic political struggles using alternative and social media throughout the world from 2011 into the present.

From the 2011 Arab Uprisings Through Occupy and the Trump Resistance

In 2011, the Arab Uprisings, the Libyan revolution, the UK Riots, the Occupy movements and other political insurrections cascaded through broadcasting, print, and digital media, seizing people's attention and emotions, and generating complex and multiple effects that may make 2011 as memorable a year in the history of social upheaval as 1968 and perhaps one as significant (Kellner 2012). In September 2011, a movement "Occupy Wall Street" emerged in New York as a variety of people began protesting the economic system in the United States, corruption on Wall Street, and a diverse range of other issues. The project of "Occupy Wall Street" was proposed by *Adbusters* magazine on July 13, 2011 and on August 9 Occupy Wall Street supporters in New York held a meeting for "We, the 99%." On September 8 a "We are the 99 Percent Tumblair" was launched, and on September 17 Occupy Wall Street protesters began camping out and demonstrating Zuccoti Park in downtown New York close to Wall Street, setting up a tent city, that would be the epicenter of the Occupy movement for some months.

Using social media, more and more people joined the demonstrations which received wide-spread media attention when police attacked peaceful demonstrators, yielding pictures of young women being pepper-gas sprayed by police. Mainstream media attention and mobilizing through social media brought more people to demonstrate and by the first weekend in October, there was a massive protest in lower Manhattan that marched across the Brooklyn Bridge and blocked traffic, leading to over 700 arrests.

The idea caught on and during the weekend of October 1–2, similar "Occupy" demonstrations broke out in San Francisco, Los Angeles, Chicago, Boston, Denver, Washington and several other cities. On October 5 in New York, major unions joined the protest and thousands marched from Foley Square to the Occupy Wall Street encampment in Zuccotti Park. Celebrities, students and professors, and ordinary citizens joined the protest in support, and daily coverage of the movement was appearing in U.S. and global media.

As it has come to own many major political stories of 2011, the British *Guardian* was initially the place to go for Occupy Wall Street in the global media, with a Live Blog documenting news and actions related to the movement, and a web-page collecting their key stories with links to other stories at https://www.guardian.co.uk/world/occupy-wall-street (accessed on October 3, 2011). As the Occupy movement came to London, the *Guardian* focused special attention on their local occupation that involved dramatic clashes with the City of London and Catholic Church when occupiers set up a camp outside the venerable St. Paul's Cathedral; church debates over how to deal with the occupation led high-ranking officials to resign.

In the U.S., police violence against the movement appeared to intensify its support and *Al-Jazeera* television had telling footage on October 5, 2011, of demonstrators videotaping police beating up their colleagues, calling attention to the fact that the participants were using media to organize, to document violence against them, and to circulate their message globally, and that the Occupy Wall Street was traversing the globe as a major media spectacle of the moment.

During the weekend of October 8 and 9, large crowds gathered in Occupy sites throughout the country, and it appeared that a vibrant protest movement had emerged in the United States that articulated with the global struggles of 2011. Like the movements in the Arab Uprising, the Occupy movements were using digital media and social networking to both organize their movement and specific actions, as well as to document police and government assaults on the movement—documentation used to recruit more members and to intensify the commitment and resolve of its participants.

Occupy Wall Street was focused against financial capitalism and the corruption of the political class in the U.S., just as the 1990s anti-corporate global capitalism movement focused on the WTO, World Bank, IMF, and other instruments of global capital. In Greece, Spain, and Italy, people were demonstrating against these same institutions of global capitalism, as well as their own national governments. Like the Arab Uprisings, the Occupy Wall Street and other anti-corporate movements were outside of the domain of old-fashioned party politics, embraced diversity and tended to be leaderless. Although after meeting with Egyptian and other militants, some members of Occupy Wall Street indicated that they were going to search for specific issues that could lead to particular actions, no agreed-upon specific demands were made to define the movement as a whole, although specific actions were undertaken by different Occupy groups, retaining their autonomy.

Slogans such as "We Are the 99 percent" and "Banks Got Bailed Out, We Got Sold Out," and critiques of economic inequality and greed were becoming characteristic of the movement, which was producing a great diversity of slogans, including humorous ones like "We Demand Sweeping, Unspecified Change!" and "One Day the Poor Will Have Nothing to Eat but the Rich." Momentum continued, the protests spread globally, and by mid-October there were over 1000 Occupy sites in over 80 countries. Activism in these movements was taking place simultaneously on-line and in the streets, and activists circulated information, planned events, and mobilized for action. Indeed, by mid-October, there were over 1.2 million followers of the Occupy Wall Street movement on Facebook and hundreds of pages all over the world; during the global protests on October 15–16, 2011, the overall volume of Twitter doubled, as an analysis from Trendrr indicated; see https://blog.trendrr.com/2011/10/21/trendrr-occupy-wall-street-press-recap/ (accessed October 22, 2011).

Interestingly, many of the tactics and goals of the Occupy movement replicated the politics and vision of Guy Debord and the Situationist International,[9] creating situations, demonstrating outside of organized party or movement structures,

---

[9] Guy Debord's *The Society of the Spectacle* (1967) was published in translation in a pirate edition by Black and Red (Detroit) in 1970 and reprinted many times; another edition appeared in 1983 and a new translation in 1994. The key texts of the Situationists and many interesting

using slogans and art of different forms to raise consciousness and inspire revolutionary movements. 2011 was looking more and more like 1968 with eruptions of struggle, police and establishment brutality, and renewed protest and actions. Yet digital media and social networking were creating new terrains of struggle. In using digital media and social networking, the Occupy movements had the same decentralized structure as the computer networks they were using, and the movement as a whole had a virtual dimension as well as people organized in specific spaces. Hence, even if people were not occupying the spaces where the organizing and living were taking place they could participate virtually and be mobilized to participate in specific actions.

While the rightwing Tea Party movement which had helped the Republicans win control of the U.S. Congress in 2010 and block all and any progressive and even mildly ameliorative initiatives during the Obama era, were hierarchical and top-down, the Occupy movements were genuinely bottom-up. The Occupy movement exemplified Deweyean strong democracy, was highly participatory, and experimental in its ideas, tactics, and strategies.[10] While the Tea Party was financed by rich rightwing Republicans like the Koch brothers and had a national television network in *Fox News* to promote their goals and fortify their troops, the Occupy movements produced their own media including their own website, news media, videos, and Livestream that broadcast live action taking place in Occupy sites (see the Occupy Wall Street website at https://occupywallst.org/ [accessed on January 3, 2012] and Livestream at https://www.livestream.com/occupywallstnyc [accessed on January 3, 2012]).

As Michael Greenberg points out, by the middle of October, polls indicated that more than half of Americans polled had a positive view of the movement:

> By mid-October, according to a Brookings Institution survey, 54 percent of Americans held a favorable view of the protest. Suddenly, or so it seemed, there was less talk of budget cuts that would limit, if not dismantle, social insurance programs such as Medicare while extending Bush's tax cuts, and more talk about how to deal with economic inequality.

---

commentaries are found on various Web-sites, producing a curious afterlife for Situationist ideas and practices. For further discussion of Debord and the Situationists, see Steven Best and Douglas Kellner, *The Postmodern Turn*. New York and London: Guilford Press and Routledge, 1997, Chap. 3. On Debord's life and work see also Kaufmann 2006. On the complex and highly contested reception and effects of Guy Debord and the Situationist International, see Marcus 1990; McDonough, editor (2002); and Wark (2008).

[10] On Dewey and participatory democracy, see Antonio and Kellner (1992). On Occupy Wall Street, see Bauer 2012, Chomsky 2013, and Kellner and Satchel 2020.

Several events pointed to an altered political climate. In New York, Governor Andrew Cuomo partially reversed his opposition to extending the so-called millionaire's tax, pushing through legislation for a higher tax rate for the wealthiest New Yorkers. Bank of America, Wells Fargo, and JPMorgan Chase abandoned plans to charge a monthly fee to use their debit cards after an outpouring of indignation from customers—a minor event in the larger picture, but indicative of the public's rapidly shifting mood.

More significantly, in Ohio 61 percent of voters rejected a referendum favored by Republican Governor John Kasich that would have severely restricted the collective bargaining rights of 360,000 public employees. And in Osawatomie, Kansas, on December 6, President Obama gave a speech that echoed almost verbatim what I had been hearing from protesters in Zuccotti Park. Obama deplored "the breathtaking greed of a few" and called the aim to "restore fairness" the "defining issue of our time."[11]

By the end of October, establishment violence against the Occupy movements intensified, and on October 25, 2011 police brutality was used to forcefully remove Occupy Oakland militants, causing a concussion and hospitalization of Scott Olsen, a young Iraq war veteran. Olsen became a cause célèbre and the Oakland movement organized a general strike on November 2 that closed down much of the inner city and first slowed down and then shut down the Port of Oakland, the country's fifth biggest as thousands of marchers descended on the Port. The same day in New York, demonstrators ascended on Lehman Brothers where George W. Bush was allegedly meeting, shouting "Arrest George Bush" and calling for a citizen's arrest that apparently kept Bush imprisoned in the Lehman Brothers building until he was spirited out in a limousine after the demonstrators left for other destinations. Henceforth, demonstrators could be assembled in flash mobs that could occupy any site at a moment's notice and submit corrupt businessmen, politicians, and others to the wrath of the people.

The Occupy movements had generated a critical political discourse widespread through the young and disseminated through media and technology networks that focused on economic inequalities, greed and the corruption of Wall Street and financial institutions, and the need for people to organize and demonstrate to force government to meet their needs. As evidence that the Occupy movements were constituting a threat to the established system of power in November 2011, police and city governments closed down some of the biggest Occupy tent sites, sometimes violently, yet people continued to rally to the cause of the movement and demonstrations, occupations, and actions continued through the year. The brutality

---

[11] See Michael Greenberg, "What Future for Occupy Wall Street?" *New York Review of Books*, February 9, 2012 at https://www.nybooks.com/articles/archives/2012/feb/09/what-future-occupy-wall-street/?pagination=false&printpage=true#fn-1 (accessed on February 10, 2012).

pictured in the closing down of the Occupy Wall Street site on December in Zuccotti park presented images of a fascist police state as images documented police beating up demonstrators, tearing apart and bull-dozing their camp-sites, and throwing their possessions in garbage trucks, including the Occupy Wall Street library that had collected over 5,000 books, thus presenting a frightening image of a fascist police state.

One of the main features of the Occupy movements was having media on hand to document their activities and those of police brutality, and the spectacle of police throughout the United States brutally tearing down Occupy camps made the U.S. look like the thug regimes overthrown in the Arab Uprisings. The documentation accumulated of brutal police power provided material to radicalize new members and harden the resolve of the experienced ones that made possible a continuation of radical Occupy movements in the future.

After the political establishment shut down some of the major Occupy sites, like Occupy Wall Street, members began taking specific actions, transforming public spaces into "temporary autonomous zones" occupied temporarily by flash mobs of protestors. As Michael Greenberg indicates:

> On December 1, for instance, protesters gathered in front of Lincoln Center to await the end of the final performance of Philip Glass's opera *Satyagraha*, about the life of Gandhi. The idea was to dramatize their affinity with Gandhi's method of nonviolent resistance. The following day, occupiers launched twenty-four hours of dance, "radical theater," and "creative resistance" near Times Square meant "to educate tourists and theater-goers about OWS" and to demonstrate "a more colorful image of what our streets could look like." December 6 was the day to "reclaim" selected bank-owned vacant homes in poor neighborhoods, reinstalling a handful of willing families that had been foreclosed upon and evicted. On December 12 there was a march on Goldman Sachs's offices in Manhattan. On December 16 there was a rally at Fort Meade in Maryland where Private Bradley Manning, a hero to the movement, was standing trial for allegedly releasing classified government documents to WikiLeaks. The next day, more rallies were scheduled in New York and elsewhere, this time for immigrants' rights. And so on.[12]

On December 16, the third month anniversary of the Occupy Wall Street movement happened to correspond to the first anniversary of the death of the vegetable vendor Mohamed Bouazizi in Tunisia who had set himself on fire and burned to death in protest, a media spectacle that was frequently taken as the spark that ignited the Arab Uprisings. As argued above, the Occupy Wall Street and Occupy

---

[12] Greenberg, op. cit.

Everywhere! Movements were inspired by the Arab Spring, creating an American Autumn and Winter that guaranteed that 2011 would long be remembered in history books and popular memory as a time in which media spectacle took the forms of political resistance and insurrection.

As 2012 began to unfold, Occupy movements continued to undertake actions throughout the U.S. and the globe. In the U.S. and other countries, the movement had morphed from being primarily located in tent cities and occupations of specific sites to groups focused on particular actions. The Movement's base was expanding to include individuals who had not participated in the first wave of occupations and to make coalitions with varying groups for targeted actions.

Occupy groups in the U.S. also began focusing on politicians, heckling candidates for the Republican presidential nomination in the primaries which began in earnest in early 2012. Those affiliated with the Occupy movement demonstrated against various and sundry politicians of both parties, and carried out protest actions at various politicians' offices in Washington or locally. How the Occupy movements would participate in the 2012 presidential election was of interest to both parties and those participating in or sympathizing with the movement. Indeed, it was the very nature of the multiplicity and complexity of the Occupy movements that they could not fit into standard political models and were thus spontaneous and unpredictable in nature.

The Occupy groups and their allies could point to specific victories in early 2012, to which their movements had partially contributed. On January 18, 2012, major Internet industry web-sites went black in a day of protest against a proposed Congressional bill Stop Online Piracy Act (SOPA) and a Protect Identity Property Act, which opponents claim could lead to online censorship and force some websites out of business. By midday, Google officials asserted that 4.5 million people had signed its petition against SOPA,[13] while Wikipedia claimed that 5.5 million people had accessed the site and clinked on a link that would put them in touch with local legislators to register their opposition to the act. Evidently, the action had an impact as politicians who had been for the bill, suddenly indicated opposition to it, and the bill's sponsors withdrew it for further consideration.

On January 18, 2012, the Obama administration announced it would temporarily deny a permit for the building of the highly toxic Keystone XL Pipeline

---

[13] There are a variety of on-line petitions against SOFA including the ACLU's "Sign the Pledge: I Stand With the ACLU in Fighting SOPA" at https://secure.aclu.org/site/SPageS erver?pagename=sem_sopa&s_subsrc=SEM_Google_Search-SOPA_SOPA_sopa%20bill_ p_10385864662 (accessed on February 9, 2012) and Broadband for America's "Hands off the Internet" at https://www.broadbandforamerica.com/handsofftheinternet?gclid=COqHzp uska4CFQN8hwod0GBVew (accessed on February 9, 2012).

which would have transported extremely dirty oil from a vast oil deposit in Alberta, Canada, to refineries on the Texas Gulf Coast.[14] And on the same day, activists were celebrating in Wisconsin having received over one million signatories to have a recall election to potentially unseat Governor Scott Walker who was financed with ultra-right wing Tea Party movement money and had attacked union bargaining rights in a highly publicized affair that led union workers, students, activists and their supporters to occupy the Madison Wisconsin state capital in protest in May 2011,[15] linking Occupy movements in the Middle East with the U.S. and anticipating the Occupy Wall Street movement by some months.

Hence, new politics and subjectivities were emerging from specific sites of the Occupy movement, which were global in inspiration, tactics, and connections, leading to an new era of global, national, and local political struggle with unforeseeable outcomes in the Time of the Spectacle. These movements were inspired and connected in certain ways with the North African Arab Uprisings that began an intense year of struggle throughout the world in 2011. History and the future are open and depend on the will, imagination and resolve of the people to create their own lives and futures rather than being passive objects of their masters. Media spectacle is a contested terrain upon which the key political struggles of the day are fought and 2011 was a year rich in examples of media spectacle as insurrection.

Media and Cultural Activism in the Trump Era

The Trump era exhibits the further expansion of youth and political resistance movements paralleling new social movements of the 1960s. Following the year of upheaval in 2011, new movements such as Black Lives Matter emerged, accompanied by Dreamers and Latino youth struggling for basic rights to immigration and to stay in a country to which their parents brought them and the #MeToo movement opposing sexual harassment and violence against women. These struggles were followed by Bernie Youth during the 2016 Democratic primaries where an army of youth mobilized behind progressive candidate Bernie Sanders in his struggle against Hillary Clinton for the Democratic Party Presidential nomination.

---

[14] There are multiple web-sites devoted to blocking the construction of the Keystone XL pipeline such as the Natural Resources Defense Council (NRDC)'s site Stopping the Keystone XL Pipeline at https://www.nrdc.org/energy/keystone-pipeline/?gclid=CMX6o7Gtka4CFQV ahwodkAwofQ (accessed on January 9, 2012).

[15] There are many Recall Scott Walker sites such as "United Wisconsin to Recall Walker" at https://www.unitedwisconsin.com/onedaylonger (accessed on February 8, 2012).

Bernie Youth remained active during and after the election and were part of a Trump resistance movement consisting of black, brown, white, and a rainbow of youth of all ages struggling against the Trump administration, along with highly organized groups of older women and men. After a series of mass shootings in the Trump era where the Trump administration and Republican controlled Congress refused to take any action on rational gun control, youth in Parkland, Florida, after a school shooting on February 14, 2018, which left 17 dead and 17 wounded, the students of Parkland mobilized a national pro-gun control movement under the hashtag #NeverAgain, a movement inspired in part by the ground broken by the #MeToo movement and the 2018 Women's March. #NeverAgain demanded legislative action to prevent similar mass shootings, and vowed to organize and defeat lawmakers who received political contributions from the National Rifle Association in the upcoming 2018 Congressional elections and beyond.[16] The group rallied on February 17 in Fort Lauderdale planned to focus on legislative action and rallies in support took place all over the world. The Women's March Network organized a school walkout that took place on March 14, and the Parkland youth and their supporters helped organize demonstrations named "March for Our Lives" on March 24 which included a march in Washington, D.C.

In all these cases, insurrectionary movements used media spectacle and digital and social media to mobilize resisting youth, people of color, women, and members of the Trump resistance. Yet all of these movements were overshadowed by the explosion of Black Lives Matter which constitutes one of the most important resistance movements in U.S. history.

Race and racism has been a highly contested feature of US life throughout US history and the color line between Black and White has been a defining feature of life in the US. Since the Civil Rights movement emerged in the 1960s, racism has been sharply opposed and resistance and movements opposing racism have been a recurrent phenomenon in US politics. In 2013, after the acquittal of George Zimmerman, who admitted to shooting a young unarmed Black man named Trayvon Martin, activist Alicia Garza turned to Twitter to express her devastation at the verdict. She wrote, "Our Lives Matter, Black Lives Matter," which her friend

---

[16] During this period, the NRA went from being one of the most powerful lobby groups in the country which blocked rational gun safety and control issues and elected rightwing politicians who would embrace the agenda of the weapons industry to a disgraced organization in financial crisis, facing major financial and other scandals. See Jane Coaston, "The NRA's big, bad financial mismanagement crisis, explained. The NRA is low on cash and embroiled in complex lawsuits," *Vox.com,* May 14, 2019 at https://www.vox.com/2019/4/30/18510946/nra-finances-lapierre-north-accusations-corruption (accessed on December 3, 2020).

Fig. 5.1 Protests in May
2020 after George Floyd's
death. (source: https://en.
wikipedia.org/wiki/Black_
Lives_Matter)

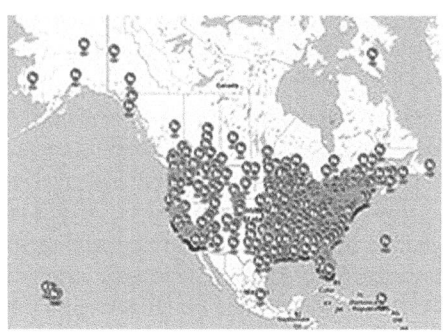

and fellow organizer, Patrisse Cullors, quickly adapted into #BlackLivesMatter.[17]
The hashtag is commonly believed to have sparked a movement (NPR Staff, "The
#BlackLivesMatter Movement"). While the hashtag, and subsequently, the official
Black Lives Matter network, provided a visible slogan and cause around which
groups could coalesce and those new to activism could identify, the movement
itself is built on a broader foundation established by those working for social
justice in intersecting areas of American life.

As it evolved, the Black Lives Matter movement came to encompass a broad
array of people and organizations. The slogan "Black Lives Matter" became a
rallying cry used against police violence towards black people as well as a protest
slogan against other injustices, as well as a call for reform progressive social
change that would mitigate the racism that is an endemic stain on U.S. politics
and culture.

Throughout 2020, there were increasing intense #BlackLivesMatter demonstrations with 450 major protests held by the end of May 2020, incited by a stream
of police murder of unarmed blacks and unceasing police brutality against people
of color (see Fig. 5.1 and 5.2). As Wikipedia summarizes: "The breaking point
was due primarily to the killing of George Floyd by Minneapolis police officer Derek Chauvin, eventually charged with second degree murder after a video
circulated showing Chauvin kneeling on Floyd's neck for nearly nine minutes
while Floyd pleaded for his life, repeating, "I can't breathe". Following protesters' demands for additional prosecutions, three other officers were charged with
aiding and abetting second degree murder."[18]

---

[17] On Black Lives Matter see Kellner and Satchel 2018.

[18] Wikipedia has an extremely extensive and well-documented Black Lives Matter entry which
I am drawing on for the dramatic events of the movement in 2020 and have pasted in some

Fig. 5.2 "Black Lives Matter" on the facade of the Washington National Cathedral, June 10, 2020. (source: https://en.wikipedia. org/wiki/Black_Lives_ Matter)

In the following weeks, Black Lives Matter organized rallies in the United States and worldwide to protest George Floyd's death and calling for the end of police brutality, the end of police funding which ranged from calls to defund the police, reinvest in social services for communities of color, halting the construction of new jails, and other demands which varied from community to community. On June 5, Washington, D.C.'s Mayor Muriel Bowser announced that part of the street outside of the White House had been officially renamed to Black Lives Matter Plaza, and from May 22, to August, 22, 2020, there were more than 10,600 BLM protest events in the United States.[19]

Hence, we see that during the Trump era, there were a vast array of political insurrections as media spectacle in contemporary U.S. politics that used digital technologies and social media to combat manifold forms of oppression and to battle for social change across gender, race, class, regional, political, and other

---

pictures from the site for illustration; see https://en.wikipedia.org/wiki/Black_Lives_Matter (accessed December 3, 2020).

[19] "Demonstrations & Political Violence in America: New Data for Summer 2020". ACLED. September 3, 2020 at https://en.wikipedia.org/wiki/Black_Lives_Matter (accessed December 3, 2020).

divides. The progressive forces of these movements coalesced to defeat Trump in the U.S. Presidential Election and now the U.S. has a future, with possibilities for social change if individuals stay or become committed to democracy, social justice, and a better life for all. Technopolitics will be part of this struggle, as it has been for the past decades, and so individuals and activists need to devise ways that digital technologies and social media can become emancipatory forces in the struggle for the future. I will expand on this argument in the next section.

Concluding Remarks

Critical media and cultural studies has been especially negligent of developing strategies and practices for media intervention and the production of alternative media and devising strategies for technopolitics. There has been little discussion within cultural studies circles concerning how radio, television, film, and other media could be transformed and used as instruments of social enlightenment and progress. Likewise, the Frankfurt School seemed inherently skeptical of media technologies and viewed them as totally controlled by capitalist corporations. Indeed, when the classical theories of the cultural industries were being formed, this was more or less the case. The failure of critical social theory and media and cultural studies today to engage the issue of alternative media is more puzzling and less excusable since there are today a variety of venues for alternative film and video production, community radio, computer bulletin boards and discussion forums, and other forms of communications within which progressives can readily intervene.

Thus, critical social and media theory today should discuss how the media and culture can be transformed into instruments of social enlightenment and progress. This requires more focus on alternative media than has previously been evident and reflections on how media and digital technology can be reconfigured and used to empower individuals. It requires developing activist strategies to intervene in public access television, community radio, digital activism, social media, and other domains that will emerge in the future. To genuinely empower individuals requires giving them knowledge of digital technology and media and becoming proficient in using media and technoculture to produce texts and messages that are then disseminated to the public and used for political mobilization and struggle for democracy and social justice. Increasing media activism could significantly enhance democracy, making possible the proliferation of voices and allowing those voices that have been silenced or marginalized to speak.

A first step in this process, which I shall address later in more detail in Chap. 10, involves cultivating, teaching, and disseminating critical media and digital literacies so that individuals can become literate in engaging media and digital culture and develop new technoliteracies as new media and technologies evolve. Technoliteracies can be deployed in technopolitics which uses technologies to advance political goals such as I've discussed in this chapter and will also engage in the next chapter on "Globalization, Technopolitics, and Revolution."

As I argued in this chapter, emergent public spheres and technologies produce new roles and functions for intellectuals and for intellectuals to deploy digital technologies in the new public spheres. Media and the technoculture are producing complex cyberspaces to explore and map, and new terrains of political struggle and intervention. The emergent cyberintellectuals of the present may not be the organic intellectuals of a class, but we can become technointellectuals of digital technologies and new cultural spaces, charting and navigating through the brave new worlds of media culture and technoculture and deploying and inventing technopolitics as instruments of progressive social transformation. These technologies can be used as instruments of domination or liberation, of manipulation or social enlightenment, and it is up to the activist intellectuals of the present and future to determine which way digital technologies and social media will be used and developed and whose interests they serve.

A democratic media politics will accordingly be concerned that the media and computer technologies will be used to serve the interests of the people and not corporate elites. A democratic media and technopolitics will strive to see that media are used to inform and enlighten individuals rather than to manipulate them. A democratic media and technopolitics will teach individuals how to use the media and digital technologies, to articulate their own experiences and interests, and to promote democratic debate and diversity, allowing a full range of voices and ideas to become part of the cyberdemocracy of the future.

Globalization, Technopolitics and Revolution

6

> *'The coincidence of the changing of circumstances and of human activity can be conceived and rationally understood only as revolutionary practice.'*
>
> *– Karl Marx*
>
> *'A community will evolve only when a people control their own communication.'*
>
> *– Frantz Fanon*

Abstract

Chapter 6 on "Globalization, Technopolitics and Revolution," I analyze in more detail the global economy, politics, and culture from the standpoint of technology and a networked world. I will engage some issues involving technopolitics, and the alleged rise of a new economy, networked society, and cyberspace, in relation to the problematic of revolution and the prospects for a radical democratic or progressive transformation of society. Globalization and

---

This study and the concepts of globalization and technological revolution developed here are grounded in the studies of Cvetkovich and Kellner 1997; Best and Kellner 2001; Kahn and Kellner 2007; and my earlier articles on globalization, including Kellner 1998, 1999, 2002, 2003, 2005, 2014. By "revolution," I am assuming a concept of fundamental economic, political, social, and cultural transformation, such as was developed in the works of Herbert Marcuse. See Kellner 1984 and the six volumes of Marcuse's collected and largely unpublished papers that I have published with Routledge; Marcuse 1994–2004.

© The Author(s), under exclusive license to Springer Fachmedien Wiesbaden GmbH, part of Springer Nature 2021
D. Kellner, *Technology and Democracy: Toward A Critical Theory of Digital Technologies, Technopolitics, and Technocapitalism*, Medienkulturen im digitalen Zeitalter, https://doi.org/10.1007/978-3-658-31790-4_6

153

the rise of a digital computer and information technology based economy and society are interpreted in both popular and academic literature as modes of technological revolution in which new digital technologies are transforming every mode of life from how individuals do research to how people communicate and interact socially. There is some truth in this notion, but it is also true that the technological revolution can perpetuate the interests of the dominant economic and political powers, intensify divisions between haves and have nots, and is a defining feature of a new form of global technocapitalism.

Keywords

Globalization • technopolitics • revolution • New economy • networked society • and cyberspace • Technological revolution • Technocapitalism

In this chapter, I will engage some issues involving globalization, technopolitics, and the alleged rise of a new economy, networked society, and cyberspace in relation to the problematic of revolution and the prospects for a radical democratic or progressive transformation of society. Globalization and the rise of a digital computer and information technology based economy and society are interpreted in both popular and academic literature as modes of technological revolution in which new digital technologies are transforming every mode of life from how individuals do research to how people communicate and interact socially. There is some truth in this notion, but it is also true that the technological revolution can perpetuate the interests of the dominant economic and political powers, intensify divisions between haves and have nots, and is a defining feature of a new form of global technocapitalism.

Yet while I would argue that there are novelties and discontinuities in the current configuration of economic, political, social, and cultural constellations that constitute the contemporary moment, there are also continuities with the previous forms of modern society. In particular, the so-called "new economy" exhibits crucial features of the old capitalism such as the driving forces of capital accumulation, competition, commodification, exploitation, and the business cycle. From this perspective, globalization and technological revolution are best theorized as forms of the global restructuring of capitalism in which technological development and a turbulent socio-economic transformation are intrinsically interconnected.

On the topic of whether globalization renders revolution in the classical Marxian tradition obsolete, I would argue that much significant political struggle today is mediated by technopolitics and social movements. The use of computer and information technology is becoming a normalized aspect of politics, just as the broadcasting media were some decades ago. Deploying computer-mediated technology for technopolitics, however, opens new terrains of political struggle for voices and groups excluded from the mainstream media and thus increases potential for resistance and intervention by oppositional groups. Hence, if revolution is to have a future in the contemporary era, it must incorporate technopolitics as part of its strategy, conceiving of technopolitics, however, as an arm of emancipatory struggle aimed at social justice and not an end in and of itself.

Consequently, in this chapter, I focus my discussion on the ways in which an oppositional democratic politics can use digital technologies and social media to intervene within the global restructuring of capitalism to promote democratic and progressive social movements aiming at radical structural transformation. I would argue that globalization and technological revolution are in some ways inevitable—barring an apocalyptic collapse of the global economy—but the forms that they take are not. That is, I think that the trends toward a more global economy and culture, a networked society, and the continued flow of commodities, images, cultural forms, technology, and people across the globe will continue apace, as will intense technological revolution. Both take the form of what Schumpeter (2009) called "creative destruction" and guarantee that the next decades will be highly turbulent, contested, and full of struggle and conflict.

Yet, I would argue, the forms that globalization and technological revolution will take are neither fixed nor determined. Hence, it is arguably perfectly reasonable to oppose corporate capitalist globalization and its market model of society, its neo-liberal laissez-faire ideology, and its putting profit, competition, and market logic before all other aspects of life. I will accordingly focus on the ways that technopolitics can and are being used for anti-capitalist contestation, while noting the limitations of this conception. First, however, some comments on theorizing globalization critically.

Theorizing Globalization Critically

Globalization continues to be one of the most hotly debated and contested phenomena of the past several decades. A wide and diverse range of social theorists have argued that today's world is organized by accelerating globalization, which is strengthening the dominance of a world capitalist economic system, supplanting the primacy of the nation-state by transnational corporations and organizations, and eroding local cultures and traditions through a global culture. Contemporary theorists from a wide range of political and theoretical positions are converging on the position that globalization is a distinguishing trend of the present moment, but there are hot debates concerning its origins, nature, effects, and future.[1]

For its defenders, globalization marks the triumph of capitalism and its market economy. Apologists for capitalist globalization such as Fukuyama (1993) and Friedman (1999, 2005) perceive this process as positive, while its critics portray globalization as destructive and negative (see Mander and Goldsmith 1996; Eisenstein 2004; Robins and Webster 1999). Thomas Friedman sees globalization as promoting both capitalism and democracy a la capitalist ideologue Milton Friedman.[2] For Thomas Friedman, capitalism's allegedly free markets help generate free democratic societies, and globalization for Friedman brings both markets and liberal democracies in it path and thus serves to promote democratic societies through technology and globalizing forces. Friedman stresses the integration of markets, nation states, and technologies in a friction-free way parallel to Bill Gates and both claim that an information technology propelled globalization will enable individuals, corporations, and nation states to circulate, ideas, commodities, and political forms like democracy around the world in a more encompassing, faster, deeper, and more efficient way than ever before and in a way that is enabling to business, democracy, and individual mobility within a cosmos of global integration.

For others, globalization promotes market economies at the expense of democracy and I shall attend to this debate throughout the chapter. Some theorists highlight the emergence of a new transnational ruling elite and the universalization of consumerism (Sklair 2001), while others stress global fragmentation of "the clash of civilizations" (Huntington 1996). While some argue for the novelties of globalization and even claim it constitutes a rupture in history, others stress continuities with modernity and play down differences and novelties (see Rossi 2007). Driving "post" discourses into novel realms of theory and politics, Hardt and Negri (2000, 2004, 2011) present the emergence of "Empire" as producing evolving forms of sovereignty, economy, and culture that clash with a "multitude" of disparate groups, unleashing political struggle and an unpredictable flow of novelties, surprises, and upheavals.

---

[1] The literature on globalization is vast including a pro-globalization book by Friedman (1999), the Marxist critiques by Hardt and Negri (2000) material in the readers edited by Ritzer and Dean 2015. See also books by Falk 1999; Brecher, Costello, and Smith 2000; Steger 2002; Ritzer 1993; and Stiglitz 2017.

[2] Thomas and Milton Friedman are apparently not related. See the Google page on the two Friedmans that summarizes their major ideas;

Discourses of globalization initially were polarized into pro or con "globophilia" that celebrates globalization contrasted to globophobia that attacks it.[3] For critics, "globophilia" provides a cover concept for global capitalism and imperialism, and is accordingly condemned as another form of the imposition of the logic of capital and the market on ever more regions of the world and spheres of life. For defenders, globalization is the continuation of modernization and a force of progress, increased wealth, freedom, democracy, and happiness. Its "globophiliac" champions thus present globalization as beneficial, generating fresh economic opportunities, political democratization, cultural diversity, and the opening to an exciting new world. Its "globophobic" detractors see globalization as harmful, bringing about increased domination and control by the wealthier overdeveloped nations over the poor underdeveloped countries, thus increasing the hegemony of the "haves" over the "have nots". In addition, supplementing the negative view, globalization critics assert that it produces an undermining of democracy, a cultural homogenization, hyperexploitation of workers, and increased destruction of natural species and the environment. There was also a tendency in some theorists to exaggerate the novelties of globalization and others to dismiss these claims by arguing that globalization has been going on for centuries and there is not that much that is new and different. Some imagine the globalization project—whether viewed positively or negatively—as inevitable and beyond human control and intervention, whereas others view globalization as generating new conflicts and new spaces for struggle, distinguishing between globalization from above and globalization from below (see Brecher et al. 2000).

Engaging the "dialectics of globalization," I would argue for developing a **critical theory of globalization** that will undercut the opposing globophobic and globophilia discourses in order to discuss the fundamental transformations in the world economy, politics, and culture in a dialectical framework that distinguishes between progressive and emancipatory features and oppressive and negative attributes. Such a dialectical theory of globalization is similar to the critical theory of technology that I am developing which rejects technophilia and technophobia by arguing that technology has both extremely positive and negative aspects and effects. Developing a critical theory of technology and globalization thus requires articulations of the contradictions and ambiguities of globalization and the ways

---

[3] What now appears as the first stage of academic and popular discourses of globalization in the 1990s tended to be dichotomized into celebratory globophilia and dismissive globophobia. There was also a tendency in some theorists to exaggerate the novelties of globalization and others to dismiss these claims by arguing that globalization has been going on for centuries and there is not that much that is new and different. For an excellent delineation and critique of academic discourses on globalization, see Steger 2002.

that globalization is both imposed from above and yet can be contested and reconfigured from below in ways that promote democracy and social justice. Theorizing globalization critically involves theorizing it at once as a product of technological revolution and the global restructuring of capitalism in which economic, technological, political, and cultural features are intertwined (Best and Kellner 2001, Kellner 2002).

From this perspective, one should avoid both technological and economic determinism and all one-sided optics of globalization in favor of a view that theorizes globalization as a highly complex, contradictory, and thus ambiguous set of institutions and social relations, as well as involving flows of goods, services, ideas, technologies, cultural forms, and people (see Appadurai 1996; Kellner 2002). Finally, I focus on the politics of globalization, stressing resistance and oppositional movements to corporate and neo-liberal globalization, and sketch a "cosmopolitan globalization" as an alternative model which meets the demands of democracy and provides a model of a democratic, multicultural, and cosmopolitan globalization.

As the ever-proliferating literature on the topic indicates, the term "globalization" is thus often used as a code word that stands for a tremendous diversity of issues and problems and that serves as a front for a variety of theoretical and political positions. While it can serve as a legitimating ideology to cover over and sanitize ugly realities, a critical globalization theory can inflect the discourse to point precisely to these phenomena and can elucidate a series of contemporary problems and conflicts. In view of the different concepts and functions of globalization discourse, it is important to note that the concept is a theoretical construct that varies according to the assumptions and commitments of the theory in question. Seeing the term globalization as a construct helps rob it of its force of nature, as a sign of an inexorable triumph of market forces and the hegemony of capital, or, as the extreme right fears, of a rapidly encroaching world government. While the term can both describe and legitimate capitalist transnationalism and supranational government institutions, a critical theory of globalization does not buy into ideological valorizations and affirms difference, resistance, democratic self-determination, and an alternative cosmopolitan globalization against forms of global domination and subordination.

Technopolitics and Oppositional Political Movements

Significant political struggles today against globalization are mediated by technopolitics, that is the use of digital technologies and social media to advance political goals such as democracy and social justice. To some extent, politics in the modern era have always been mediated by technology, with the printing press, photography, film, and radio and television playing crucial roles in politics and all realms of social life, as McLuhan, Innis, Mumford, and others have long argued and documented. Participation in representative democracies is mediated by technology, and the right to vote and casting of ballots is often presented as the core of democracy, whereby the people choose their representatives and government in a free, open, and non-coercive where the voters can choose the candidate of their choice. In the U.S., there were successive struggles to give workers the vote, and then women through the suffragette movements in the 1920s, followed by there were struggles to give immigrants the vote during the progressive era. And the civil rights movements has been battling from the post-World War II period until the present to secure the right to vote for people of color who had long been excluded from the ballot.

Of course, disasters can befall the voting process and to keep democracy alive requires vigilance, care, and often struggle. Voting can go wrong as it did in the disastrous failure of voting machines and the voting-counting process in the U.S. 2000 Presidential Election dramatized (Kellner 2001), as well as the Russian intervention into Facebook in the 2016 U.S. presidential election which helped Trump win a narrow victory and wreak havoc in his four-year assault on democracy (see Kellner 2017; Nance 2018; Corn and Ishikoff 2018; Hettena 2018). There have been long-time battles over suppression of votes, especially in the case of people of color and the poor, over the counting of votes, and over alleged irregularities in the voting process although through the tumultuous history of struggles over voting in the last century there is a democratic consensus that the sanctity of the vote and the voting process is a key foundation of democracy.

The computerization of voting has also promoted struggles over the role of technology and democracy in the electoral process in the United States. What is new about computer and information technology mediated politics is that information can be instantly communicated to large numbers of individuals throughout the world who are connected via computer networks. The Internet and social media are also potentially interactive, allowing discussion, debate, and online and archived discussion. Digital technology and social media are increasingly multimedia in scope, allowing the dissemination of images, sounds, video, and other cultural forms. Moreover, the use of digital technology and networks is becoming a normalized aspect of politics, just as the broadcasting media were some decades ago. The use of digital technology for technopolitics, however, opens new terrains of political struggle for voices and groups excluded from the mainstream media and thus increases the potential for intervention by oppositional groups, significantly expanding the scope of democratization.

Given the extent to which capital and its logic of commodification have colonized ever more areas of everyday life in recent years, it is somewhat astonishing that cyberspace is by and large decommodified for large numbers of people—at least in the overdeveloped countries like the United States.[4] On the other hand, using digital technology, transforming information into data-packets that can be sent through networks, and hooking up to digital networks oneself, involves a form of commodified activity, inserting the user in networks and technology that are at the forefront of the information revolution and global restructuring of capital. Yet the networks and social media are owned by giant corporations like Facebook, Google, Apple, Microsoft, and other Titans of the Digital Economy, while titans like Amazon, Walmart, and giant corporations control the rapidly expanding domain of digital commerce, given a tremendous boost during the COVID-19 pandemic of 2020.[5] Thus the Internet is highly ambiguous from the perspective of commodification, as it is from other perspectives, being both an instrument of individual and social empowerment, and corporate control, exploitation, surveillance, power, and profit.

Digital technology is also ambiguous when it comes to democracy. While digitization of information brings vast amounts of competing information to politics, it is often highly conflictual, thus the citizens must become information literate, and learn what sources of information are reliable, what sources are biased, and what sources are accurate. Just as voting is the life-blood of democracy, so do is reliable information, as the informed voter must have good information to make an intelligent vote that corresponds to her or his interests, politics, and life-situation.

---

[4] In many parts of the world, individuals must pay telephone companies for each unit of time on the Internet, giving rise to movements everywhere for an affordable flat-rate for monthly Internet access. In the United States, the first Internet access companies like CompuServe, Prodigy, the Source, America On-Line, Yahoo, and other Internet access services provided connections for a reasonable rate, and Universities and businesses started to provide free Internet access to its students, workers and others in the 1980s and 1990s, while today a diversity of services provide service, still charged on a monthly basis and not according to use.

[5] On the giant corporations who control digital networks, communication, and commerce, see Galloway 2018 and Noble 2018.

This highlights again the importance of critical digital literacies for democracy, where the informed voter can access the best sources of information, assess the biases on different sides, and can assess what information is most reliable and accurate and a sound info-base upon which to vote. A discerning voter in the U.S. knows that *Fox News* is pro-Republican, conservative, and has been a Trump propaganda apparatus over the past few years, normalizing his lies in Foxy news, readily consumed by his supporters as truth. Similar pro-Trump propaganda goes through the Internet ranging from far-right and extremist sites to popular radio hosts like Rush Limbaugh, other popular figures on talk radio and TV, and pro-Trump ideologues in a number of pro-Trump rightwing internet sites, publications, and books (see Stelter 2020).

An info-literate citizen knows, by contrast, that *MSNBC* is fiercely anti-Trump and anti-Republican and pro-Democrat. MSNBC came to national prominence criticizing the Bush-Cheney regime and then turned pro-Obama, before swiveling to fierce critique of Trump, and now they are back to being firmly pro-Biden as I've watched during the first weeks of the Biden administration as they feature interviews with their top Administration officials, and strongly defend their policies, initiatives, and actions.

Likewise, experience with newspapers, internet sites, social media, books, and other material allows one to become info-literate so that one can discern the best information on any given target. Given the ubiquity of the Internet, Google, and other key information sources it is especially important to have cultivated critical digital literacies in relation to digital technology sources as I shall argue below and again in Chapter 9 on the importance of education is cultivating critical media and digital literacies.

However widespread and common computers and digital technologies become, it is clear that they are of essential importance already for labor, politics, education, and social life, and that people who want to participate in the public and cultural life of the future will need to have computer access and literacies. Although there is a real threat that the computerization of society will intensify the current inequalities in relations of class, race, and gender power, there is also the possibility that a democratized and computerized public sphere might provide opportunities to overcome these injustices (see Chapter 4).

Cyberdemocracy and the Internet should be seen therefore as a contested terrain. Radical democratic activists should look to its possibilities for resistance and the advancement of political education, action, and organization, while engaging in struggles over the digital divide and excessive high-tech corporate power. Dominant corporate and state powers, as well as conservative and rightist groups, have been making sustained use of digital technologies and social media to

advance their agendas. If forces struggling for democratization and social justice want to become players in the cultural and political battles of the future, they
must devise ways to use new technologies to advance a radical democratic and
ecological agenda and the interests of the oppressed.

There are by now copious examples of how the Internet and cyberdemocracy
have been used within oppositional political movements. A large number of insurgent intellectuals are already making use of new technologies and public spheres
in their political projects, as I described in the last chapters. Seeing the progressive potential of advanced communication technologies in revolutionary struggle,
Frantz Fanon (1967) described the central role of the radio in the Algerian Revolution, and Lenin stressed the importance of film in spreading communist ideology
after the Bolshevik revolution. Audiotapes were used to advance the insurrection
in Iran in the late 1970s and to disseminate alternative information by political
movements throughout the world (Downing 1984, 2000). The Tiananmen Square
democracy movement in China and various groups struggling against the remnants of Stalinism in the former communist bloc used computer bulletin boards
and networks, as well as a variety of forms of communications, to promote their
movements (Béja 2010). Anti-NAFTA groups made extensive use of the new communications technology (Brenner 1994; Fredericks 1994), as did other sectors of
the anti-globalization movement. Such multinational networking and distribution
of information failed to stop NAFTA, but created alliances useful for the politics
of the future. As Nick Dyer-Witheford (1999) notes:

> The anti-NAFTA coalitions, while mobilizing a depth of opposition entirely unexpec-
> ted by capital, failed in their immediate objectives. But the transcontinental dialogues
> which emerged checked—though by no means eliminated—the chauvinist element in
> North American opposition to free trade. The movement created a powerful pedagogi-
> cal crucible for cross-sectoral and cross-border organizing. And it opened pathways for
> future connections, including electronic ones, which were later effectively mobilized by
> the Zapatista uprising and in continuing initiatives against maquilladora exploitation.

And in the last chapter I discussed how in the Trump era, the Trump resistance,
Black Lives Matter, #MeToo, and other movements deployed technopolitics to
advance their movements and struggles. Thus, using digital technologies and
social media to link information and transformative social practice and to advance
oppositional politics is neither extraneous to political battles nor merely utopian.
Even if immediate gains are not won, often the information circulated or the
alliances formed can have material effects. There are, moreover, striking examples of how digital technology-centered organizing campaigns effectively worked
against the institutions and corporations of capitalist globalization.

Successful struggles against the Multilateral Agreement on Investment (MAI) in 1995–1998 involved websites and email campaigns against the U.S.-supported effort to develop binding rules on how states treat foreign investors and list-serves linking the groups struggling against the agreement. Obviously, digital technology alone did not defeat this initiative for capitalist globalization, but it enabled the non-government organizations fighting against it to circulate information, share resources, and link their struggles (Smith and Smythe 2000).

There have been many campaigns against the excesses of capitalist global corporations such as Nike and McDonald's. Hackers attacked Nike's site in June 2000 and substituted a global justice message for Nike's corporate hype. Many anti-Nike websites and list-serves emerged, helping groups struggling against Nike's labor practices circulate information and organize movements against Nike, which have forced them to modify their labor practices.[6]

A British group that created an anti-McDonald's website against the junk food corporation and then distributed the information through digital and print media received significant attention. This site was developed by supporters of two British activists, Helen Steel and Dave Morris, who were sued by McDonald's for distributing leaflets denouncing the corporation's low wages, advertising practices, involvement in deforestation, cruel treatment of animals, and patronage of an unhealthy diet. The activists counterattacked and with help from supporters, organized a McLibel campaign, assembled a McSpotlight website with a tremendous amount of information criticizing the corporation, and mobilized experts to testify and confirm their criticisms. The three-year civil trial, Britain's longest ever, ended ambiguously on June 19, 1997, with the Judge defending some of McDonald's claims against the activists, while substantiating some of the activists' criticisms (Vidal 1997, pp. 299–315).

The case created unprecedented bad publicity for McDonald's which was disseminated throughout the world via Internet websites, mailing lists, and discussion groups. The McLibel/McSpotlight group claims that their website was accessed over fifteen million times and was visited over two million times in the month of the verdict alone (Vidal 1997, p. 326). The *Guardian* reported that the site "claimed to be the most comprehensive source of information on a multinational corporation ever assembled," and was part of one of the more successful anti-corporate campaigns up to that point (February 22, 1996; see also www.mcspot light.org).

---

[6] For an overview of Nike, see Goldman and Papson 1999. For a dossier of material assembled on Nike's labor practices and campaigns against them, see the website constructed by David M. Boje at https://web.nmsu.edu/~dboje/ (accessed on November 4, 2020).

Anti-Nike, McDonalds, and other websites critical of global capitalist corporations have disseminated a tremendous amount of information. Many labor organizations are also beginning to make use of the new technologies. Mike Cooley (1987) has written on how computer systems can reskill rather than deskill workers, while Shoshana Zuboff (1988) has discussed how high-tech can be used to informate workplaces rather than automate them, expanding workers' knowledge and control over operations rather than reducing and eliminating it, as well as providing powerful instrument of surveillance.

Indeed, the use of technologies in the workplace also has negative effects. Many jobs are eliminated as production is automated or moved to the developing world (itself an ambiguous phenomenon); technology-mediated labor also involves deskilling and can have harmful effects on workers; a new class of "net-slaves" is developing with relatively low-paid and/or temporary job status, without benefits; and computer-surveillance gives capital powerful new tools to monitor and spy on workers.[7]

Yet, digital technologies have been extensively used in labor movements and actions. The Clean Clothes Campaign, a movement started by Dutch women in 1990 in support of Filipino garment workers, has supported strikes throughout the world, exposing exploitative working conditions (see www.cleanclothes.org/1/index.html). In 1997, activists involved in Korean workers strikes and the Merseyside dock strike in England used websites to promote international solidarity (for the latter, see www.gn.apc.org/labournet/docks/ (accessed January 22, 1998)). Jesse Drew (1998) has extensively interviewed representatives of major U.S. labor organizations to see how they were making use of digital communication technologies and how these instruments helped them with their struggles; many of his union activists indicated how useful email, faxes, websites, and the Internet have been to their struggles and, in particular, indicated how such technopolitics helped organize demonstrations or strikes in favor of striking English or Australian dockworkers, as when U.S. longshoremen organized strikes to boycott ships carrying material loaded by scab workers. Technopolitics thus helps labor create global alliances in order to combat increasingly transnational corporations.[8]

---

[7] The classic text on deskilling of labor is still Braverman 1972; on "net-slaves," see Crompton and Jones 1984; Lessard and Baldwin 2000; and on computer-mediated surveillance, see Robins and Webster 1999 and Lyon 2001.

[8] For an overview of the use of electronic communication technology by labor, see the studies by Moody 1998; Waterman 1990, 1992; Brecher and Costello 1994; Dyer-Witheford 1999; Drew 1998, 2013, and https://www.jessedrew.com/ (accessed December 4, 2020) which documents Drew's later interventions in alternative media, arts, and communications research. Labor projects in 1990s using the new technologies include the U.S. based Labornet,

On the whole, labor organisations, such as the North South Dignity of Labor group, note that computer networks are useful for organizing and distributing information, but cannot replace print media, which are more accessible to many of its members, face-to-face meetings, and traditional forms of political action. Thus, the challenge is to articulate one's communications politics with actual movements and struggles so that cyberpolitics is an arm of real battles rather than their replacement or substitute. The most efficacious Internet projects have indeed intersected with activist movements encompassing campaigns to free political prisoners, boycotts of corporate projects, and various labor and even revolutionary struggles, as noted above.

The Global Movement against Capitalist Globalization

One of the more instructive examples of the use of digital technology to foster global struggles against the excesses of corporate capitalism occurred in the protests in Seattle and throughout the world against the World Trade Organization (WTO) meeting in December 1999, and the subsequent emergence of a worldwide anti-globalization movement in 2000–2001. Behind these actions was a global protest movement using the Internet to organize resistance to the institutions of capitalist globalization, while championing democratization. In the build-up to the 1999 Seattle demonstrations, many websites generated anti-WTO material and numerous mailing lists used the Internet to distribute critical material and to organize the protest.[9] The result was the mobilization of caravans from throughout the United States to take protestors to Seattle, as well as contingents of activists

the European Geonet, the Canadian LaborL, the South African WorkNet, the Asia Labour Monitor Resource Centre, Mujer a Mujer, representing Latina womens groups, and the Third World Network, while PeaceNet in the United States was devoted to a variety of progressive peace and justice issues. For contemporary perspectives on labor and politics, see Chandler 2019.

[9] In addition, while the organizers demanded that the protesters agree not to engage in violent action, there was one web site that urged WTO protesters to help tie up the WTO's Web servers, and another group produced an anti-WTO web site that replicated the look of the official site (see RTMark's website, gatt.org/; the same group had produced a replica of George W. Bush's website with satirical and critical material, winning the wrath of the Bush campaign). For compelling accounts of the anti-WTO demonstrations in Seattle and an acute analysis of the issues involved, see the documents collected in Danaher and Burbach 2000, and Cockburn, St Clair, and Sekula 2000. See Smith and Smythe 2001 for detailed analysis of the use of the Internet in the anti-WTO demonstrations; they located 4089 websites with material specific to the Seattle WTO meetings and selected 513 to examine and classify.

throughout the world. Many of the protestors had never met and were recruited through the Internet. For the first time ever, labor, environmentalist, feminist, anti-capitalist, animal rights, anarchist, and other groups organized to protest aspects of globalization and to form new alliances and solidarities for future struggles. In addition, demonstrations took place throughout the world, and a proliferation of anti-WTO material against the extremely secret group spread throughout the Internet.[10]

Furthermore, digital technologies and networks provided critical coverage of the event, documentation of the various groups' protests, and debate over the WTO and globalization. Whereas the mainstream media presented the protests as "anti-trade," featured the incidents of anarchist violence against property, and minimized police brutality against demonstrators, digital networks provided pictures, eyewitness accounts, and reports of police viciousness and the generally peaceful nature of the protests. While the mainstream media framed the Seattle anti-WTO activities negatively and privileged suspect spokespeople like Patrick Buchanan as critics of globalization, the Internet provided multiple representations of the demonstrations, advanced reflective discussion of the WTO and globalization, and presented a diversity of critical perspectives.

The Seattle protests had some immediate consequences. The day after the demonstrators made good on their promise to shut down the WTO negotiations, Bill Clinton gave a speech endorsing the concept of labor rights enforceable by trade sanctions, thus effectively making impossible any agreement during the Seattle meetings. In addition, at the World Economic Forum in Davos a month later there was much discussion of how concessions were necessary on labor and the environment if consensus over globalization and free trade were to be possible. Importantly, the issue of overcoming divisions between the information-rich and the information-poor, and improving the lot of the disenfranchised and oppressed, bringing these groups the benefits of globalization, were also seriously discussed at the meeting and in the media.

More important, many activists were energized by the new alliances, solidarities, and militancy, and continued to cultivate an anti-globalization movement. The Seattle demonstrations were followed by the April 2000 struggles in Washington, DC, to protest the World Bank and IMF, and later in the year against capitalist globalization in Prague and Melbourne; in April 2001, an extremely large and militant protest erupted against the Free Trade Area of the Americas summit in Quebec City. It was apparent that a new worldwide movement was in the making that was uniting diverse opponents of capitalist globalization throughout the world. The anti-corporate globalization movement favored globalization-from-below, which would protect the environment, labor rights, national cultures, democratization, and other goods from the ravages of an uncontrolled capitalist globalization (Falk 1999; Brecher et al. 2000).

The movement against capitalist globalization used digital technologies and networks to organize mass demonstrations and to disseminate information to the world concerning the policies of the institutions of capitalist globalization. The events made clear that the protestors were not against globalization per se, but were against neoliberal globalization, opposing specific policies and institutions that produce intensified exploitation of labor, environmental devastation, growing divisions among the social classes, and the undermining of democracy. The emerging anti-globalization-from-above movements have located these problems in the context of opposition to a restructuring of a neo-liberal market capitalism on a worldwide basis for maximum profit with zero accountability. The anti-capitalist movements, by contrast, have made clear the need for democratization, regulation, rules, and globalization in the interests of people and not profit.

The new movements against globalization-from-above have thus placed the issues of global justice, democracy, and environmental issues squarely in the center of important political concerns of our time. Hence, whereas the mainstream media had failed to vigorously debate or even to report on globalization until the recent past, and rarely, if ever, critically discussed the activities of the WTO, World Bank and IMF, there is now a widely circulating critical discourse and controversy regarding these institutions. Stung by criticisms, representatives of the World Bank, in particular, are pledging reform. Pressures are mounting concerning proper and improper roles for the major global institutions, highlighting their limitations and deficiencies, and the need for reforms like debt relief for overburdened developing countries to solve some of their fiscal and social problems.

Hence, to capital's globalization=from-above, cyberactivists have been attempting to carry out globalization-from-below, developing networks of solidarity and propagating oppositional ideas and movements throughout the planet. To the capitalist international of transnational corporate-led globalization, a Fifth International, to use Waterman's phrase (1992), of digital technology-mediated activism has emerged, that is qualitatively different from the party-based socialist and communist Internationals. Such networking links labor, feminist, ecological, peace, and other anti-capitalist and anti-imperialist groups, providing the basis for a new democratic politics of alliance and solidarity to overcome the limitations of postmodern identity politics (Dyer-Witheford 1999; Burbach 2001; Best and Kellner 2001).

Technopolitics: A Contested Terrain

A key to developing a robust technopolitics is articulation, the mediation of technopolitics with real problems and struggles, rather than self-contained reflections on the internal politics of the Internet. The Zapatista movement in Chiapis addressed problems of survival and transforming social, cultural, political, and economic conditions, using new technologies as an instrument of political struggle. Likewise, the campaigns against major capitalist corporations and the institutions of capitalist globalization are attempting to advance progressive political agendas and to engage key issues of the day, as have the movements of the Trump era like Black Lives Matter, the Dreamers, #MeToo, and the Climate Change movement struggling for the survival of the earth.[10]

There are, moreover, many other examples of how technopolitics are effectively serving as an instrument of social transformation on many national and local terrains. The movement in Serbia in the mid-1990s that led to the electoral defeat and eventual overthrow of the Milosevic regime deployed digital technologies as organs of information and organization for years. The Serbian government's annulment of a November 1996 election won by oppositional forces in 1996 sparked fierce criticism disseminated worldwide through the Internet, forcing a reversal of the results and producing what was described as the "first Internet revolution" (*New York Times*, September 8, 1998). Although the repressive forces of the Serbian government continued to exercise power, growing Internet and public criticism helped generate mobilization of the opposition against the Milosevic government, including Internet radio that was disseminated throughout the region. By the summer of 2000, opposition forces were able to generate massive anti-government rallies and peacefully vote Milosevic out of office, thanks in part to Internet-generated information and discourse, and when he refused to leave office, massive demonstrations, sparked in part by Internet critique and mobilization, forced him to resign.

Increased numbers of groups and individuals are also using technopolitics to advance local issues. A 1998 *New York Times* article discusses "Santa Monica Seeking a Return to Online Civic Forum of Yore" (September 8, 1998). In addition, the Santa Monica Public Electronic Network, or PEN, established itself, providing jobs, shelter, and amenities to the homeless and inaugurating a forum

---

[10] See the American Archive of Public Broadcasting page on "the History of the Environmental Movement" at https://americanarchive.org/exhibits/climate-change/history (accessed on December 4, 2020).

of public dialogue.[11] With the demise of Santa Monica's only local newspaper in March 1998, such forums are more essential than ever to public dialogue, and the city is attempting to reinvigorate its online civic platform, the oldest public affairs network in the U.S. Other *NYT* articles the same week discuss a lawsuit over the right of students to criticize their schools and educational authorities on websites (September 4, 1998), and how various corporations are using electronic mail and bulletin-boards to promote criticism of company policies and personnel (September 7, 1998).

Youth are also using a form of technopolitics to create new venues for culture, youth communities, and forms of communication and interaction. After cultivating varieties of websites, zines, and chat-rooms youth have been developing P2P (person-to-person, or peer-to-peer) forms of communication and interaction. These involve swapping texts, music, video, and other cultural forms. It requires leaving one's computer on so that others can access files available for sharing, generating new de-commodified cultural spaces and communities. P2P-sharing is also now being used for medical research and to share computer power and archives.[12]

By 2020, youth are making use of the Internet for a wealth of projects ranging from individual to P2P to group projects including politics. The decades in which youth have been involved in a variety to technopolitics suggest how use of digital and social media make possible a refiguring of politics, a refocusing of politics on everyday life and using the tools and techniques of digital information and communication technology to expand the field and domain of politics by supporting a variety of social movements organized around race, gender, class, and sexuality (see Kellner and Satchel 2020). In this conjuncture, the ideas of Guy Debord and the Situationist International are especially relevant with their stress on the construction of situations, the use of technology, media of communication,

---

[11] On PEN, see Kevin McKeown, "Social Norms and Implications of Santa Monica's PEN (Public Electronic Network)" on-line at https://www.mckeown.net/PENaddress.html (accessed January 1, 2021) and the study "The_Public_Electronic_Network_PEN_and_the_Homeless_in_Santa_Monica" at https://www.researchgate.net/publication/247959082_The_Public_Electronic_Network_PEN_and_the_Homeless_in_Santa_Monica (accessed January 1, 2021).

[12] See Damien Cave, "Come together, right now, over P2P," December 14, 2000. The article discusses an organization "Popular Power" which pays individuals five dollars a month to allow use of their computing power for businesses (that in turn pay the parent organization), non-profit organizations, or scientific research groups, who are given free access. P2P networks continue to be active up to the present in the 2000s, although they are now utilized by commercial networks. See, for example, the Google page on P2P that I accessed today at https://www.google.com/search?source=hp&ei (accessed December 4, 2020).

and cultural forms to promote a revolution of everyday life, and to increase the realm of freedom, community, and empowerment.[13]

To some extent, digital technologies and social media *are* revolutionary, they *do* constitute a revolution of everyday life, but it is often a revolution that promotes and disseminates the capitalist consumer society and involves new modes of fetishism, enslavement, and domination, yet to be clearly perceived and recognized by many. Clearly, rightwing and reactionary forces can and have used digital technology and social media to promote their political agendas as well. In a short time, one can easily access websites maintained by the Ku Klux Klan and myriad neo-Nazi assemblages, including the Aryan Nation and various militia groups—although during the Trump era many of the extreme rightwing hate groups are being banned from social media after the Trump-promoted assault and occupation of the Capital in January 2021 and the threats by rightwing extremist groups against federal and state government institutions and politicians.[14]

Both foreign and domestic groups have been circulating divisive hate speech and promoting domestic terrorism during the Trump era and rightwing extremists are aggressively active on many social media forums and apps, as well as radio programs and stations, public access television programs, fax campaigns, video and even rock music productions. These organizations are hardly harmless, having carried out terrorism of various sorts extending from church burnings to the bombings of public buildings, and even the occupation and discretion of the Capital of the United States with the alleged intent of some of the rioters of capturing and killing many politicians.[15] Adopting quasi-Leninist discourse and tactics for ultra-right causes, these groups have been successful in recruiting working-class members devastated by the developments of global capitalism which has resulted in widespread unemployment for traditional forms of industrial, agricultural, and unskilled labor—such aggrieved working class people have become willing shock troops for demagogues like Trump and other rightwing authoritarians throughout

---

[13] On the importance of the ideas of Debord and the Situationist International to make sense of the present conjuncture, see Best and Kellner 1997, 2001; Knabb 2006; Trier 2019.

[14] Vinopal, Courtney (2021) "How the U.S. Capitol attack highlights the challenges of thwarting online right-wing extremism," *PBS Newshour*, January 13, 2021 at https://www.pbs.org/newshour/politics/how-the-u-s-capitol-attack-highlights-the-challenges-of-thwarting-online-right-wing-extremism (accessed January 16, 2021).

[15] Godfrey, Elaine (2021) "Some Pro-Trump Rioters Wanted More Violence," *The Atlantic*, January 9, 2021 at https://www.theatlantic.com/politics/archive/2021/01/trump-rioters-wanted-more-violence-worse/617614/ (accessed January 16, 2021) and "Boogaloo Bois Prepare for Civil War," *The Atlantic*, January 15, 2021 at https://www.theatlantic.com/politics/archive/2021/01/boogaloo-prepare-civil-war/617683/(accessed on January 16, 2021).

the world. Moreover, extremist websites have influenced alienated middle-class youth as well. A 1999 HBO documentary on *Hate on the Internet* provides a disturbing number of examples of how extremist websites influenced disaffected youth to commit hate crimes.[16]

Hence, a distressing feature in the saga of technopolitics is that alleged terrorist groups throughout the globalized world are increasingly using the social media, the Internet, and apps to promote their causes. This process has evolved through the twenty first century. An article in the *Los Angeles Times* (February 8, 2001: A1 and A14) reports that groups like Hamas use their website to post reports of acts of terror against Israel, rather than calling newspapers or broadcasting outlets. A wide range of groups around the globe labelled as "terrorist" reportedly use digital technology, mailing and email lists, and websites to further their struggles, including Hezbollah and Hamas, the Maoist group Shining Path in Peru, and a variety of other groups throughout Asia and elsewhere. The Tamil Tigers, for instance, a liberation movement in Sri Lanka, offers position papers, daily news, and free email service. According to the *Times*, experts are still unclear "whether the ability to communicate online worldwide is prompting an increase or a decrease in terrorist acts (ibid.)."

After the January 6, 2021 assault on the Capital by pro-Trump and extreme right insurrectionary forces officials worried that anti-government demonstrations would be carried out in all 50 state capitals to protest Biden's inauguration, although only a handful actually showed up. There are daily media reports in early 2021 that document how participators in the insurrection in the riot organized over the internet and social media and continue to use digital technologies to recruit new members, propagandize, and organize new insurrections.

Different political groups in various countries of the globe are in fact engaging in cyberwar[17] as adjuncts of their political battles. Israeli hackers have repeatedly attacked the websites of Hezbollah, while pro-Palestine hackers have reportedly placed militant demands and slogans on the websites of Israel's army, foreign ministry, and parliament.[18] Likewise, Pakistani and Indian computer hackers have

---

[16] For a contemporary take on hate speech on the Internet globally, see the Council on Foreign Relations site and the study by Zachary Laub, "Hate Speech on Social Media: Global Comparisons. Violence attributed to online hate speech has increased worldwide. Societies confronting the trend must deal with questions of free speech and censorship on widely used tech platforms." June 7, 2019 at https://www.cfr.org/backgrounder/hate-speech-social-media-global-comparisons (accessed December 4, 2020).

[17] I discuss variants and theorists of cyberwar in more detail in Chapters 6 and 7 below.

[18] A google search for "terrorism israel palestine" listed "Israel Ministry of Foreign Affairs" as the top two sources, followed by the US State Department, and then more largely Israeli

waged similar cyberbattles against opposing websites in the bloody struggle over Kashmir, while rebel forces in the Philippines taunt government troops with cell-phone calls and messages and attack government websites. This trend continued into the 2000s with every oppositional and government political group waging cyberwar of various sorts against their opponents.

In addition, it is now *de rigueur* for mainstream politicians to run their political campaigns as a critical part of their overall strategy, as websites can provide information on the candidate, citizen feedback, and, of course, links for volunteer efforts and donations. It is widely held that without the Internet, former wrestler and independent candidate Jesse "The Body" Ventura would have lost his bid for governor of Minnesota, since his funding and influence grew primarily through a plain Internet site and a burgeoning email list.[19] Unlike one-way transmission TV ads, the sites of Bush and Gore in the 2000 presidential race featured interactive links for citizens to get involved. Gore had special links for students, African Americans, Asians, Hispanics, and gays and lesbians. Such links were conspicuously absent on Bush's site, except for an "en espanol" link and a "Just For Kids!" page that likened a presidential campaign to a baseball game (Kellner 2001). And, as I argued in previous chapters and studies, both the Obama and Trump campaigns benefitted significantly for their use of digital technologies and technopolitics.

And yet despite the terrorist groups who have used technopolitics to promote hate and terror, global media and technopolitics provide for the possibilities of better understanding, communication, and a more cosmopolitan globalization. Hence, I will conclude that despite the global horror and upsurge in domestic and foreign terrorism and the increasing dangers of cyberwar, there are also trends toward a more positive globalization that provides an alternative from the horrors of the contemporary moment.

---

sources before getting to Palestinian sources; see https://www.google.com/search?source=hp&ei=l2HvX960FcjK-gSUv5mwAg&q=terrorism+israel+palestine&oq=terrorism+israel+palestine&gs_lcp=CgZwc3ktYWIQDDIFCAAQyQNQqhFYqhFg_B1oAHAAeACAAYIBiAGCAZIBAzAuMZgBAKABAqABAaoBB2d3cy13aXo&sclient=psy-ab&ved=0ahUKEwie3P7WpvvtAhVIpZ4KHZRfBiYQ4dUDCAg#spf=1609523609999 (accessed January 1, 2021).

[19] Rebecca Fairley Raney, "Former Wrestler's Campaign Got a Boost" From the Internet November 6, 1998 at https://archive.nytimes.com/www.nytimes.com/library/tech/98/11/cyber/articles/06campaign.html (accessed January 1, 2021).

Toward a Cosmopolitan Globalization

The first stage of the anti-corporate globalization movement was largely negative and produced a form of globophobia toward corporate globalization and neoliberalism. Yet pursuing the need for an alternative vision and an answer to TINA (There Is No Alternative, i.e. to corporate globalization), in the past years the search has been for alter or other globalizations, providing positive visions of what a more democratic, just, ecological, and peaceful globalization could be and how to attain it, or at least move beyond the disastrously flawed and largely failed neo-liberal vision.

There are certainly reasons to develop a robust critique of globalization as I have been arguing and documenting. Indeed, writing in 2020 as a global COVID-19 virus pandemic has swept the globe the entire year and despite lockdowns continues to ravage communities like Los Angeles, from where I write, the pandemic continues to witness appalling levels of death, overstressed hospitals, and a population sick and tired of lockdown and quarantine. It is clear that only a global solution to a global pandemic like COVID-19 can help manage and defeat the virus and in the first days of his administration Biden has put out many constructive initiatives to fight COVID-19 on a global scale,[20] going back into the WHO and other global health organizations that Trump with his demagogic "America First" pulled out of, betraying our allies and U.S. engaged democratic global traditions.

As I write in early 2021 we also are learning that for months the Russians have infiltrated U.S. security systems throughout the government,[21] reminding us that the Russians hacking of the Democratic Party helped elect the horrific Trump

---

[20] The Council of Councils affirms multiple groups supporting Biden's returning the U.S. to global institutions and politics, see Global Memo by ISS, IAI, CEPS, CARI, PISM, Genron, SVOP, CIGI, RSIS, Lowy, COMEXI, GRC, SAIIA, SWP, INSS, GRF, IISS/PKU "Biden and the World: Global Perspectives on the U.S. Presidential Election," November 11, 2020 at https://www.cfr.org/councilofcouncils/global-memos/biden-and-world-global-perspectives-us-presidential-election-0 (accessed January 27, 2021).

[21] Ben Fox, "Hack against US is 'grave' threat, cybersecurity agency says," *AP News*, December 18, 2020 at https://apnews.com/article/technology-malware-hacking-russia-software-b3f 993fb7bc9390302f0df26ecb6c10e (accessed December 19, 2020) and Karl Paul, "What you need to know about the biggest hack of the US government in years. Russian agents are suspected in the Orion breach, which affected the treasury and commerce departments—and perhaps others," *The Guardian*, December 15, 2020 at https://www.theguardian.com/technology/2020/dec/15/orion-hack-solar-winds-explained-us-treasury-commerce-department (accessed December 19, 2020).

whose denial and appallingly inadequate national response to the COVID-19 pandemic allowed it to ravage the country. While Putin's poodle Trump initially failed to respond to news reports of massive Russian hacking of U.S. security and infrastructure and then tried to minimize it, President-elect Biden has promised to retaliate and U.S. national security hawks are urging extreme measures.[22]

Hence, just as we have lived through the threat of Nuclear Armageddon, we now face Network Armageddon and the dangers of a world in which global computer networks are hacked and perhaps destroyed. Globalization has thus exhibited global health pandemics and now confronts us with global attacks on the networked economy which should give pause to ideologues of globalization who have long told us how wonderful it is.

Yet we are condemned to living in a global world and we must attempt to deal with its problems and challenges and develop a positive vision of the global world we are now living in. A critical theory of globalization and dialectical emancipatory vision thus needs to not only develop a critique of neoliberal or corporate globalization and analyze its contradictions, but needs to project a positive ideal of alternative globalizations, focusing on democracy and a cosmopolitan globalization. Resistance and struggle against corporate globalization need to have a positive ideal of what kind of globalization to struggle for since we are fated to live in a global world. Different societies and groups will, of course, have different alternative versions and strategies in mind but in conclusion I want to stress that corporate and neoliberal globalization could be opposed by alternative globalizations that are multipolar and multilateral, involving autonomous partners and alliances, and that are radically democratic and ecological. Such a **cosmopolitan globalization** would include NGOs, social movements, and popular institutions, as well as states and global institutions like the UN. A democratic and multipolar globalization would be grounded philosophically in Enlightenment cosmopolitanism, democratic theory, human rights and ecology, drawing on notions of a cosmos, eikos, global citizenship, and genuine democracy.[23]

The need for cosmopolitan globalization shows the limitations of one-sided anti-globalization positions that dismiss globalization out of hand as a form of capitalist or U.S. domination. Taking this position is admitting defeat before you've started, conceding globalization to corporate capitalism and not articulating contradictions, forms of resistance, and possibilities of democracy grounded in globalization itself. Rather, an U.S.-dominated or corporate globalization

---

[22] David E. Sanger and Nicole Perlroth, "Range of Tools in a Vast Hack Elevates Fear. Biden Says Retribution Will Be 'Substantial,'" *New York Times*, December 18, 2020: A1; see also Thomas P. Bossert, "We're Being Hacked. The Damage is Already Done," *New York Times*, December 18, 2020: A19.

[23] On cosmopolitanism, see Cheah and Robbins (1998) *Cosmopolitics,* and special issue of *Theory, Culture & Society* on Cosmopolis, Vol. 19, Nrs. 1–2 (February-April 2002).

represents a form of neoliberal globalization which, interestingly, Wallerstain claims is "just about passé" (2004, p. 18). The argument would be that Bush administration unilateralism has united the world against U.S. policies, so that the U.S. can no longer push through whatever trade, economic, or military policies that they wish without serious opposition. Wallerstein points to the widely perceived failures of IMF and WTO policies, the collapse of Cancun and Miami trade meetings that ended with no agreement as strongly united so-called southern countries opposed U.S. trade policy, and, finally, global opposition to the Bush administration Iraq intervention. He also points to the rise of the World Social Forum as a highly influential counterpoint to the Davos World Economic Forum, which has stood as an organizing site for a worldwide anti-neoliberal globalization movement (see Hardt 2002).[24]

A cosmopolitan globalization would overcome the one-sidedness of a nation-state and national interest dominant politics and recognize that in a global world the nation is part of a multilateral, multipolar, multicultural, and transnational system. A cosmopolitan globalization driven by issues of multipolar multilateralism, democratization and globalization from below, would embrace women's, workers', and minority rights, as well as strong ecological perspectives. Such cosmopolitan globalization thus provides a worthy way to confront challenges of the contemporary era ranging from inequalities between haves and have nots to global warming and environmental crisis.

Some Concluding Remarks on Technopolitics and Globalization

Digital technologies and social media are thus a contested terrain, used by the Left, Right, and Centre to advance their own agendas and interests. Technopolitics is becoming more important in recent years on the local, national, and global scale. Major political battles of the past decades on a global scale have been fought through technopolitics as I have argued in this chapter, and local, national,

---

[24] Trump also attacked globalization constantly and tended to unite U.S. allies and other forces and institutions against Trump and his unhinged nationalism. See Pete Buttigieg and Philip H. Gordon, "Present at the Destruction of U.S. Power and Influence. Four years of neglect, unilateralism, and failed diplomacy have left America's alliances in tatters. It's time to rebuild them." Foreign Policy, July 14, 2020 at https://foreignpolicy.com/2020/07/14/trump-biden-foreign-policy-alliances/ (accessed January 1, 2021). Fortunately, in 2021, the Biden administration and globalists like Pete Buttigieg who is in Biden's cabinet are working to restore the multilateral global alignments that characterize the Obama and Clinton eras.

and global battles of the future will be fought in the streets, factories, parliaments, and in digitally mediated cyberspaces. The Internet is global so cyberspace reaches to wherever there are devices that can access communications and information and deploy this information in a technopolitics that scans the globe and embraces struggles on many levels. Virtual politics have included calls to revolution and insurrection (on both the left and right!), have organized democratic movements around race, ecology, gender, religion, sexuality, and many other key issues of the contemporary era.

Hence, today political struggle is now mediated by media, computer, and information technologies and increasingly will be so. Those interested in the politics and culture of the future should therefore be clear on the important role of the new public spheres and intervene accordingly.

Active citizens thus need to acquire new forms of technological literacy to intervene in the new public spheres of the media and information society. In addition to traditional literacy skills centered upon reading, writing, and speaking, engaged citizens and public intellectuals need to learn to use the new technologies to engage the public and participate in democratic discussion and debate.[25] Computer and digital technologies thus expand the field and capacities of the intellectual as well as the possibilities for political intervention (see Chapter 4).

During the Age of the Big Media, critical-oppositional intellectuals were by and large marginalized, unable to gain access to the major sites of mass communication. With the decentralization of the Internet, social media, and digital technology, however, new possibilities for public intellectuals exist to reach broad audiences and expand the terrain and impact of democracy. It is therefore the responsibility of the active citizen to creatively engage these technologies, as well as to critically analyze the diverse developments of the cyberculture, and explore the connections between technology and democracy. This requires dialectical thinking that discriminates between the benefits and the costs, the upsides and downsides, of digital technologies and social media, while devising ways in which the technological revolution can be used to promote positive values like education, democracy, enlightenment, and ecology. Active citizens thus face new challenges, and the future of democracy depends in part on whether digital technologies and social media will be used for domination or democratization, and whether each individual will sit on the sidelines or participate in the development of digitally-mediated democratic public spheres.

---

[25] For the new forms of critical media and digital literacies needed to use the new technologies for education, communication, and politics, see Kellner and Share 2020 and Chapter 9 below.

I have not discussed the ways that technopolitics could be used to struggle not only against capitalism, but for socialism. I would argue that socialist ideas are still relevant to the politics of the contemporary era and that in particular Karl Marx's ideas, far from being obsolete, are still relevant in developing critical theories of globalization, technology, and capitalism in the current conjuncture. It could be that only a socialist politics could overcome the digital divide, making accessible to all the benefits of the technological revolution. A socialist government could provide wireless communications in underdeveloped societies making possible access to the Internet and use of new communications and information technology even to societies that are not yet wired, or whose telephone systems only extend to the privileged. Interestingly, societies like Korea, Japan, and the Philippines make more extensive use of wireless communications than the U.S.,[26] with wireless messaging systems and Internet connections made use of by the working classes as forms of popular communication.

This study has suggested that in the era of globalization and the Internet political struggles are at once local and global, that there are continuities and discontinuities with struggles and movements of the past, and that we can therefore continue to draw on the most progressive ideas of the modern tradition while also developing new concepts of politics and new strategies for social transformation. A revolution of the future needs to articulate models and ideals of a post-capitalist economy, a radical democratic polity, an egalitarian and socially just multicultural society, a diverse, free, and open culture, and a sustainable environment in the era of climate change. Ideals of the past can and no doubt will enter into revolutionary thought of the future, but new ideals, values, and forms of everyday life will no doubt emerge. The future of revolution is thus open and requires new theory and practice as well as appropriation of the best progressive heritages of the past.

---

[26] Kenji E. Kushida, "Correspondencekkushida@stanfordalumni.orgWireless Bound and Unbound: The Politics Shaping Cellular Markets in Japan and South Korea," *Journal of Information Technology & Politics*, Volume 5, 2008 - Issue 2, pp. 231–254 and published online October 11, 2008 at https://www.tandfonline.com/doi/abs/10.1080/193316808022 94461 (accessed December 19, 2020). In the Covid-19 era, it is more essential than ever to deploy wireless communications to unserved communities, see Catherine Sbeglia, "Globe Telecom to expand 5G coverage to 17 cities in the Philippines." *RCR Wireless News*, November 18, 2020 at https://www.rcrwireless.com/20201118/5g/globe-telecom-to-expand-5g-coverage-to-17-cities-in-the-philippines (accessed January 1, 2021).

Virilio, War, and Technology

7

Abstract

Chapter 7 presents French theorist and technology critic Paul Virilio in an analysis of "Virilio, War, and Technology," which explores destructive and unsettling aspects of digital technologies. I follow Virilio in pursuing what he calls the "riddle of technology" and interrogate his attempts to elucidate this conundrum. I probe Virilio's perspectives on technology to determine the extent of his insight and use-value, and to indicate what I see as the limitations of his perspectives. In this reading, Virilio emerges as one of the major critics of war, technology, and vision machines in our time, albeit with excessively negative and even technophobic proclivities.

Keywords

Paul Virilio • "riddle of technology" • War • Military • Speed • "Dromomatics" • Vector • City • Time and space • Disappearance and loss

Paul Virilio is one of the most prolific and penetrating critics of the drama of technology in the contemporary era, especially military technology, technologies of representation, and computer and information technologies. For Virilio, the question of technology is *the* question of our time and his life-work constitutes a sustained reflection on the origins, nature, and effects of the key technologies that have constituted the contemporary world. In particular, Virilio carries out a radical

---

This article was first published as Douglas Kellner, "Virilio, War, and Technology: Some Critical Reflections," *Theory, Culture and Society,* Vol. 16(5–6), 1999: 103–125. It was revised and updated for this chapter.

D. Kellner, *Technology and Democracy: Toward A Critical Theory of Digital Technologies, Technopolitics, and Technocapitalism*, Medienkulturen im digitalen Zeitalter, https://doi.org/10.1007/978-3-658-31790-4_7

critique of the ways that technology is transforming culture, society, politics, war, and even the human species, and for the purposes of our studies of technology and democracy, Virilio provides a striking account of how technology can be turned against democracy and provides obstacles to democracy.

While Virilio has a flawed conception of technology that is excessively one-sided and that misses the emancipatory and democratizing aspects of digital and media technologies, it is precisely his laser-like focus on how technology endangers democracy and helps produce destructive effects that are detrimental to the future of humanity that is unsettling, but valuable for a critical theory of democracy and technology. My argument is that Virilio's vision of technology is overdetermined by his intense focus on war and military technology and that this optic drives him to predominantly negative and technophobic perspectives on technology per se. However, precisely the one-sidedness and extremely critical discourse on war and military technology, as well as his reflections on cinema, technologies of representation, and vision machines, constitute some of the most valuable aspects of his work. I will also suggest that Virilio's analysis of the rapidly approaching future can be read as a science fiction story about the world just ahead of us.

Consequently, in the following pages I will follow Virilio in pursuing what he calls the "riddle of technology" and interrogate his attempts to elucidate this conundrum. Nowhere, however, does Virilio directly theorize technology in any systematic or sustained way, although reflections on it permeate his analyses. Thus, I want to probe Virilio's perspectives on technology to determine the extent of his insight and use-value, and to indicate what I see as the limitations of his perspectives. In this reading, Virilio emerges as one of the major critics of war, technology, and vision machines in our time, albeit with excessively negative and even technophobic proclivities.

Speed, Politics, and Technology

> We must take hold of the riddle of technology and lay it on the table as the ancient philosophers and scientists put the riddle of Nature out in the open, the two being superimposed (Virilio and Lotringer 1983, p. 30).

> Totalitarianism is latent in technology (Virilio 1995a).

In *Speed and Politics* (1986 [1977]), Paul Virilio undertakes his first sustained attempt to delineate the importance of accelerated speed, of the impact of technologies of motion, of types of mobility and their effects in the contemporary era. Subtitled "Essay on Dromology," Virilio proposes what he calls a "dromomatics," which interrogates the role of speed in history and its important functions in urban and social life, warfare, the economy, transportation and communication, and other aspects of everyday life. "Dromology" comes from the Latin term, *dromos*, signifying race, and *dromology* studies how innovations in speed influence social and political life. The "dromocratic revolution" for Virilio involves means of fabricating speed with the steam engine, then the combustion engine, and in our day nuclear energy and instantaneous forms of warfare and mass communication.

Virilio was initially an urbanist who suggests that the city is a dwelling place organized by channels of communication and transportation, penetrated by roadways, canals, coastlines, railroads, and later airports. Each crossing has its speed limits, its regulations, and its systematic enclosure and spaces within a system of societal organization. The city itself is a conglomeration of these roads, a stopover for travel, and a system of "habitable circulation" (Virilio 1986, p. 6). City life unfolds in the spectacle of the street with its progressions and movements, its institutions and events, mobilizing and moving flows of traffic and people. Likewise, politics unfolds in the streets and urban sites of demonstration, debate, revolt, and revolutionary insurrection.

For Virilio, the city and its institutions have military origins. In his view, the medieval cathedral and early modern fortified cities were military camps. In Virilio's words: "Before it became the throne of totality, the Christian sanctuary was a stronghold, a bunker, a fortified church for those who remained within it; all their powers and capacities were deployed and strengthened in, through and as combat" (1986, p. 38). Likewise, although Virilio himself does not make this point, the early Christian missions in the Americas were military fortifications used by the colonizing powers as defense and control mechanisms (Tinker 1993).

The *vector* is a key term for Virilio that indicates the trajectory of various technologies along a fixed length and direction, but from no fixed point. It refers to any trajectory along which goods, money, information, or military apparatuses can flow, including roadways, airwaves, and communication and military circuits. *Territory* is the space across which speed, technology, politics, economics, and urban and everyday life flow through vectors of transportation, commerce, war, social interaction, communication, and information.

From a political and military perspective, "territory" is the space of human habitation, it is a space to be defended and secured, and to be invaded and colonized. Within modern societies, the nation-state was the territory that defined politics and the city, with its public spaces, institutions, and fortifications, serving as its privileged site. In the contemporary world, however, the city has been displaced by technologies of speed and power.

In the military sphere, the city no longer serves as a break against military conquest and as a site of protection of its citizens when instantaneous military violence can assault it from hidden spaces (airplanes, nuclear submarines, and missiles shot from afar and even outer space). With politics occurring through media and information circuits, the time of deliberation and consensus is obliterated. As I argued in the last chapter, democracy requires adequate information to make intelligent political and voting decisions, but it also requires deliberation and consensus. The speed of transmission and overload of information through media and digital technologies makes democratic debate and reaching consensus more difficult. Although Virilio doesn't make this point, the cities that he excavated in his analyses were an origin of democracy. The agora in Athens was a place that citizens could meet to debate issues concerning their urban lives—including war and peace as Thucydides tell us in his history, as was the Forum in Rome. Today political battles are often fought out in media and digital spaces, frequently in the form of spectacle in which image and narrative overshadows discourse and democratic debate, which requiring a careful weighing of alternative positions.

In technological societies, space and time are thus overwhelmed by technologies that travel at ever faster speeds, and digital media and communication technologies instantaneously circulate images and information across space, replacing one story, scandal, and spectacle after another making it difficult to analyze and appraise rapidly shifting position positions, alliances, and debates. This situation is evident in both the Trump and Biden eras as citizens are overloaded with information as fast moving spectacles shift the fate of the nation and overwhelm, confuse, and alienate citizens who do not have adequate training in information literacy and the forms of contemporary politics, requiring, in the spirit of Dewey and Freire, a reconstruction of education to create informed and active citizens as I argue in Chapter 9.

Dromology also involves analysis of the forces that brake or diminish speed, as well as those forces that accelerate it. War, for instance, involves both offensive attempts to rapidly control space and territory contrasted to defensive efforts to slow down the attack, to decelerate the offensive, just as laws and rules brake or slow down certain actions deemed destructive to the community. Virilio claims that, strictly speaking, there was no production of speed until the nineteenth century with the combustion engine to the electric telegraphy, in which, first, transportation is greatly speeded up and then communication takes place instantaneously over great distances, thus obliterating traditional barriers of time and space. Consequently, the generation of modernity involves transition from the age of the brake to that of the accelerator (Virilio and Lotringer 1983, p. 44 f.), as intensification of speed generates new economic, political, social, and cultural forms.

Virilio argues that the role of speed had been previously overlooked in the organization of civilizations and politics, and that speed is crucial to the production of wealth and power. Resolutely rejecting the forms of economic determinism associated with Marxism, Virilio's dromology focuses on those instruments that accelerate and intensify speed and that augment the wealth and power of those groups who control them. In his vision, the military comes to control speed and thus to become a dominant societal power. This situation produces an accelerating decline of the state and politics and primacy of the military, which, for Virilio, becomes a key force in politics and society whose importance he believes is usually underestimated. It also creates crises of democracy, as we will explore in the following sections.

War, Technology, and Civilization

From the beginning of his studies, Virilio was concerned to theorize the interconnection between speed, technology, and war. On Virilio's view, the importance of warfare in understanding human history had been grossly underestimated. Initially an urbanist and specialist in architecture, Virilio came to the view that war was at the center of civilization, that the city, for instance, was formed as a garrison for warfare, that need for defense and the preparation for war was at the origins of the foundation of cities.[1] For Virilio, war involved the organization of space, through preparing and undertaking the conquering of territory, and thus in terms of logistics, offensive tactics, strategy, and defense, there was a unique spatial organization for war. Defense required slowing down the enemies' military assault, and cities provided walls, ramparts, fortresses, and enclosed areas that could repel invasion, that could protect individuals gathered within its spaces.

For Virilio, logistics, the preparation for war, is the beginning of the modern industrial economy, fuelling development of a system of specialized and mechanized mass production. War and logistics require increased speed and efficiency, and technology provides instruments that create more lethal and effective instruments of war. The acceleration of speed and technology, in turn, create more dynamic

---

[1] Virilio's text *Bunker Archeology* (1994 [1975]) explores this theme; see also Virilio 1986, p. 3 ff. and Virilio and Lotringer 1983, p. 2 f.

industry, and an industrial system that obliterates distances in time and space through the development of technologies of transportation, communication, and information. The fate of the industrial system is thus bound up with the military system which provides, in Virilio's vision, its origins and impetus.

Thus, on Virilio's optic, cities, cathedrals, the economy, politics, and other key aspects of the modern world are products of military mobilization and deployment, thus war serves as the motor of history, culminating in what Virilio calls "pure war" (Virilio and Lotringer 1983, 2008). In Virilio's view, the system of deterrence in the Cold War nuclear stalemate created a situation in which technological development channels technology into military forms and technocratic political domination. In this situation, "Weapons and armor constantly need to be strengthened. Technological development thus leads to economic depletion. The war-machine tends toward societal non-development" (Virilio and Lotringer 1983, p. 5). With more and more resources going to the military and military imperatives dominating production, government, and the evolution of science and technology, societal development is undermined and social underdevelopment becomes a defining mark of the contemporary world.

Many individuals have noted that war and the overdevelopment of the military sector of the state can provide impediments to democracy and President Eisenhower's farewell address warning against the military-industrial complex and its threats to democracy are relevant in this context:

> American makers of plowshares could, with time and as required, make swords as well. But now we can no longer risk emergency improvisation of national defense; we have been compelled to create a permanent armaments industry of vast proportions.... This conjunction of an immense military establishment and a large arms industry is new in the American experience.... Yet we must not fail to comprehend its grave implications.... In the councils of government, we must guard against the acquisition of unwarranted influence, whether sought or unsought, by the military-industrial complex. The potential for the disastrous rise of misplaced power exists and will persist.[2]

In addition, for Virilio, the acceleration of events, technological development, and speed in the current era designates "a double movement of implosion and explosion," so that "the new war machine combines a double disappearance: *the disappearance of matter in nuclear disintegration and the disappearance of places in vehicular extermination*" (Virilio 1986, p. 134). The increased speed of destruction in military technology is moving toward the speed of light with laser

---

[2] President Dwight D. Eisenhower's Farewell Address (1961) at https://www.ourdocuments.gov/doc.php?flash=true&doc=90 (accessed on January 27, 2021).

weapons and computer-controlled weapons systems constituting a novelty in war-fare in which there are no longer geo-strategic strongpoints since from any given spot we can now reach any other, producing what Virilio calls "a strategy of Brownian movement through geostrategic homogenization of the globe" (Virilio 1986, p. 135). Thus, "*strategic spatial miniaturization* is now the order of the day," with micro-technologies transforming production and communication, shrinking the planet, and preparing the way for what Virilio calls "pure war," a situation in which military technologies and an accompanying technocratic system come to control every aspect of life.

In Virilio's view, the war machine is the demiurge of technological develop-ment and an ultimate threat to humanity, producing "a state of emergency" in which nuclear holocaust threatens the very survival of the human species. This involves a shift from a "geo-politics" to a "chrono-politics," from a politics of space to a politics of time, in which whoever controls the means of instant infor-mation, communication, and destruction is a dominant socio-political force. For Virilio, every technological system contains its specific form of accident and a nuclear accident would, of course, be catastrophic. Hence, in the contemporary nuclear era, in which weapons of mass destruction could create an instant world holocaust, we are thrust into a permanent state of emergency that enables the nuclear state to impose its imperatives on ever more domains of political and social life.

Politics too succumbs to the logic of speed and potential holocaust as increased acceleration in military violence, instantaneous information and communication, and the flow of events diminishes the time and space of deliberation, discus-sion, and the building of consensus that is the work of democratic politics. Speed and war thus undermine democracy, with technology replacing democratic par-ticipation and the complexity and rapidity of historical events rendering human understanding and control ever more problematical. Ubiquitous and instantaneous media communication in turn makes spin-control and media manipulation diffi-cult, but essential, to political governance. Moreover, the need for fast spin control and effective media politics further diminishes the space and role of democratic political participation and interaction.

Propaganda and spin control in politics raises the issue of the politics of truth that is currently of central importance in an era when the politics of lying had nurtured authoritarian regimes throughout the world. Truth is the antidote to the poisons of lies and deception, the cleanser of falsehood and mendacity, and the light which brings truth into the world. The *Washington Post* has on its masthead "Democracy Dies in Darkness" and it is the role of the media, public intellectuals, and citizens interested in truth who must combat the lies, deceptions, and false-hoods that are spread by autocratic and authoritarian forces on the local, national and global level. A politics of truth is essential to democracy but as Virilio warns us the loss of political certitude and rectitude in an out-of-control infodemic in which disinformation and lies become prevalent, thus creating great challenges to especially a transparent democracy based on truth, virtue, and a solid foundation of fact and information.

Disappearance and Loss: Virilio's Complaint

> We now have the aesthetics of the disappearance of a numerical, unstable image of fleeting nature, whose persistence is exclusively retinal (Virilio 1991a, p. 36).

Throughout his works, Virilio describes the loss of key human capacities and powers in the contemporary world under the influence of always accelerating technology. While from the 1970s to the present, Virilio discusses the decline of politics in a technological world in which individuals are losing control over their technology, society, and polity, *The Lost Dimension* (1991b [1984]) deals with the decline of the city, its decentering and displacement in the information and postindustrial society and, crucially, the loss of the object, of the very concreteness of lived experience in a new world of technologically-generated representations and modes of vision.

For Virilio, the city is decentered in relation to the rise of suburbs and then tele-communications and new sites of work and interaction in a postindustrial society. Virilio's "overexposed city" is penetrated by media and advertising, information technology, and what Debord (2002 [1968]) called "Society of the Spectacle," overwhelming urban space and life. In Virilio's vision (1991b, p. 9 ff.), the urban wall and gateways have given way to a plethora of openings to media channels, information and communication networks, and diverse new technologies. Each technology is a window to the outside world, obliterating urban boundaries and spaces to the geopolitical channels of the global world and the world of ato-pic cyberspace. Exposed to global culture and communication, the city loses its specificity and city life gives way to technological cyberlife, an aleatory, heteroge-neous and fractured space, and a world-time that enables individuals to experience events simultaneously from every time zone in the world.

Henceforth, fragmentary images derived from diverse sources constitute one's "image of the city," rather than the grids of maps or personal experience. Virilio is analyzing a momentous shift in the image and imaging of the world, of what he calls a "morphological irruption," of an "iconological disruption" mutating from qualitative perception to quantitative representation and then to digitization. This shift in experience progressively volatilizes the real and obliterates the object of lived experience into technological modes of representation, that constitute a derealization and dematerialization of the object. That is, whereas the object of lived experience was once an object of perception, an object seen and handled by the bodily subject, the objects of cyberspace and virtual reality—as well as the objects of contemporary scientific theory—are abstract and immaterial, generating a new form of technological idealism.

As I revise this study in 2020, I am struck by observations I have made of people living in Los Angeles where I have resided the past twenty-five years and how their wanderings or driving through the city, public spaces, or Universities where I work and spend much time are mediated by digital devices. When I walk in my neighborhood, more than half the people seem to be talking to a digital device, checking messages, listening to headphones, and thus traversing through a technologically-mediated space rather than strolling through a vibrant city in the mode of Baudelaire or Benjamin's *flaneur*. At home and work, likewise much time is taken by media and digital devices, meaning that much of our live is lived in cyberspace, where a la Virilio, we speed from one country, or author, or topic, or friend, or TV show or movie, to another, speeding through our lives with increasing velocity until the end-point when all is still.

And as for the effect of speed, digital devices and information overload in the context of democracy, just like the city dweller is overwhelmed by speed and digital technology, so that, for instance, to get to school on time requires driving fast through back neighborhoods and avoiding big boulevards crowded with traffic, or the freeways that become jammed up like a parking lot during rush hour, so too is the channel surfer or web cruiser overwhelmed with information and information sources and is forced to speed through multiple sites while checking email, messaging, and lecture and zoom schedules in the virtual world of the pandemic and lockdown.

For Virilio, theories of light and speed replace time and space, as a new immateriality and "new illuminism" comes to dominate contemporary scientific thinking. Virilio believes that as with the notions of critical mass or temperature, when states of affairs break up and become radically other, space too becomes "critical" (Virilio 1997b, p. 9 ff.). The notion of "critical space" refers to the breaking up and dissolution of previous configurations of space under the impact of technology. For Virilio, telecommunication that eradicates all duration and extension of time in the transmission of messages and images, as well as mass transportation and interactive computer technologies that decenter urban or lived space, all constitute threats and dissolutions of previous configurations of experience as space becomes virtual and takes on new modalities. Previous configurations of space and time are replaced by time-light (i.e. the time of the speed of light) and a new "lumiocentrism" (1997b, pp. 5 f., 14 f.), in which the instantaneous flow of information ruptures previous configurations of time and space, requiring new concepts to describe the parameters and processes of the new worlds of technology and technological experience.

For Virilio, developments in science and technology are obliterating both modern and common sense views of the world and producing new objects and spaces that cannot be explained by current conceptual schemes. The "physics of the infinitesimally small" and the cosmological speculations on outer space produce novelties and puzzles that put in question the facts of perception, the realm of experience, and that point to new, unperceived and imperceptible entities, that confound common sense and current scientific schemes (Virilio 1991b, p. ff.). Moreover, new technologies are producing both new objects (i.e. cyberspace, virtual reality, etc.) and new modes of perception and representation (i.e. fractal geometry, computer-generated representations of external and internal realities, etc.) that themselves require new modes of thought and cognition. Such shifts in modes of perception and representation began with cinematic photography that captured motion and phenomena not visible to the naked eye, increased with developments in microscopes and telescopes, and proliferated new modes of perception and representation with computers and virtual technologies.

In short, Virilio is mourning the loss of the object of ocular perception in the emergent forms of technological perception and representation, the displacement of the dimension of direct observation and common sense (1991b, p. 111), and thus the loss of the materiality and concreteness of the objects of perception, and the realm of appearance and lived experience. Virilio mourns the loss of the phenomenological dimension that privileged lived experience. Always a phenomenologist, as he affirms in his interview with John Armitage (1999), Virilio roots his thought in concrete experience of objects, people, and processes in the observed and experienced worlds of everyday life and the natural and social worlds. The new technological worlds, for him, constitute a break and rupture with ordinary experience and thus shift the locus of truth, meaning, and validity to, for Virilio, an abstract and enigmatic virtual realm.

Contemporary science and technology for Virilio are thus producing new forms of experience, new modes of perception and representation, and new objects of experience that decenter the human subject, that replace human cognition with technological vision, and displace human labor power in favor of automated technological production. Losing control over its world, the human subject becomes a mere recording device and the human body is reduced to functions in a technological system. Material reality is decentered and a new technological idealism generates concepts increasingly distant from common sense, the body and material world, the conceptual systems of the past, and lived experience.

In addition to the loss of the concrete object of perception, of the realm of appearance and material reality, of the body, Virilio mourns the disappearance of the city, the state, and the end of politics in the new globalized technopolis. Just as computer-aided production and a new virtual form of automation displace human labor power, so too does "flexible accumulation" (David Harvey), the new global division and organization of production, and international financial markets, data bases, and simultaneity of information transmission, communication, and video representation obliterate previous experiences and concepts of time and space, producing a grave new world of transnational global corporations, political organizations, and cities, and virtual spaces, displacing the national firm, the city, the nation-state, and previous forms and sites of modern politics and culture.

Indeed, for Virilio part of the "lost dimension" is the end of politics in a world of increased speed and virtualities. This is most evident in the realm of military technology in which the complexity of weapon systems create ever shorter response times for humans to react to frightening computer-generated information concerning military threats and in which military technology itself can autonomously generate catastrophes ranging from "friendly fire" incidents that could disable computer networks to nuclear apocalypse that could end human existence. Further, the loss of stable referents of the political—the city, state, nation—in the deterritorialized and volatilized virtual and global spaces of the new information economy and polity, also render human participation in politics perplexing and perhaps futile.

This vision of technological domination, of technology displacing human beings, has echoes of the theories of "autonomous technology" (Winner 1978) developed by Heidegger, Ellul, and other totalizing critics of contemporary technology. Virilio does positively cite Heidegger on technology, though he suggests critiques of Heidegger and totalitarianism, specifically his affiliation with German National Socialism (1986, pp. 90, 113 f., passim, and Virilio and Lotringer 1983, p. 23 f.). Thus, while Virilio is quasi-Heideggerian in his perspective on technology, seeing technology as the enframing demiurge of the modern world, as the matrix in which human practice unfolds, he is clearly anti-totalitarian, and might be seen perhaps as a left-Heideggerian.[3]

---

[3] For Heidegger on technology, see Heidegger 2013 and the studies in Idhe 2010.

Further, in the light of Virilio's Christian religious beliefs, he has certain affinities with Jacques Ellul's radical critique of technology, that sees technology as an autonomous force that is coming to dominate the contemporary world,[4] effacing human freedom and meaning. When asked if Ellul or Christian existential philosopher Gabriel Marcel influenced his thought, Virilio affirmed the influence of Ellul while denying the impact of Marcel.[5]

Certainly, there are echoes of Ellul's technique, of a totalitarian tendency toward domination and destruction from technological development, running throughout Virilio's work, although he uses more concrete models of war machines, or vision machines, to characterize technology, is less overtly totalizing than Ellul, and is more muted in his religious perspectives. Yet there are similar themes of the demise of human autonomy and creativity in a world in which technique and technological development imposes its imperatives on human beings and both have a predominantly negative and critical take on what they see as the totalitarianism of modern technology. Like Ellul, Virilio denies the technological imperative and affirms the dignity and sovereignty of human beings over things.

Yet, against all forms of economic determinism and idealist humanism, Virilio posits an autonomous force and power of technology and describes the ways that it constrains economic and social life. In place of Marxian economic determinism, Virilio arguably substitutes a form of military-technological determinism. On his view, the military organization and deployment of people is the origin of proletarianization and predates capitalism; military mobilization is exploited by political, economic, and military forces to augment their power; and the result is the ever-more sophisticated and lethal development of a war machine, a destructive apparatus that is increasingly automated, lethal, fast, effective and removed from human control or values, producing "a state of emergency" and totalitarian threat in which the very fate of the earth, democracy, and humanity is at stake.

To the totalitarianism of technology, in which technology inhabits and dominates ever more forms of life, there is the threat of political totalitarianism, such as we've seen from fascism, communism, and absolute state governments that have ruled much of the world over the past century. Democratic societies face ever present threats of political totalitarianism from authoritarian parties, movement, and politicians who have become dominant in Russia, China, and many parts of the world and under Trump the U.S. faced its worst crisis of democracy in

---

[4] For Ellul on technology, see Ellul 1964 and the commentaries in Van Vleet and Rollison 2020.

[5] See Virilio's interview with John Armitage in *Theory, Culture and Society*, Vol. 16(5–6), 1999, and Virilio 1997b, p. 139 f.

facing a totalitarian leader and movement which sought to overthrow democracy and establish a totalitarian dictatorship. For now, in 2021, the attempted coup against U.S. democracy failed but as the Trump era showed, the totalitarian danger can appear everywhere and democratic societies must gird themselves against totalitarian forces by strengthening their democracy.

Virilio, Baudrillard and the Present Moment

> There is a nihilistic dimension in Baudrillard's writing that I cannot accept. It is quite clear to me that Baudrillard has totally lost faith in the social. To me, this is sheer nihilism. I have not at all lost faith in the social (Virilio 1998).

For Virilio, it is technology that accelerates speed, that intensifies war, that creates totalitarian modes of domination and so something like a technological military determinism is present in Virilio's thought. His displacement of the primacy of economics and focus on the key constituent role of technology brings his thought into dialogue with his contemporary Jean Baudrillard. In the 1970s, Baudrillard and Virilio were two of the world's most advanced theorists of technology, both focusing on the new technologies that were creating the *novum* of the contemporary. Both were concerned to grasp the nature of technology and the present age with Virilio theorizing contemporary societies in terms of technology and speed, and its impact on war, politics, and modes of representation, with new configurations of space and time emerging in pure war. Baudrillard, by contrast, theorized the end of modernity in a postmodern turn moving toward a society of simulation, hyperreality, implosion and other postmodern technological and social novelties (Kellner 1989a).

Baudrillard began in the field of social theory and his early and to some extent middle works provided aspects of a sociology of new media, information, and biogenetic technology, while Virilio centered on technologies of war and representation. By contrast, Virilio eschewed sociology, preferring to focus on war and politics. In his later work, Baudrillard too moved beyond conventional social theory and sociology, moving into a new type of philosophical discourse and cultural metaphysics of the present age.[6] While the post-1980s Baudrillard engaged in abstract theorizing and increasingly obscure metaphysical discourse, Virilio undertook extremely detailed empirical and historical research, albeit presented

---

[6] For the twists and turns of Baudrillard's tortured theoretical trajectory, see Kellner 1989b; Best and Kellner 1991; Kellner 1994, 1995; Best and Kellner 1997; Gane 2008.

in an often cryptic and fragmentary style. Both engage in comprehensive historical analysis, though Virilio arguably develops more penetrating historical and political analysis, somewhat in the mode of Foucault who, however, he says he respects more than he likes, claiming that his own work is more fragmentary and disruptive, deploying collage methods of assembling fragments and quickly moving from one topic to another in contrast to Foucault's more classical style (see Virilio and Lotringer 1983, p. 38 f.).

There is a strong convergence on some themes with Baudrillard and his French contemporaries concerning the radical breaks and ruptures in the contemporary technological world with past modes of social organization, as well as significant differences in theorizing this rupture. For Baudrillard, postmodernity means the end of reality, the end of being able to distinguish between the real and unreal, the end of being anchored in and living in a real material world (see Kellner 1989a). On Baudrillard's optic, we dwell increasingly in the realms of hyperreality—broadcast media, the cyberspace of computer interaction, video and computer games, or a range of mass-mediated worlds—film, music, multimedia, VR devices, and, we would add, digital devices and social media. Moreover, it becomes increasingly difficult to distinguish between the real and hyperreality, leading to a dissolution of the real. Thus, as Virilio notes, "The question of modernity and postmodernity is superseded by that of reality and post-reality" (1994, p. 84).

Yet Virilio differs from Baudrillard in his theorizing of contemporary technological society. In an interview with John Armitage (1999), Virilio says that he disagrees with Baudrillard over the issue of simulation, seeing simulation not as an obliteration of reality, but instead as substitution, in which a technological reality replaces a human one, as photography substitutes itself for real life, or film substitutes the static representation of the real with "moving pictures," or, in our day, when virtual reality substitutes itself for "real life." Consequently, unlike Baudrillard, Virilio believes reality does not disappear, but is rather displaced by another mode of reality, a virtual reality: "Thus, there is no simulation, but substitution. Reality has become symmetrical. The splitting of reality in two parts is a considerable event which goes beyond simulation" (Virilio 1997a, p. 43). Thus, whereas for Baudrillard reality disappears in hyperreality, for Virilio new technologies provide a substitute reality, a virtual reality which becomes more powerful and seductive than ordinary reality.

Virilio theorizes speed, dynamics, and the simultaneous eruption of a dialectic of implosion and explosion, while Baudrillard theorizes inertia, implosion, and the crisis of the political. Both, however, evoke the end of history and politics in the contemporary moment. More than Virilio, who often articulates political and religious passions, Baudrillard more neutrally describes, accepts, perhaps even affirms, the end of politics, history, in the "catastrophe of modernity." Virilio, by contrast, wants to preserve and expand the social and politics against pure war and the military, opposing a transpolitics which denies the continued relevance of modern politics.[7]

In terms of concrete political analysis, Baudrillard has had a particularly poor record as a social and political analyst and forecaster. As a political analyst, Baudrillard has often been superficial and off the mark. In a essay "Anorexic Ruins" published in 1989, Baudrillard read the Berlin wall as a sign of a frozen history, of an anorexic history, in which nothing more can happen, marked by a "lack of events" and thus the end of history, taking the Berlin wall as a sign of a stasis between communism and capitalism that would endure for the millennium. Shortly thereafter, in 1989 rather significant events destroyed the wall that Baudrillard took as eternal, the two Germanies split by the Cold War reunited, and the Cold War itself was in the process of dissolving, opening up a new historical era.

The Cold War stalemate was long taken by Baudrillard as establishing a frozen history in which no significant change could take place. Virilio, by contrast, perceived the beginning of the break-up of the Soviet Empire and the opening up of a new era. Already in *L'Insecurite du territiore* (1976), Virilio cited Helene Carriere de Encausse's *Decline of an Empire*, and returned to this theme in *Pure War*, noting that the Soviet empire is "breaking apart" (Virilio and Lotringer 1983, p. 155). Of course, no one anticipated the extent and suddenness of the breakdown of the Soviet empire and collapse of the Soviet Union itself, and in the 1980s Virilio tended to exaggerate the continuing power of the Soviet military machine and operated with the Cold War model of a bipolar world as the key constituent of contemporary history, much as Baudrillard.

With the collapse of the Soviet Union and end of the Cold War, we are, arguably, in a new historical era, in which technology drives us and impels us into new modes of speed and motion, as it carries us along into an unknown future. Virilio suggests that:

---

[7] In an interview with Sylvere Lotringer, Virilio states that: "For me, trans-politics is the beginning of the end. That's where my understanding of it radically differs from Jean Baudrillard's; for him it's positive. For me, it's totally negative. I fight against the disappearance of politics. I'm not saying that we should revert to ancient democracy, stop the clock and all that. I'm saying that there's work to be done... in order to re-establish politics" (Virilio and Lotringer 1983: 28). One might argue, however, that Baudrillard does not see the end of politics, or "trans-politics," as "positive," or "negative," but rather as inevitable in an era in which politics, aesthetics, sexuality, and culture implode; see Baudrillard 1992 and 1994.

the question, "Can we do without technology?" cannot be asked as such. We are forced to expand the question of technology not only to the substance produced, but also to the accident produced. The riddle of technology we were talking about before is also the riddle of the accident (Virilio and Lotringer 1983, pp. 31–32).

Virilio claims that every technology involves its accompanying accident: with the invention of the ship, you get the ship wreck; the plane brings on plane crashes; the automobile, car accidents, and so on. For Virilio, the technocratic vision is thus one-sided and flawed in that it postulates a perfect technological system, a seamless cybernetic realm of instrumentality and control in which all processes are determined by and follow technological laws (Baudrillard also, to some extent, reproduces this cybernetic and technological imaginary in his writings; see Kellner 1989b). In the real world, however, accidents are part and parcel of technological systems, they expose its limitations, they subvert idealistic visions of technology. Accidents are consequently, in Virilio's view, an integral part of all modes of transportation, industrial production, war and military organization, and other technological systems. He suggests that in science a Hall of Accidents should be put next to each Hall of Machines: "Every technology, every science should choose its specific accident, and reveal it as a product–not in a moralistic, protectionist way (safety first), but rather as a product to be 'epistemo-technically' questioned. At the end of the nineteenth century, museums exhibited machines: at the end of the twentieth century, I think we must grant the formative dimensions of the accident its rightful place in a new museum" (Virilio and Lotringer 1983).[8]

Virilio is fascinated as well by *interruptions* to active life ranging from sleep to day dreams to maladies like picnolepsy or epilepsy to death itself (1991a and Virilio and Lotringer 1983, p. 33 ff.). Interruption is also a properly cinematic vision in which time and space are artificially parceled and is close to the microscopic and fragmented vision that Lyotard identifies with "the postmodern condition" (Virilio and Lotringer 1983, p. 35). For Virilio, the cinema shows us that "consciousness is an effect of montage" (Virilio and Lotringer 1983, p. 35), that perception itself organizes experience into discontinuous fragments, that we are aware of objects and events in a highly discontinuous and fragmented mode.

Virilio further argues that new technologies alter our mode of perception and experience, change the way we see and experience the world, and that in particular

---

[8] Of course, the English novelist J.G. Ballard actually staged an exhibition of wrecked cars in London in 1966, described in his 1970 novel *The Atrocity Exhibition*, and the apocalyptic vision of western civilizations systems of control and circulation erupting in proliferations of car wrecks informed Ballard's 1973 novel *Crash* and the 1996 film directed by David Cronenberg based on the novel.

technologies of speed have produced an increasingly fragmented, discontinuous, and transhistorical mode of experience that grasps instances and partial relations rather than whole fields. In his view, technological time has thus invaded the time and space of the city and other sites of habitation, creating new rhythms, experiences, and modes of interaction that dramatically transform social and everyday life. Virilio describes what he calls "endo-colonization" in which the state colonizes its own urban spaces and then global institutions colonize the entire world. Concretizing this vision, Mike Davis writes: "No wonder that the contemporary American inner city resembles nothing so much as the classical colonial city, with the towers of the white rulers and colons militarily set off from the Kasbah or indigenous city" (1990, p. 111).

Yet more striking, as we shall see in the following section, is the eruption of new "vision machines" that create autonomous realms of experience and perception.

War, Cinema, and Representation: Vision Machines

> These new technologies try to make virtual reality more powerful than actual reality, which is the true accident. The day when virtual reality becomes more powerful than reality will be the day of the big accident. Mankind never experienced such an extraordinary accident (Virilio 1997a, p. 43).

With *War and Cinema* (1989 [1984]) and his subsequent writings such as *The Vision Machine* (1994 [1988]) and *Open Sky* (1997b [1995]), Virilio focuses more on the relation between war, speed, technology, and the means of representation, particularly vision machines and the logistics of perception embodied in that pre-eminent vision machine—cinema. War, Virilio suggests, has long been dependent on the logistics of representation, on providing accurate representations of the enemy's troop and weapon deployment. As military surveillance progressed, cinematic representation became more and more salient to military strategy, although, more recently, informatics, computer simulation, and satellite imaging have become more central.

Virilio claims that from approximately 1904, accelerating in the First World War, and until the recent high tech explosion, the apparatus of cinema was deployed as part of military strategy, involving lighting the terrain of battle and enemy forces, accurately representing their strength and movement, and instantaneously perceiving the actual battlefield itself as a dynamic field of motion, all of which was crucial to military strategy. Cinema too followed a certain military logic with great directors serving as dictators and authoritarian orchestrators of cinematic illusion spectacle, leading Virilio to conclude that: "War is cinema, and cinema is war" (1989, p. 26).

Cinema has long been part of Virilio's imaginary and his reflections on cinema and war cover a vast expanse of modern history, providing a unique take on the history of cinema and the ways that modes of cinematic representation are also crucial to war. Virilio's theme is the progressive dematerialization of warfare in high tech and virtual war, in which technologies progressively replace human beings:

> What the video artist Nam June Paik calls the triumph of the electronic image over universal gravity has carried this [dematerialization] still further. The sense of weightlessness and suspension of ordinary sensations indicates the growing confusion between 'ocular reality' and its instantaneous, mediated representation. The intensity of automatic weaponry and the new capacities of photographic equipment combine to project a final image of the world, a world in the throes of dematerialization and eventual total disintegration, one in which the cinema of the Lumiere brothers becomes more reliable than Junger's melancholy look-out who can no longer believe his eyes (1989, p. 73).

This passage refers to the tendency of technology to displace modes of human perception and representation in military planning and execution, as computer programs replace military planners and computer simulations replace charts and map of the territory. On the level of the battlefield itself, human power is replaced by machines, reducing the soldier to a cog in a servomechanism. Virilio comments:

> The disintegration of the warrior's personality is at a very advanced stage. Looking up, he sees the digital display (opto-electronic or holographic) of the windscreen collimator; looking down, the radar screen, the onboard computer, the radio and the video screen, which enables him to follow the terrain with its four or five simultaneous targets; and to monitor his self-navigating Sidewinder missiles fitted with a camera of infra-red guidance system (1989, p. 84).

With *The Vision Machine* (1994 [1988]) and Virilio's subsequent reflections on information and computer technologies, the epistemological turn to focus on modes of representation and the logistics of perception decenter to some extent Virilio's intense focus on war, though his interconnection of the themes of war, technology, and representation produce a unity and coherence to his otherwise highly fragmentary and elusive thought.

In the concluding chapter of *The Vision Machine* (1994), Virilio distinguishes between painting as the age of the image's *formal logic*, contrasted to photography and film as the age of the image's *dialectical logic*, and video recording, holography, and computer graphics as the beginning of an age of *paradoxical logic*: the latter emerges "when the real-time image dominates the thing represented, real time subsequently prevailing over real space, virtuality dominating actuality and turning the very concept of reality on its head" (1994, p. 63). In this situation, images and representations replace the real, the object of representation declines in importance, and a domain of images and digital representation replaces reality.

Culturally, this involves the proliferation of new vision machines that proliferate an artificial realm of data, images, and information that constitute a novel realm of experience. In war, it involves new modes of weapons based on the annihilation of time (just as nuclear technology involved the disintegration of matter and space). Just as computers and new image machines dramatically transform the nature of culture, so too do new laser technologies, modes of surveillance, and new modes of image warfare, disinformation, and high tech military spectacle change the nature of war (1994, p. 66 f.), such as was evident in the Gulf war of 1991.

So we see that for Virilio new modes of visual representation provide new ways of seeing and experiencing the world. Virilio's book *The Vision Machine* did not, however, explore in any great detail the new forms of information technology, multimedia, or cyberspace, providing instead an introduction to the ways that new vision machines influence perception and representation. It is in *Open Sky* (1997b [1995]) and many interviews and articles of the 1990s that Virilio interrogates the new information technology. His central insight is that emergent information, communication, and transportation technologies are taking us out of this world, beyond the limits of space and time, outside of nature and the material world into a new dimension with its own temporality, spatiality, and modes of being. Virilio fears that this journey will take us out of our bodies, minds, nature, and world as we have experienced and known it into a terrifying new sphere that will cause disastrous, possibly fatal, mutations of mind, body, and experience.

For Virilio, the astronauts are harbingers of a new experience beyond the familiar space and time coordinates of material existence. Shot into outer space beyond the laws of gravity and earth's spatial and temporal coordinates, the astronauts found themselves in a no place and no time continuum without fixed coordinates or dimensions. In this new dimension, some experienced the vertigo of disorientation and collapsed into madness after their return, or into strange metaphysical musings. Virilio' comments here, however, are somewhat anecdotal and serve more as metaphorical and rhetorical devices to dramatize the strangeness of outer space travel and the displacement of our scientific and conceptual schemes in this new dimension than a serious scrutiny of the effects of space flight on human beings.

Cyberspace, Virilio claims, supplies another space without the usual coordinates of space and time that also produces a disorienting and disembodying form of experience in which communication and interaction takes place instantaneously in a new global time, overcoming boundaries of time and space. It is a disembodied space with no fixed coordinates in which one loses anchorage in one's body, nature, and social community. It is thus for Virilio a dematerialized and abstract realm in which cybernauts can become lost in space and divorced from their bodies and social world.[9]

Although Virilio does not spell this out in any detail, information technology and cyberspace may produce an erosion of politics as individuals spend more time in virtual and cyberworlds they have less time and interest in the real world and thus less time and interest in their communities, societies, and politics. In my view, there is an ambiguity concerning technology and cyberspace because one can gain political information and engage with political communities in cyberspace, although one is prey to disinformation as well, and it is becoming increasing obvious how rightwing terror groups recruit their commandos on Internet hate sites,[10] as al Qaeda recruited members on Jihad sites.[11]

In addition, Virilio analyzes and denounces what he calls "a pernicious industrialization of vision" (1997b, p. 89) and what he fears is a displacement of vision by machines. Virilio is afraid that increasingly vision machines are seeing for us, ranging from cameras to video to satellite surveillance to nanotechnology which probes the body (and next the mind?). For Virilio, we are increasingly subjected to bombardment by images and information and thus by "a discreet pollution of our vision of the world through the sundry tools of communication" (1997b, p. 96). Moreover, he fears, media, like cinema and television train and constrain vision, leading to degradation of vision and experience: "If, according to Kafka, cinema means pulling a uniform over your eyes, television means pulling on a

---

[9] The cyberexplorer, by contrast, revels in explorations of cyberspace and trips into its depths and surprises; see Kelly 1995 and the critique in Best and Kellner 1999.

[10] Rightwing hate groups are also recruiting members from video gamers, see Anya Kamenetz, "Right-Wing Hate Groups Are Recruiting Video Gamers," *National Public Radio*, November 5, 2018 at https://www.npr.org/2018/11/05/660642531/right-wing-hate-groups-are-recruiting-video-gamers (accessed January 28, 2021).

[11] Andrew Dornbierer for The Diplomat, "How al-Qaeda Recruits Online. The internet has been a boon to terrorist recruitment. And extremists are finding some surprising mediums for spreading their message." *The Diplomat*, September 13, 2011 at https://thediplomat.com/2011/09/how-al-qaeda-recruits-online/ (accessed January 28, 2021).

straitjacket, stepping up an eye training regime that leads to eye disease, just as the acoustic intensity of the walkman ends in irreversible lesions in the inner ear" (1997b, p. 97).

Shrilly technophobic and at times quasi-hysterical, Virilio demonizes modern information and communication technologies, suggesting that they are doing irreparable damage to the human being. Sometimes over-the-top rhetorical, as in the passage just cited, Virilio's 1990's comments on new information technology suggest that he is deploying the same model and methods to analyze the new technologies that he used for war technology. He speaks regularly of an "information bomb" that is set to explode (1995a, 1995b, 1995c, 1997a, 1997b), evoking the specter of "a choking of the senses, a loss of control of reason of sorts" in a flood of information and attendant disinformation.

Deploying his earlier argument concerning technology and the accident, Virilio argues that the information superhighway is just waiting for a major accident to happen (1995a, 1995b, 1997a, 1997b), which will be a new kind of global accident, effecting the whole globe, "the accidents of accidents" (Epicurus):

> The stock market collapse is merely a slight prefiguration of it. Nobody has seen this generalized accident yet. But then watch out as you hear talk about the 'financial bubble' in the economy: a very significant metaphor is used here, and it conjures up visions of some kind of cloud, reminding us of other clouds just as frightening as those of Chernobyl... (1995b).[12]

In a 1995 interview with German media theorist Friedrich Kittler (1995c), titled "The information Bomb," Virilio draws an analogy between the nuclear bomb

---

[12] Now that Cloud-computing has become widespread and many individuals and organizations store their data and entire computer systems in a cloud—as we do at UCLA—Virilio's warning of an "accident of accidents" and catastrophic event in the technosphere becomes ever more frightening and compelling. Indeed, as I noted in an earlier chapter, as I write in December 2020 revelations of a Russian hack of multiple institutions of the U.S. National Security System create a potential dangerous calamity of the infosphere and beyond. See David E. Sanger, Nicole Perlroth and Julian E. Barnes, "As Understanding of Russian Hacking Grows, So Does Alarm," *New York Times*, January 2, 2021 at https://www.nytimes.com/2021/01/02/us/politics/russian-hacking-government.html (accessed on January 2, 2021). And just as I was working on this book yesterday, I heard of a big Internet outage on the East Coast; see Rachel Lerman, "Big Internet outages hit the East Coast, causing issues for Verizon, Zoom, Slack, Gmail. Internet outages and slowed services hit many areas of the East Coast just as the work and school day was ramping up Tuesday," *Washington Post*, January 26, 2021 at https://www.washingtonpost.com/technology/2021/01/26/internet-outage-east-coast/ (accessed January 28, 2021).

and the "information bomb," talking about the dangers of "fallout" and "radiation" from both. In contrast to the more dialectical Kittler, Virilio comes off as exceedingly technophobic in this exchange and deploys an amalgam of military and religious metaphors to characterize the world of the new information technologies. In one exchange, Virilio claims that "a caste of technology-monks is coming up in our times," and "there exist monasteries of sorts whose goal it is to pave the way for a (kind of) 'civilization' that has nothing to do with civilization as we remember it." These monks are avatars of a "technological fundamentalism" and "information monotheism," a world-view that replaces previous humanist and religious worldviews, displacing man and god in favor of technology.

> [This world-view] comes into being in a totally independent manner from any controversy. It is the outcome of an intelligence without reflection or past. And with it goes what I think as the greatest danger (of all), the derailment, the sliding down into the utopian, into a future without humanity. And that is what worries me. I believe that violence, nay hyperviolence, springs out of this fundamentalism.

Virilio goes on to claim that fallout from the "information bomb" will be as lethal for the *socius* as nuclear bombs, destroying social memory, relations, traditions, and community with an instantaneous overload of information. And we would add, as we suggested above, that a robust democracy would be another victim of the information bomb, as people would be so involved in media, virtual, and information technology that their relations to other people, society and politics would be atrophied and democracy weakened.

Thus, the technological "monks" who promote the information revolution are guilty of "sins in technical fundamentalism, of which we witness the consequences, the evil effects, today." One wonders, however, if the discourse of "sin," "evil," and "fundamentalism" is appropriate to characterize the effects and uses of information technologies which are, contrary to Virilio, hotly and widely debated, hardly monolithic, and, in my view, highly ambiguous, mixing what might be appraised as positive and negative features and effects.

Yet Virilio is probably correct that the dominant discourse is largely positive and uncritical and that we should be aware of negative aspects and costs of the new digital information technologies and debate their construction, structure, uses, and effects. Virilio is also right that they constitute at least a threat to community, social relations, and democracy, as previously established, though one could argue that the new communities and social relations generated by use of the digital technologies have positive dimensions as well as potentially negative ones that could strengthen democracy and social bonds although I am sure Virilio would vigorously argue that the negative effects of information technology far outnumber the positive ones.

Virilio notes as well the ways that digital technologies are penetrating the human body and psyche, taking over previous biological, perceptual, and creative functions of human beings, making humans appendages of a technological apparatus. He writes: "I am a materialist of the body which means that the body is the basis of all my work" (Virilio 1997a, p. 47). In his early writings, Virilio spoke of the body as "a vector of speed" and "metabolic vehicle" in which increased speed and velocity overwhelmed the human sensorium and empowered controllers of technologies of speed over other humans (1986).

In later work, Virilio has described the body as a planet, as a unique center around which objects gravitate, and criticizes increasing derealization of the body in cyberspace and virtual technologies (1997a, 1997b). Virilio is thus in part a materialist humanist and phenomenologist who is disturbed by the invasion of the human body by technology and the substitution of the technological for human and lived experience. We noted above Virilio's disagreement with Baudrillard over the issue of simulation which Virilio prefers to interpret in terms of substitution of one mode of experience or representation for another. Virilio's project is to describe the losses, the disappearances, of the substitution, describing now technology displaces human faculties and experience, subjecting individuals to ever more powerful modes of technological domination and control.

Thus, Virilio describes the effects of digital information technologies in terms of an explosion of information as lethal as a nuclear explosion and warns of the ubiquity of new types of accident that will require new modes of deterrence and dissuasion. He also envisages progressive derealization and dematerialization of human beings in the realm of virtual reality which may come to rule every realm of life from war to sex. From this perspective, technology emerges as the major problem and threat of the contemporary era, as a demonic force that threatens to erase the human. Much as his predecessors, Heidegger and Ellul, Virilio warns of the totalitarian threat in technology and calls for a critical discourse on technology, recognition of its possible negative effects, and regulation of technological development, subjecting technology to human and political control.

Yet Virilio uses the same model and categories to analyze war technology to characterize information technology. Thus, he has not really unraveled the riddle of technology which would have to interrogate the fascination, power, and complexity of digital technology and digital worlds, and not just its negativity. Virilio criticizes the discourses of technophilia, that would celebrate technology as salvation, that are totally positive without critical reservations, but he himself is equally one-sided, developing a highly technophobic and negative discourse that fails to articulate any positive aspects or uses for new technologies, claiming that negative and critical discourses like his own are necessary to counter the overly optimistic and positive discourses. In a sense, this is true and justifies Virilio's predominantly technophobic discourse, but raises questions concerning the adequacy of Virilio's perspectives on technology as a whole and the extent to which his work is of use in theorizing the digital information technologies with their momentous and dramatic transformation of every aspect of our social and everyday life.

Summing Up: Virilio, War, and Technology

> I don't claim to define the situation, I try to reveal tendencies. And I think I've revealed a number of important ones: the question of speed; speed as the essence of war; technology as producer of speed war as logistics, not strategy; war as preparation of means and no longer as battles, declaration of hostilities (Virilio and Lotringer 1983, p. 157).

This seems like a fair summary of Virilio's lasting contributions and I would suggest that the power of his work resides in his sustained interrogation of the virulence and power of military technology, but his works' limitation in turn results from using the model of military technology to interrogate technology as such and particularly digital information technologies. Virilio was justly distressed by the specter of total war, by the forces of military-technological domination, by the inexorable growth of power and the danger of the military-industrial complex during the Cold War and in particular the era between the Vietnamese and Gulf wars. During the era of the Cold War, the propagation and growth of the military-industrial complex and military state capitalism was the fundamental project and the organizing force behind the development of science, technology, and the allocation of public resources. More money was spent on this project than any other domain of existence and military priorities helped determine the mode of science, technology, and industry that developed in the Cold War period.

Military capitalism helped produce Big Government, Big Corporations, and a Big Military that deployed a tremendous array of manpower, weapons, and resources. Computers were largely developed from military imperatives, producing large, centralized calculating machines and information machines, including the so-called "information superhighway" which had its origins in the defense industry (see Edwards 1996; Turner 2008). The military, big government, and giant corporations also controlled scientific and technological research and development, with the military-industrial complex dominating the post-World War Two Cold War economies (see Melman 1965, 1974; Boggs 2004, 2016).

Yet while there are still threats to world peace and even human survival from the dark forces of military capitalism, one of the surprising events of the past decade is the emergence of a new form of global and computerized techocapitalism, with less lethal and more decentralized information and communication technologies, and new modes of peaceful connection and communication, as well as destruction. The project of this new form of technocapitalism is the development of an information-entertainment society based on the expansion of technoculture further into leisure and everyday life. This form of capitalism is a softer capitalism, a less violent and destructive one, a more ecological mode of social organization, based on more flexible, smaller-scale, and more ludic technologies.[13]

The differences between hard military capitalism and a softer computerized technocapitalism are evident in the transformation of the computer from a top-down, highly centralized, specialized machine controlled by big organizations, like the military or corporation, to the smaller scale, more flexible, and more ludic personal computer (see Turkle 1996 and Poster 1991 for elaboration of this distinction). Moreover, the surprising development of the Internet opens up new public spheres and the possibility of political intervention by groups and individuals excluded from political dialogue during the era of Big Media, controlled by the state and giant corporations, as I argued in the last chapters.

Of course, global technocapitalism has its own dangers ranging from economic worries about near-monopoly control of economic development through software and technological domination to the dangers of individuals getting lost in the proliferating terrains of cyberspace and the attendant decline of individual autonomy and initiative, social relations and interaction, and community. Ecological crises and worries continue to exist as an urgent problem and so far there has emerged no high-tech solution to solving ecological problems.[14]

Defenders of high-tech capitalism argue that the infotainment society promises more connections, interactions, communication, and new forms of community.

---

[13] For my earlier analysis of technocapitalism, see Kellner 1989a and for more detailed analysis of its current forms see Kellner 1998. On the infotainment media complex see Matthew Jordan, "For the 'political-infotainment-media complex,' the Mueller investigation was a gold mine," *The Conversation*, April 8, 2019 at https://theconversation.com/for-the-political-infotainment-media-complex-the-mueller-investigation-was-a-gold-mine-114417 (accessed December 5, 2020).

[14] For a discussion of capitalism, technology and ecological crises, see John Bellamy Foster, "Capitalism's Environmental Crisis—Is Technology the Answer?" *Monthly Review December* 1, 2000 at https://monthlyreview.org/2000/12/01/capitalisms-environmental-crisis-is-technology-the-answer/ (accessed January 17, 2021).

The project is in far too early stages to be able to appropriately evaluate so for now we should rest content to avoid the extremes of technophobia which would reject emergent information and entertainment technologies out of hand as new forms of alienation or domination, contrasted to technophilic celebrations of the information superhighway as the road to a computopia of information, entertainment, affluence, and democracy.

Virilio misses a key component of the drama of technology in the present age and that is the titanic struggle between national and international governments and corporations to control the structure, flows, and content of information and communication technologies in contrast to the struggle of individuals and social groups to use the information technologies for their own purposes and projects. This optic posits technology as a contested terrain, as a field of struggle between competing social groups and individuals trying to use digital technologies for their own projects contrasted to giant corporations, the state, and global bodies struggling over rules, regulations, and control of the powerful technoculture.

Despite his humanism, there is little agency or politics in Virilio's conceptual universe and he does not delineate the struggles between various social groups for the control of digital technologies and the new forms of politics that they will produce. Simply by damning, demonizing and condemning new technologies, Virilio substitutes moralistic critique for social analysis and political action, reducing his analysis to a lament and jeremiad rather than an ethical and political critique a la Ellul and his tradition of Catholic critique of contemporary civilization, or critical social theory. Virilio has no theory of justice, no politics to counter, reconstruct, reappropriate, or transform technology, no counterforces that can oppose technology. Thus, the increasing shrillness of his lament, the rising quasi-hysteria, and a sense of futility in the face of the technological juggernaut.

While Virilio's take on technology is excessively negative and technophobic, his work is still of importance in understanding the great transformation currently underway. Clearly, speed and the instantaneity and simultaneity of information are more important to the information economy and military than ever before, so Virilio's reflections on speed, technology, politics, and information technology and culture are extremely relevant. Yet he seems so far to have inadequately conceptualized the enormous changes wrought by an infotainment society and the advent of a new kind of multimedia information-entertainment technology. If my hunch is correct, his view of technology and speed is integrally structured by his intense focus on war and the military, while his entire mode of thought is a form of military-technological determinism which forces him not only to overlook the important role of capital, but also the complex ambiguities, the mixture of positive and negative features, of the new technologies now proliferating and changing every aspect of society and culture in the present era.

Virilio thus emerges as a highly useful theorist of the post-World War Two and Cold War era of the military with the domination of military technology and military capitalism, but he never analyses the complicity of capitalism and those economic forces that deploy technology for power and profit, instead putting all blame for contemporary problems on technology and its deployment by the military and perhaps the state. Thus, against Virilio, it should be recognized that digital information and entertainment technologies and social media are part of the capitalist project, that capital recognizes, along with Marx, that surplus value is gained by productive deployment of information technologies and social media, and that technology provides powerful weapons of profit and social control.

By eschewing critical social theory, Virilio does not have the resources to theorize the complex relations between capital, technology, the state, and military in the present age, substituting a highly elusive and evocative method for systematic theoretical analysis and critique. Virilio himself acknowledges his allusive and suggestive approach to writing, noting: "I don't believe in explanations. I believe in suggestions, in the obvious quality of the implicit. Being an urbanist and architect, I am too used to constructing clear systems, machines that work well. I don't believe it's writing's job to do the same thing. I don't like two-and-two-is-four-type writing. That's why, finally, I respect Foucault more than I like him" (Virilio and Lotringer 1983, pp. 38–39).

Indeed, Virilio's style is extremely telescopic, leaping from topic to topic with alacrity, juxtaposing defuse elements and themes, proliferating images, quotes, and ideas which rapidly follow each other, often overwhelming the reader and making it difficult to grasp the thrust of Virilio's argument. Sometimes, however, the trip with Virilio through the matrix of technology, war, representation, and dromology is exhilarating and illuminating. Virilio drives his text fast and covers a lot of ground quickly, although there are diversions and occasionally it appears the text-machine is getting lost. The Virilio-machine often encounters accidents but keeps on going and his followers are continuing his quest after his death in 2018. And unlike Camus who died young in a car crash, Virilio died of cardiac arrest at 86.

Hence, the speed which Virilio so well theorizes enters into the very fabric and substance of his writings. Virilio's texts move along quickly, they catch their topics on the run, they overwhelm with detail, but rarely develop a topic in systematic and sustained fashion. His style thus reflects his themes with speed, fragmentation, and complexity the warp and woof of his writing and texts. One wonders, however, whether a critic of speed, war, and technology should not occasionally slow down and more carefully and patiently delineate his theoretical position and books.

To some extent, Virilio exemplifies Walter Benjamin's theory of illuminations, fragments, and dialectical images in constellations of ideas and images which can illuminate specific phenomena and events. Like Benjamin, Virilio circles his prey with images, quotes, often startling and original ideas, and then quickly moves on to his next topic. Virilio believes in the virtue of breaks and interruptions, of gaps and absences, eschewing systematic theorizing. Yet although Virilio pursues some of the same themes as Benjamin, deploys a similar method, and cites him frequently, there are major differences. Whereas Benjamin (1969), in the spirit of Brecht, wanted to "refunction" new technologies to make them instruments of progressive social change and developed political strategies to exploit the potentially progressive features of new technologies, Virilio is relentlessly critical, eschews developing a technopolitics, and nowhere speaks of using or refunctioning technology to serve positive ends.

Thus, Virilio is highly one-sided and does not develop a dialectical conception of technology or a progressive technopolitics. Virilio produced no master oeuvre that would pull together his ideas and perspectives, or provide a synthetic overview. His long interviews with Sylvere Lotringer (1983) and John Armitrage (1999) contain the best overviews of what I take to be some of his most valuable work, but he never develops a critical theory of technology for the present age. In addition, as a critical philosopher, Virilio is quite ascetic, never articulating his normative position from which he carries on such a sustained and ferocious critique of technology. He seems to assume something like a religious humanism, that human beings are significant by virtue of their capacity for speech, reason, morality, political deliberation and participation, creative activity, and perhaps spiritually, although Virilio is ascetic in articulating his religious views. Technology is seen in Virilio's writings as undermining these human capacities, taking over human functions and rendering humans subservient to technological rationality.

Thus Virilio does not adequately articulate the humanist or religious dimension of his critique and, as noted, describes himself as a materialist and abstains from developing the normative perspective from which he carries out his critique. As noted, Virilio's reflections on technology, speed and war, recall Walter Benjamin who pointed out that the human body could simply not absorb the speed and lethality of modern war. Yet first and foremost Virilio's critique of technology has echoes of Heidegger's and Ellul's complaints concerning the totalitarian ethos of technology and modernity, and the ways that its instruments and instrumentality dominate human beings and create a novel world in which things and objects increasingly come to rule human beings. To the extent that Virilio's works illuminate the great transformation that we are currently undergoing and warn us of its dangers, too often ignored by the boosters and digiterati of information technologies and social media, he provides a useful antidote to the uncritical celebrations of the much-overhyped virtues of digital technologies and social media. Yet to the extent that he fails to provide critical perspectives which delineate how new technologies can be used for democratization, human empowerment, and to create a better world he remains a one-sided critic rather than a philosopher of technology who grasps the full range and import of the dramatic developments of the contemporary era.

In the next chapter, however, we will pursue in more detail the Virilian theme of the vicissitudes of high-tech war, since wars continue to threaten the human race. And in Chapter 9 we will explore how Stanley Kubrick's *2001* delineates a view of the future in line with Virilio's vision of technological totalitarianism coming to dominate the human species.

The Vicissitudes of High-Tech War

8

Abstract

Chapter 8 on "Vicissitudes of High-Tech War" discusses the origins of high-tech cyberwar and information war in the 1990s and some examples that played out in the Clinton, Bush, and Obama administrations, and then will discuss in more detail Russian cyberwar against the U.S. in the Trump administration. Throughout, focus will be on the increasingly central role of technology in military affairs and the increasingly perilous nature of high-tech war that has been apparent throughout the nuclear age. I argue that developments of the past decades intensify the dangers of nuclear Armageddon and create dangers of new technological Armageddon.

Keywords

High-tech cyberwar and information war • High tech war in Clinton • Bush • and Obama administrations • Russian cyberwar and the Trump administration • Cyberwarriors and Cyberwar • Infowar and Technological Armageddon

High-tech warfare emerged in the 2000's and under the Bush-Cheney administration involved wars in Afghanistan and Iraq as part of a global "War on Terror,"

This study draws upon work with Steven Best published in *The Postmodern Adventure. Science Technology, and Cultural Studies at the Third Millennium* (Best and Kellner 2001), and on my books *The Persian Gulf TV War* (1992); *From September 11 to Terror War: The Dangers of the Bush Legacy* (2003) and *Media Spectacle and the Crisis of Democracy* (2005) on the Bush II 2005 Iraq war.

D. Kellner, *Technology and Democracy: Toward A Critical Theory of Digital Technologies, Technopolitics, and Technocapitalism*, Medienkulturen im digitalen Zeitalter, https://doi.org/10.1007/978-3-658-31790-4_8

while the Obama administration continued the War on Terror and was involved in a war in Libya, using drone warfare as its weapon and tactic of choice. Trump's "victory" in the 2016 election was enabled in part through a cyberwar intervention by the Russians in which the Clinton campaign's emails were hacked, distributed to the press and public by WikiLeaks, whose release became a global media spectacle that harmed Clinton's candidacy and enabled Trump to squeeze out a narrow victory.[1] Further, in the last days of the Trump administration, there were reports that Russia had hacked many U.S. government, national security, and corporate sites for months in 2020 without the Trump administration acknowledging the hack or responding, although the Biden administration has promised to respond.[2]

In the late 1990s, The U.S. Commission on National Security for the twenty first Century maintained that "Outer Space and cyberspace are the main arteries of the world's evolving defense systems. Through technical and diplomatic means, the U.S. needs to guard against the possibility of 'breakout' capabilities in space and cyberspace that would endanger U.S. survival or critical interests."[3] In his first speech on military affairs after announcing his candidacy for President in 1999, George W. Bush affirmed the concept of a "Revolution in Military Affairs" (RMA) and was soon touting the virtues of a National Missile Defense (NMD) shield. Upon obtaining the presidency through a highly controversial electoral process (see Kellner 2001), Bush called for a dramatic increase in defense spending, and pushed for development of the missile defense program, popularly known as Star Wars II. The Bush administration also undermined collective security based on multilateral negotiations and treaties over weapons control by renouncing nuclear weapons treaties and attempts to regulate nuclear testing, biological and chemical weapons, small arms trading, land mines, and environmental treaties, which had been carefully nurtured by decades of diplomacy.

---

[1] On Trump and Russia in the 2016 U.S. presidential campaign and beyond, see Kellner 2017, Nance 2018, Corn and Ishikoff 2018; and Hettena 2018.

[2] See David E. Sanger and Nicole Perlroth, "More Hacking Attacks Found as Officials Warn of 'Grave Risk' to U.S. Government. Minutes after the government statement, President-elect Joseph R. Biden Jr. warned that his administration would impose "substantial costs" on those responsible. President Trump has been silent on the hacking. The Commerce, Treasury and Defense Departments, as well as other federal agencies, were the targets of Russian hackers," *New York Times*, December 17, 2020: A1 and Thomas P. Bossert, "We're Being Hacked. The Damage is Already Done," *New York Times*, December 18, 2020: A19. See also notes 19 and 20 in Chapter 6.

[3] See Mark Steel, *The Independent*, "The Secret Plans of the World's Most Dangerous Rogue State," June 19, 2001 at https://archive.commondreams.org/views01/0719-03.htm (accessed on December 7, 2020).

While the Obama administration returned to the collective security arrangements of rational states signing nuclear weapon treaties, negotiating treaties to limit Iran's nuclear capacity and shoring up relations with NATO allies, Donald Trump eliminated almost all of Obama's rational national defense policies and undermined U.S. national security in ways it will take years to rectify.[4] In this chapter, I first chart the genealogy and development of new trends in high-tech warfare which have emerged in the past decades in the 2000s and note its challenges and dangers. I will then discuss the Bush-Cheney administration's military program and foreign policy moves, highlighting the ways that the Bush II cabal intensified the dangers of high-tech war, while undermining efforts at collective security, environmental protection, and global peace. My argument is that the volatile mixture of a highly regressive and unilateralist and militarist administration with the development of high-tech weapons provides a clear and present danger of a protracted and frightening period of war. The mixture of rightwing unilateralism and militarism dramatically erupted in the Bush administration's military response to the terrorist attacks of September 11 and intensified the dangers to world peace in the Bush military campaign against terrorism which the hawks in the administration were labeling "World War III"—a campaign taken up by the Obama and Trump administrations that is still ongoing. Then, I conclude by looking at how U.S. national security was undermined by Trump's subservience to Russia and how Biden will likely reverse this submission to Russia, but may adopt a more aggressive policy that might reignite Cold War tensions and rivalries.

In the next section, I will discuss the origins of high-tech cyberwar and information war in the 1990s and some examples that played out in the Clinton, Bush, and Obama administrations, and then will discuss in more detail Russian cyberwar against the U.S. in the Trump administration. Throughout, focus will be on the increasingly central role of technology in military affairs and the increasingly perilous nature of high-tech war that has been apparent throughout the nuclear age. I argue that developments of the past decades intensify the dangers of nuclear Armageddon and create dangers of new technological Armageddon.[5]

---

[4] There are reports, however, that Biden will overturn Trump's worst assaults on national security, the environment, and U.S. democracy in a series of executive orders which he signed in his first ten days in office. See "Biden inauguration: Executive orders to reverse Trump policies," *BBC NEWS*, January 17, 2021 at https://www.bbc.com/news/world-us-canada-556 94415 (accessed on January 18, 2021).

[5] A Google search for "technological Armageddon" revealed a strange array of TED talks by techies, books by Christian "end-timers" talking of how developments in technologies correspond to predictions of end-times in the Bible to experts on war and weapons systems,

Cyberwarriors and Cyberwar

> On the battlefield of the future, enemy forces will be located, tracked and targeted almost instantaneously through the use of data links, computer-assisted intelligence evaluation, and automated fire control... I am confident [that] the American people expect this country to take full advantage of its technology—to welcome and applaud the developments that will replace wherever possible the man with the machine. General William Westmoreland, July 1970.

As the quote from General Westmoreland (who was head of U.S. forces in Vietnam during the period of significant U.S. intervention) indicates, the military has long anticipated a mode of high-tech war that would produce an electronic battlefield and eventually replace soldiers with machines. This would constitute a new stage of warfare in which cyborg warriors themselves would be part of a cybernetic-military apparatus marked by the merging of humans and technology and appearance of increasingly autonomous weapons systems, independent of human control. In de Landa's words (1991, p. 1):

> The image of the 'killer robot' once belonged uniquely to the world of science fiction. This is still so, of course, but only if one thinks of humanlike mechanical contraptions scheming to conquer the planet. The latest weapons systems planned by the Pentagon, however, offer a less anthropomorphic example of what machines with 'predatory capabilities' might be like: pilotless aircraft and unmanned tanks, 'intelligent' enough to be able to select and destroy their own targets.

In 1983, the Defense Advance Research Projects Agency (DARPA), responsible for development of the Internet, published a document outlining a "Strategic Computing Program" (STC) (Gray 1997). The SCP was a five year, $600,000,000 plan to produce a new generation of military applications for computers. The proposal included a thousand-fold increase in computing power and an emphasis on artificial intelligence. It envisioned "completely autonomous land, sea and air vehicles capable of complex, far-ranging reconnaissance and attack missions." These vehicles would have human abilities, such as sight, speech, understanding natural language, and automated reasoning. The SCP promoted the view

---

and a number of technophobes; see https://www.google.com/search?source=hp&ei=-fvgX5H WI5TZ9APjzZvwCg&q=technological+armaggedon+&oq=technological+armaggedon+&& gs_lcp=CgZwc3ktYWIQDDIJCAAQyQMQFhAeUP0WWP0WYKclaABwAHgAgAF1iA F1kgEDMC4xmAEAoAECoAEBqgEHZ3dzLXdpeg&sclient=psy-ab&ved=0ahUKEwjR ttTm69_tAhWULH0KHePmBq4Q4dUDCAg#spf=1608580095968 (accessed on December 21, 2020).

that the human element in many critical decision-making instances could be largely or totally taken over by machines (see the critique of the SCP by Gray 1997, p. 53 ff.). In this momentous process, just as humans are becoming like machines, machines are ever-more taking on human qualities (see Best and Kellner 2001, Chapter 4).

The Persian Gulf TV war indicated the extent to which computer and information systems were of primary importance in the planning and execution of the war and the ways that new fusions of humans and technologies engendered a cyberwarrior (see Kellner 1992; Best and Kellner 2001). The development is part of a process of creating soldiers better able to integrate themselves into technological systems and to fight increasingly complex battles. This involves cultivating high-tech skills in future soldiers. It requires disciplinary training to fit into technical apparatuses and using psych-technologies and drugs to enhance human abilities, while providing prostheses and implants that will produce technological amplification of human powers and abilities (Gray 1989, 1997, 2001).

During the 1990s and into the 2000s, reflections proliferated on the transformation of war with the incorporation of information technologies in the warfare state and the development of more de-centralized forms of social organization and a networked society. The first issue of *Wired* magazine featured a cover story by cyberpunk writer Bruce Sterling (1993) on high-tech war and during the same year cybertheorist Alvin Toffler (1993) published a book on the modes of "war and anti-war" that were unfolding in the supposed era of "Third Wave" civilization. By 1995, such views were evident in media culture with *Time* magazine publishing a cover story on "Cyberwar" (August 21, 1995), and with a cycle of films presenting technowarriors (i.e. *The Terminator* series, the *Cyborg Cop* series, *Universal Soldier*, *Cyborg Soldier*, and the like; see Kellner 2010b).

The accelerated role of information technologies in high-tech war has led some theorists to talk of new "Network-Centric Warfare" and "network-centric software systems" which are systems that focus on their communications elements of high-tech war. These changes have been produced "by the co-evolution of economics, information technology, and business processes and organizations" (Ignatieff 2000). They are, in the words of military authorities, linked by three themes: shifts from platform to network; a change from viewing actors as independent to viewing them as "part of a continuously adapting military-techno ecosystem"; and the "importance of making strategic choices to adapt or even survive in such changing ecosystems."[6]

---

[6] See the account by Vice Admiral Arthur K. Cebrokswky and John J. Garistka at www.usni. org/Proceedings/Articles98/PROcebrowski.htm (accessed January 4, 2021). Ignatieff (2000,

The evolution of high-tech war thus pertains to the increasing displacement of humans by technology and the next phase of technowar will probably reveal more "smart machines" supplementing and even replacing human beings. The 1991 U.S. Gulf War intervention, 1999 NATO war against Serbia, 2001 Afghan war, and Obama's limited military interventions in Afghanistan, Iraq, and elsewhere saw a widespread exploiting of drones, pilotless planes engaged as decoys and as instruments of surveillance, in addition to Cruise missiles and other "smart" weapons.[7] The U.S. military is developing "unmanned" technologies for ground, air, and undersea vehicles.[8] Smart tanks are already under production and as Gray notes:

> There are projects to create autonomous land vehicles, minelayers, minesweepers, obstacle breachers, construction equipment, surveillance platforms, and anti-radar, anti-armor and anti-everything drones. They are working on smart artillery shells, smart torpedoes, smart depth charges, smart rocks (scavenged meteors collected and then 'thrown' in space), smart bombs, smart nuclear missiles and brilliant cruise missiles. Computer battle-managers are being developed for AirLand battle, tactical fighter wings, naval carrier groups, and space-based ballistic-missile defense.... the Army even hopes to have a robot to 'decontaminate human remains, inter remains, and refill and mark the graves' (Gray 1989, p. 54).

By now the concept of high-tech war is widespread in the media and public sphere like the Internet and social media. For instance, an *ABC News* program on "Postmodern War" indicates a profound reorganization process in the military that is undergoing changes from heavy, slow, and large-scale machinery, such as 70 ton tanks, to smaller, lighter, faster, and more flexible vehicles. These are equipped with more accurate "smart" weapons and better mapping and sensor technologies which demand less "manpower" (see abcnews.com, 11/03/99). Exotic high-tech military devices include MEMS (Micro Electrono-Mechanical Systems) that will produce tiny airplanes or insect-like devices that can gather intelligence or attack enemies. MARV (Miniature Autonomous Robotic Vehicle) technologies

---

p. 164 ff.) describes the "revolution" in terms of the deployment of precision targeting at a distance and use of computers, also noting conservative military resistance to calls for dramatic transformation of the military (p. 171 f.).

[7] See Kellner 1992; Ignatieff 2000; and Edward Delman, "Obama Promised to End America's Wars—Has He? The president's military record, by the numbers." *The Atlantic*, March 30, 2016 at https://www.theatlantic.com/international/archive/2016/03/obama-doc trine-wars-numbers/474531 (accessed December 21, 2020).

[8] See "Pilotless Plane Pushes Envelope for U.S. Defense," *Los Angeles Times*, May 14, 2000: A1 and A-30, and "Robots with the Right Stuff," *Wired* (March 1996). See also Gunther et al. (1994) and Adams (1998).

and various other automated military systems would guide robot-ships, disable land-mines and unexploded arms, and provide more effective sensors, stabilization, navigation, control, and maintenance devices. These technologies would ultimately construct cyborg soldiers who will incorporate such devices into their own bodies and equipment (see Adams 1998, p. 122–137). Such miniature machines and cyberwarriors would be capable of gathering information, processing it, and then acting upon it, thus carrying through a technological revolution based on new intelligent machines.

Indeed, military spokespeople claim that the next generation of Army vehicles will be "Net-ready." The Army has developed and deployed a battlefield digitization project while it fields a new family of lightweight, easily deployable combat vehicles, which will have digital technology built into them, rather than bolting it on as the Army has had to do with older tanks and Bradley Fighting Vehicles.[9] Cyborg soldiers are also utilizing the Global Positioning Satellite system (which can be accessed from a computerized helmet) for precise mapping of the "enemy" and terrain. With the complex communications systems now emerging, all aspects of war—from soldiers on the ground and thundering tanks to pilotless planes overhead—are becoming networked with wireless computers providing information and exact locations of all parties. Robot scouts can roam the terrain sending back data instantaneously to commanders. SIPE (Soldier Integrated Protection Ensemble) is an army software program designed to merge all military digital technologies into one integrated data system. Even the physical state of the soldier can be monitored by computers, and one can imagine surgeons operating on wounds from continents away by using robots and the technology of "telemedicine."

Hence, phenomenal new military technologies are being produced in the Third Millennium, envisaged earlier by Philip K. Dick and other SF writers, which are changing the nature of warfare and are part of a turbulent technological revolution. They are helping to engender a new type of highly intense "hyperwar" where technical systems make military decisions and humans are put out of the loop, or are forced to make instant judgments based on technical data. As computer programs displace military planners and computer simulations supplant charts and maps of the territory, technology supersedes humans in terms of planning, decision making and execution. On the level of the battlefield itself, human power is replaced by machines, reducing the soldier to a cog in a servomechanism. French theorist Paul Virilio comments:

---

[9] See Gray 1997; Adams 1998; and www.cnn.com/2ready.combat.vehicle.idg/index.html.

> The disintegration of the warrior's personality is at a very advanced stage. Looking up, he sees the digital display (opto-electronic or holographic) of the windscreen collimator; looking down, the radar screen, the onboard computer, the radio and the video screen, which enables him to follow the terrain with its four or five simultaneous targets; and to monitor his self-navigating Sidewinder missiles fitted with a camera of infra-red guidance system (1989, p. 84).

The autonomization of warfare and ongoing displacement of humans by technology creates the specter of technology taking over and the possibility of military accidents, leading to catastrophe. There is a fierce argument raging in military circles between those who want to delegate more power and fighting to the new "brilliant" weapons opposed to those who want to keep human operators in charge of technical systems (see Arnett 1994; Adams 1998; Ignatieff 2000; Brose 2020). Critics of cyberwar worry that as technology supplants human beings, taking humans out of decision-making loops, the possibility of accidental firing of arms at inappropriate targets and even nuclear war increases. De Landa (1991) fears nuclear accidents and technology out of control in fully-automated cyberwar and calls for the manufacture of weapons over which humans maintain authority and interact creatively with technology, rather than being its object and servomechanism. Rochlin (1997) also cites dangers of accidents that emerge from automated battlefield and cyberwar where humans are forced to react ever more quickly to high speed systems. To support his arguments, Rochlin presents case studies of accidents that have happened in automated milieux over the past decade, thus warning that humans must attempt to maintain control over their technology.

The Center for Strategic and International Studies (CSIS) website list "Significant Cyber Incidents" of the past years at https://www.csis.org/programs/strategic-technologies-program/significant-cyber-incidents (accessed January 4, 2021). It claims that "This timeline records significant cyber incidents since 2006. We focus on cyber attacks on government agencies, defense and high tech companies, or economic crimes with losses of more than a million dollars." The last such incidents listed include:

> **December 2020.** Facebook found that two groups of Russians and one group of individuals affiliated with the French military were using fake Facebook accounts to conduct dueling political information operations in Africa.

> **December 2020.** More than 40 Israeli companies had data stolen after Iranian hackers compromised a developer of logistics management software and used their access to exfiltrate data from the firm's clients.

> **December 2020.** Unknown state-sponsored hackers took advantage of territory disputes between China, India, Nepal, and Pakistan to target government and military

organizations across South Asia, including the Nepali Army and Ministries of Defense and Foreign Affairs, the Sri Lankan Ministry of Defense, and the Afghan National security Council and Presidential Palace.

December 2020. Facebook announced that its users had been targeted by two hacking campaigns, one originating from state-sponsored Vietnamese hackers focused on spreading malware, and the other from two non-profit groups in Bangladesh focused on compromising accounts and coordinating the reporting of accounts and pages for removal.

December 2020. Suspected Chinese hackers targeted government agencies and the National Data Center of Mongolia as part of a phishing campaign.

December 2020. Hackers accessed data related to the COVID-19 vaccine being developed by Pfizer during an attack on the European Medicines Agency.

December 2020. Multiple U.S. agencies and private firms were breached by Russian hackers who compromised the software provider SolarWinds and exploited their access to monitor internal operations.

December 2020. A criminal group targeted the Israeli insurance company Shirbit with ransomware, demanding almost $1 million in bitcoin. The hackers published some sensitive personal information after making their demands and threatened to reveal more if they did not receive payment.

December 2020. CISA and the FBI announced that U.S. think tanks focusing on national security and international affairs were being targeted by state-sponsored hacking groups.

December 2020. Suspected state-sponsored hackers from an unknown country conducted a spear phishing campaign against organizations in six countries involved in providing special temperature-controlled environments to support the COVID-19 supply chain.[10]

In this unprecedented situation, cyberwar technologies are creating frightening types of technowar that require rethinking the very nature of military conflict and the viability of military solution to current problems in the face of such dangers. Theorists of novel modes of war focus both on the transformations of conventional warfare due to the implementation of computer technologies into the warfare state, as well as phenomena like information war; threats of hacker disruption of the economy, transportation, and communication systems; and exotic modes of biological, genetic, and chemical warfare (which are easier and cheaper to obtain than nuclear weapons).

---

[10] "Significant Cyber Incidents." Center for Strategic and International Studies (CSIS) at https://www.csis.org/programs/strategic-technologies-program/significant-cyber-incidents (accessed January 4, 2021).

There has been growing apprehension for decades concerning evolving types of chemical and biological warfare, which pertains to virulent and deadly forms of mass destruction.[11] Lethal weapons, such as a stolen nuclear bomb, or weapons using radiation exposure alone, could destroy vast urban areas, or poison water and food supplies. Biological weapons, such as anthrax, the plague, and many other disease-carrying biological agents, can be readily produced and distributed and could be extremely toxic,[12] as films such as *Outbreak* (1995) have warned and the anthrax attacks of Fall 2001 have demonstrated. Chemical weapons, which some think were deployed, or released by U.S. bombing, in the 2003 war against Iraq, are also extremely toxic and relatively easy to procure and deploy.[13] In fact, bomb-manufacturing information from the Internet makes the production of such armaments available to large numbers of groups and individuals, as well as facilitating the spread of more conventional bombs and munitions.

As noted, cyberwar was previewed in the 2003 war against Iraq and was an important component of the 1999 Kosovo war that was planned, programmed, and orchestrated through computer networks, as well as the 2001 war against the al Qaeda network and Taliban in Afghanistan. While the 1990–1991 Persian Gulf TV war was arguably the most spectacular military campaign of the TV Global Village (Kellner 1992), the Kosovo war was perhaps the first Internet war.[14] Not only was computerization deployed to plan and execute high-altitude bombing, but the Internet was a primary source of information and debate for the public. The volatile situation on the ground in Kosovo, with heavy NATO bombing, brutal retaliation against the Kosovian Albanians by the Serbs, swarms of refugees in the region, and the ensuing lawlessness made it extremely difficult for the major

---

[11] See, for example, "Weapons of Mass Destruction" in *Popular Mechanics* (June 1998, p. 80 ff.), and Peter Pringle, "Bioterrorism. America's Newest War Game," *The Nation* (November 9, 1998, p. 11–17). This article concludes that threats of the inevitability of catastrophic bioterrorism are exaggerated and are being hyped to promote another arena for military expansion—claims put in question by the terrorist attacks in Fall 2001. On bio warfare, see Miller, et al. 2001 and Boyle 2005.

[12] Michael A. Hayoun and Kevin C. King, "Biologic Warfare Agent Toxicity," *StatPearls*. Last update May 5, 2020 at https://www.ncbi.nlm.nih.gov/books/NBK441942/ (accessed on January 4, 2021).

[13] Chemical weapons were also used in the civil war in Syria. See Clare Lombardo, "More Than 300 Chemical Attacks Launched During Syrian Civil War, Study Says," *National Public Radio*, February 17, 2019 at https://www.npr.org/2019/02/17/695545252/more-than-300-che mical-attacks-launched-during-syrian-civil-war-study-says (accessed January 28, 2021).

[14] On the Kosovo war, see Walter L. Perry et al. (2002) *Disjointed War: Military Operations in Kosovo*. Washington, D.C.: Rand Publishing and Wesley K. Clark (2002) *Waging Modern War: Bosnia, Kosovo, and the Future of Combat*. Washington: PublicAffairs.

broadcasting and news corporations to bring in their employees to the battlefield. Instead, freelance reporters wrote on-the-ground testimonials and accounts, sent to leading web-based 'zines like *Salon* and *Slate*,' or in some cases newspapers. In addition, there was a tremendous amount of information from the region transmitted over the Internet via list-serves, e-mail, and websites. The NATO war was intensely debated over the Internet, if not the mainstream broadcasting and print media, bringing the Internet to the fore of political communication and debate.

Infowar and Technological Armageddon

The Afghan and Iraq wars fought by the U.S. after the September 11, 2001 terror attacks against the U.S. by the Al Qaeda networks combined high-tech surveillance, drone and missile attacks, and secret missions run by covert operations to supplement U.S. troop deployments.[15] Emergent forms of high-tech cyberwar war would include modes of Netwar fought in cyberspace wherein warring nations, or terrorists, attempt to destroy information and communications systems.[16] This type of Netwar was previewed in what might be called the "hacker wars." The term "hacker" initially meant someone who made creative innovations in computer systems to facilitate the exchange of information and construction of new communities (see Levy 1984; Hafner and Lyons 1996, p. 189 f.). Yet it came to refer to a mode of "terrorism" whereby malicious computer nerds either illegally invade closed computer systems, or breed viruses or worms that will disable computers and even entire computer networks (see Hafner and Markoff 1991; Sterling

---

[15] A site run by "The Bureau of Investigative Journalism" on "Drone warfare" states that "Between 2010 and 2020 the Bureau tracked US drone strikes and other covert actions in Pakistan, Afghanistan, Yemen and Somalia. The comprehensive reporting on civilian deaths helped lead to greater official transparency on targeted killing, and provided the data needed to hold the White House to account" at https://www.thebureauinvestigates.com/projects/drone-war (accessed on December 21, 2020).

[16] On "Netwar," see the studies in Mahnken, Editor, 2020. Some analysts use "information war" (Schwartau 1996) to cover all the modes of new high-tech war, while Rand theorists David Arquilla and David Ronfeldt (1996) distinguish, unsuccessfully in my opinion, between "netwar" and "cyberwar." I am conceiving cyberwar, as defined above, as a component of technowar, that includes "information war" in a specific sense of using information as a form of warfare, and more generally as a mode of warfare that is governed by information technology. "Netwar" in this sense is thus a form of war within or against computer networks that could include use of computer viruses, logic bombs, worms, and so on against enemy communications networks, or defense of one's systems against enemy attack or intrusion. It might also involve destroying communications satellites with nuclear weapons to disable the networks, or protecting one's own systems against such attack.

1992). During the 1990s and into the Third Millennium, panic emerged whenever a new computer virus was discovered and the national security apparatuses are preparing for information war that might disable important computer systems, disrupting the world economy, a nation's defense establishment, or any aspects of the system of production, transportation, and communication (Schwartau 1996; Adams 1998).

Such new modes of military conflict have evoked much discussion of "cyberwarriors," an "electronic Pearl Harbor," and dire threats to the world economy and individual security from a "technological Armageddon" (see Note 17). In this scenario, information guerilla warriors could disrupt or dismantle vital infrastructure systems of the military and civilian sectors of a country or region, creating problems ranging from power outages and airline crashes to the shutdown of banks, the stock market, the growing realm of electronic commerce, and the use of computers, phones, and electronic communication.

Quite unlike hand-to-hand combat, cyberwarriors can attack a nation from continents away. Dispensing with guns, tanks, and airplanes, cyberwar takes place through computers and high-tech devices. While cyberwar may unfold as abstract and bloodless, it too can have deadly "collateral damage" by effecting institutions like hospitals, emergency services, and air traffic control systems. Hackers and infowarriors employ new weapons such as viruses, logic bombs, trojan horses, and worms, all designed to replicate within and destroy the systems they penetrate.[17]

"Infowar" has been subjected to a variety of different analyses, referring on one hand to a new form of combat waged in the virtual and digital realms (making

---

[17] The Department of Defense estimates that its 2.1 million computer networks were infiltrated 250,000 times in 1995 (www.fas.org/irp/eprint/snyder/infowarfare.htm), and that the most significant breach of military computers occurred in 2008; see William J. Lynn III, "Defending a New Domain. The Pentagon's Cyberstrategy," United State Cybercommand Cybersecurity at https://archive.defense.gov/home/features/2010/0410_cybersec/lynn-article1.aspx (accessed December 7, 2020). The National Security Association (NSA) calculates that more than 120 countries now have "computer attack capacities" that could overtake Pentagon computers in a way that would "seriously degrade the nation's ability to deploy and sustain military forces" (www.govexec.com/dailyfed/0497/042297b1.htm). Moreover, teenage hackers, or "script kiddies," can develop programs that will disable electronic commerce and invade computer systems and destroy programs, as has happened regularly in recent years (see Best and Kellner 2001, Chapter 4). With summer 2000 virus attacks of the "Lovebug" and "resume," and summer 2001 Code Red and Siricam attacks, netwar began spreading throughout all domains of society, targeting citizens as well as businesses and governments—a danger that has continued up to the present. See Singer and Friedman 2014.

it a "third wave war" in Toffler's terms), as well as covering geopolitical struggles over information and communication. Yet it also refers to everything from
mischievous parahacker attacks on military sites to menacing assaults on communication systems by "terrorists." The fast-evolving technologies of infowar waged
by hackers and cyberactivists are causing a merging between military and civilian
targets, in that it targets civilians and noncombatants.[18]

After the bombing of the Chinese Embassy in Belgrade by NATO forces in
May 1999, hackers broke into the NATO website protesting the action and there
are many examples of hackers breaking into Pentagon and Defense Department
sites to post critical messages and deface government bulletin boards. Hacker
campaigns have also been organized against the governments of Mexico, Indonesia and others, protesting against unpopular policies by defacing official websites
or bombarding government sites and servers with spam or logic bombs, attempting
to shut them down. One of the more spectacular set of hacker attacks against commercial e-business sites occurred in February 2000. Hackers temporarily blocked
access to the popular Internet sites Yahoo, Amazon.com, CNN, and Buy.com, an
e-business retail site. Attacks followed on the news site ZDNet and E-Trade, an
online brokerage. Demonstration of the ease with which commercial Internet sites
can be disabled set jitters through the stockmarket, put the FBI and law enforcement agencies in motion, and set off a flurry of discussions of the need for better
cybersecurity.

Thereafter, cyberattacks against the stockmarket and banks became frequent,
including in September 2012 denial of service attacks that were carried out against
the New York Stock Exchange and a number of banks including J.P. Morgan
Chase.[19] In August 2014, "gigabytes" of sensitive data were reported stolen from
JPMorgan Chase, and the company's internal investigation was reported to have
found that the data was sent to a "major Russian city."[20] The FBI was said to be
investigating whether the breach was in retaliation for sanctions the United States

---

[18] See John Markoff, "Ideas & Trends: Blown to Bits; Cyberwarfare Breaks the Rules of
Military Engagement," *The New York Times*, Oct. 17, 1999 at https://www.nytimes.com/
1999/10/17/weekinreview/ideas-trends-blown-to-bits-cyberwarfare-breaks-the-rules-of-mil
itary-engagement.html (accessed December 21, 2020).

[19] See Singer and Friedman 2014.

[20] Elizabeth Weise, "JP Morgan breach affected 76 million households," USA TODAY, October 2, 2014 at https://www.usatoday.com/story/tech/2014/10/02/jp-morgan-security-breach/
16590689/ (accessed December 16, 2020).

had imposed on Russia in relation to the 2014 Russian military intervention in Ukraine.[21]

U.S. Government agencies have also been the object of cyberwar. In July 2001, the White House website was attacked by the "Code Red" virus that allegedly infected more than 225,000 computer systems around the world, and in the U.S. during the 2000s, up to 2020, there have been frequent cyberattacks on the White House and U.S. government.[22] Attorney General John Ashcroft announced that the government was forming nine special units to prosecute hacking and copyright violations in 2001.[23] Shortly thereafter, a Sircam virus infected countless computer systems and personal computers and once again hacking, worms, and viruses were being perceived as a serious threat to the digital economy and culture. And in May 2001, after a Chinese plane crashed in a skirmish with a U.S. intelligence plane, Chinese hackers launched several days of attacks at U.S. websites.

During this period computer hacker groups like Anonymous carried out multiple hacks against targets that they deemed worthy of attack and became a media sensation and referenced in a graphic novel and film *V for Vendetta* (2005) with their Guy Fawkes masks and mystique. Anonymous presents itself as a decentralized international activist/hacktivist collective/movement that has carried out various cyberattacks against government institutions, corporations, and organizations, such as Project Chanology in 2008 which consisted of protests, pranks, and hacks targeting the Church of Scientology.

Individuals all over the world aligned themselves with Anonymous and undertook protests and actions in their name in copyright-focused campaigns against motion picture and recording industry trade associations, and later Anonymous hacked government agencies of the United States, Israel, Tunisia, Uganda and others; the Islamic State of Iraq and the Levant; child pornography sites; copyright protection agencies; the Westboro Baptist Church; and corporations such as PayPal, MasterCard, Visa, and Sony (see Olson 2013). Anonymous was a precursor of and later supported WikiLeaks and the Occupy movement (which I discus

---

[21] Michael Riley and Jordan Robertson "FBI Examining Whether Russia Is Tied to JPMorgan Hacking". *Bloomberg*. August 27 2014 at https://www.bloomberg.com/news/articles/2014-08-27/fbi-said-to-be-probing-whether-russia-tied-to-jpmorgan-hacking (accessed December 23, 2020).

[22] Zoe Thomas, "Data breach hits agency overseeing White House communications," *BBC News*, February, 20, 2020 at https://www.bbc.com/news/technology-51580925 (accessed December 6, 2020). See Singer and Friedman 2014.

[23] See "Ashcroft Aims at Cyber-Criminals," *Associated Press*, July 20, 2001, and "Inept Virus Hits White House," *Wired News*. July 20, 2001.

in other parts of this book) and continue to be active, most recently in the George Floyd Murder protests.[24]

The possibility of new forms of cyberwar, and terrorist threats from chemical, biological, or nuclear weapons, creates new vulnerabilities in the national defense of the overdeveloped countries and provides opportunities for weaker nations or groups to attack stronger ones. Journalist William Greider (1998), for instance, author of *Fortress America: The American Military and the Consequences of Peace*, claims that: "A deadly irony is embedded in the potential of these new technologies. Smaller, poorer nations may be able to defend themselves on the cheap against the intrusion of America's overwhelming military strength" (abcsnew.com, 11/01/99). So-called asymmetrical war also allows smaller countries or groups to exercise deadly terrorism against civilian populations. Conversely, it is becoming clear that the more technologically advanced a society is, the more vulnerable it is to cyberwar.

Realizing the dangers, the Pentagon has been in the process of assembling something like a digital Manhattan Project with multi-billion dollar investments. Alarmed by threats to the national information infrastructure, the U.S. has been organizing a Federal Intrusion Detection Network, or Fidnet, to monitor computer networks and attempt to block intrusions and other illegal acts.[25] Jeffrey Hunter, the U.S. National Security Council director of information who is in charge of the initiative stated that: "Our concern about an organized cyberattack has escalated dramatically. We do know of a number of hostile foreign governments that are developing sophisticated and well-organized offensive cyber attack capabilities, and we have good reason to believe that terrorists may be acquiring similar capabilities." Civil libertarians, however, have been concerned that the project might compromise privacy and threaten civil liberties while increasing exponentially the power of the state.[26]

---

[24] Andrew Griffin, "'Anonymous' is back and is supporting the Black Lives Matter protests". *The Independent*. June 1, 2020 at https://flipboard.com/article/anonymous-is-back-and-is-supporting-the-black-lives-matter-protests/a-5sv-SL-ZQPqZRJb2k4DCbg%3Aa%3A2205 07260-bea9a90465%2Fco.uk (accessed December 23, 2020).

[25] Brian Fuller, "Federal Intrusion Detection Cyber Early Warning Federal Response," *Sans Institute*, 2020 at https://www.sans.org/reading-room/whitepapers/warfare/federal-intrusion-detection-cyber-early-warning-federal-response-1095 (accessed on December 7, 2020).

[26] For discussion of the earlier October 1997 Marsh report for the President's Commission on Critical Infrastructure Protection and subsequent U.S. policy initiatives to protect the information infrastructure, see Adams 182ff. The October 2001 USA Patriot Act gave the

Ever scarier, theorists are worrying about new biotechnology and nanotechno-logy military instruments that might miniaturize weapons of mass destruction in as yet unforeseeable forms (Joy 2000). Following the logic of miniaturization that is characteristic of advanced bio and information technology, some imagine that weapons could become near invisible and release destructive forces in unimagina-ble ways. The coming stage of military technology could thus involve microscopic nanotechnologies in which what were envisaged as "engines of creation" could become "engines of destruction" (Joy 2000; see the discussion of nanotechnology and Joy's critique in Best and Kellner, 2001, Chapter 4).

Hence, technowar is increasing global insecurities and the possibility of world destruction. High-tech cyberwar thus exhibits a continuation of the worst features of modernity, and threatens to take the development of military technologies to a catastrophic end-game. Yet within the global restructuring of capital, the form of military capitalism which had been dominant since World War Two appea-red during the Clinton administration to have been overshadowed by a more user-friendly digital capitalism. In this mode, new entertainment and informa-tion technologies would reproduce an infotainment society where war would be irrelevant and even harmful to the pursuit of profit and human well-being. The Internet itself, originally conceived and funded as a multi-point communications system for the military after a nuclear attack, was restructured into an instrument for communication and information-sharing, commerce, and politics.

Indeed, the information and communications technologies produced by the military could be refunctioned and restructured to be employed for peace, human purposes and empowerment, and not destruction and war. Cyborg systems can perform dangerous industrial labor, or simple household labor, as well as gene-rate electronic battlefields. Conversion from a warfare state to a welfare state was a rational expectation of the end of the Cold War, although the conti-nued proliferation of a military-technowar establishment run counter to such expectations.

In fact, developments within the Bush-Cheney administration and the ensuing militarization of the world after the terrorist attack on the United States on Sep-tember 11, 2001 generated decades of a War on Terror that continued into the Obama and the Trump administrations. Yet the origins of the Trump Presidency, still not fully documented or understood, involved an act of cyberwar by the

---

U.S. government the right to survey without a warrant e-mail and Internet usage, telephone and cell phone communication, and other restrictions on privacy and civil liberties that greatly alarmed civil libertarians; see Declan McCullagh, "Terror Law Foes Mull Strategies," (www. wired.comm, Nov. 3, 2001). The Patriot Act surveillance continues to be debated up to the present.

Russians during the 2016 U.S. presidential election that constitutes one of the more fraught and consequential acts of the era of cyberwar that threatened U.S. democracy with a regime that attacked its institutions and was subservient to the interests of Vladimir Putin and the Russians in a way not fully grasped or understood. Hence, in the next section, I discuss the Russian cyberwar against the 2016 U.S. Presidential Election that helped elect Donald Trump, followed by a discussion of dangers of a new Cold War in the Biden administration.

Donald Trump, Cyberwar, and the Russian Intervention

In the nineteenth century, Karl Marx described a rapidly expanding form of market capitalism in which economic forces were coming to control more and more domains of the state, the legal and judicial system, the educational and cultural system, and even forms of consciousness. In the twentieth century, a group of theorists known as "The Frankfurt School" described the rise of monopoly state capitalism in which powerful corporations and the capitalist state came to control more and more dimensions of society and the polity, culminating in the fascist regimes, which during World War II ignited a global war in which the survival of freedom and democracy were at stake. A hero of World War II, General Dwight D. Eisenhower became president of the United States from 1953 to 1960, and in his oft-cited "Farewell Address" warned of the growth of a military-industrial complex controlled by corporations, the military, and authoritarian politicians.

During the Cold War both the U.S. and U.S.S.R. built up tremendous military apparatuses and weapons of mass destruction producing a nuclear stalemate. The result was ever more powerful military institutions and a stronger role for the military in the U.S. government. Ironically, in putting together his transition team, cabinet, and administration, Donald J. Trump went further than any previous U.S. president in confirming Marx's view of capitalism and embodying Eisenhower's warning against the military-industrial complex, choosing an assortment of generals, billionaires, rightwing ideologues, and cronies for top positions in his government, often without qualifications in the area in which they were chosen to serve. They also included some of the worst racists, Islamophobes, sexists, homophobes, and creatures of the swamp imaginable, suggesting that rather than draining the swamp, Trump was constructing a morass of swamp creatures who were likely to create an era of unparalleled disruption, nastiness, conflict, crisis, precarity, and warfare that would put U.S. democracy and global organizations to their more severe tests in its history (Kellner 2017).

What was also totally bizarre was the number of strongly pro-Russian figures who Trump chose for the inner circle of his Government, and how Trump himself spoke so positively of Russian dictator Vladimir Putin and the Russians during his campaign and presidency going against Republican Cold War orthodoxy that villainized Russia, the "Evil Empire," and The Enemy Red Menace during the Cold War. My generation grew up in a Cold War culture which included copious films and TV shows with evil Russians, and absorbed broadcast and print media, schooling, and other institutions that presented the Cold War as a battle between Good and Evil with the U.S. representing Good, Democracy, and Freedom, while the Soviet Union represented Evil, with its' authoritarian communism, dictators, and collectivism. In particular, Republicans vilified the Soviet Union with Ronald Reagan decrying the "evil empire," and every Republican presidential candidate in my life-time has taken a hard anti-communist and anti-Soviet, or anti-Russian, line.

Enter Donald Trump with his famously friendly words toward Vladimir Putin during the election and a strikingly Russian-friendly inner circle of his campaign and administration and pro-Russian policies throughout his presidency. Trump initially chose as National Security Director, General Michael Flynn, who was one of a cadre of close Trump associates who had fond relations with Putin and the Russians, as pictures circulated throughout the media of Flynn next to Putin in Moscow at an event celebrating Russian Television (RT); Flynn was paid for this event but did not report Russian contacts on security forms and was fired, only belatedly admitting and signing documents which confirmed he had close ties and business interests with the Russia government and was a "foreign agent" for a Russian-friendly group in Turkey. After Flynn was exited from the Trump administration as these stories circulated in the press, Trump fired FBI Director James Comey who confirmed in Congressional hearings that Donald Trump, not known for friendship or loyalty, went out of his way to try to get Comey to lay off of investigating the Flynn-Russian connections, and Trump told the FBI Director that he expected "loyalty" to himself, leading to charges of possible obstruction of justice and impeachment.[27]

---

[27] Late in his presidency, after he lost the 2020 election to Joe Biden, Trump pardoned Comrade Flynn, facing years in jail for lying to the FBI. See Kyle Cheney and Josh Gerstein, "Trump pardons former national security adviser Flynn. Michael Flynn pleaded guilty in 2017 to lying to FBI agents investigating Russian interference in the 2016 election." Politico, November 25, 2020 at https://www.politico.com/news/2020/11/25/trump-pardons-former-national-sec urity-adviser-flynn-440649 (accessed on December 21, 2020). Flynn repaid Trump by going to a demonstration with him and the Proud Boys and denouncing the stolen election, a false

In addition, there wasn't adequate focus on how former ExxonMobil CEO Rex Tillerson, Trump's choice for Secretary of State, was one of several powerful positions within the Trump administration who had especially close relations with Vladimir Putin and the Russians. In over 30 years' service with ExxonMobil, Tillerson had particularly warm relations with Putin and the Russians, cutting big business deals, becoming a personal Friend of Vladimir, and even receiving Russia's Order of Friendship.[28] As president of ExxonMobil, Tillerson was attempting to negotiate with Putin and the Russians a major oil deal to explore areas of the Artic believed to contain vast mineral wealth, when President Obama imposed sanctions on Russia because of their intervention in the Ukraine and Crimea.[29] If the Trump administration could eliminate these sanctions, Tillerson and his cronies could profit immensely, but such conflicts of interest did not bother Donald J. Trump, himself a walking and talking cauldron of conflicted interests, ranging from his hotel near the Capital in Washington D.C. to the outposts of his far-flung and largely mysterious business empire.

Trump's pro-Russian cabinet was unnerving to many because not only was Trump himself excessively well-disposed toward Putin and Russia, but Secretary of State Tillerson was also close to Putin and the Russians, as were many of his associates. Intensely pro-Russian figures in Trump's campaign included Paul Manafort, his campaign manager for six months; Carter Page, who Trump described as a key foreign policy advisor and who US intelligence claimed was the target of a Russian intelligence operation;[30] Donald Trump Jr., who had made contact with a Russian lawyer who claimed she had "dirt" on Hillary Clinton and arranged a meeting in Trump Tower with the Russians, Manafort, and Trump's son-in-law Jarod Kushner, himself under investigation for Russian connections

---

claim Flynn then made on Trump's behalf on TV in the days leading up to the infamous January 6, 2021 White Riot and storming of the U.S. capital that I describe below.

[28] On the Tillerston/Russian connections, see Steve Coll, *Private Empire: ExxonMobil and American Power* (Baltimore: Penguin Books, 2013).

[29] On Russia's intervention into the Ukraine and Crimea and ensuing global controversy, see Stephen Lee Myers, *The New Tsar. The Rise and Reign of Vladimir Putin*. New York: Vintage Books, 2016.

[30] Stephanie Kirchgaessner, "Russia 'targeted Trump adviser in bid to infiltrate campaign.' CNN claims investigators have intelligence suggesting Russians may have used Carter Page to try to access Trump campaign," *The Guardian*, April 23, 2017 at https://www.theguardian.com/us-news/2017/apr/23/russia-tried-to-use-trump-advisers-to-influence-us-election-report (accessed July 10, 2017).

that he did not disclose on security forms necessary for White House government positions.[31]

Tillerson's nomination was especially unsettling because the major story of the week of Tillerson's designation to lead the State Department was an uproar over the Russian's hacking the 2016 election, and a story in the *Washington Post*, that the Russians had intervened to help Donald Trump get elected.[32] Trump himself denied that the Russians had hacked the Democrats, and had released selected information to help Trump and hurt Clinton, whereas major figures in both parties, the U.S. intelligence services, and sectors of the media were all convinced that Russia had intervened in the U.S. election. Moreover, President Obama had announced in the last weeks of his presidency a commission that would put out a report on the Russian intervention as soon as possible.[33]

Of course, it would be wrong to claim that the United States was an innocent who had never intervened in foreign elections in the light of an entire history of the U.S. incursions in foreign elections, starting in a post-War Italy where the CIA did everything in its power to make sure the Christian Democrats beat the Italian Communist Party.[34] During the Reagan era, William Casey's CIA intervened in

---

[31] The Donald Trump Jr revelations were accompanied by a release of emails documenting Trump's connections with Russian associates who claimed to have compromising materials on Clinton, a bombshell that dominated cable news on July 11, 2017 and that was documented in the *New York Times and Washington Post* with copious articles and opinion pieces on July 11 and 12, 2017. The following days there were revelations that further individuals, who might have connections to Russian intelligence, were also present; these events are being further investigated as I write, raising serious political and legal issues for the Trump inner circle, already under attack for their Russian connections.

[32] Adam Entous, Ellen Nakashima and Greg Miller, "Secret CIA assessment says Russia was trying to help Trump win White House," *Washington Post*, December 9, 2016 at https://www. washingtonpost.com/world/national-security/obama-orders-review-of-russian-hacking-dur ing-presidential-campaign/2016/12/09/31d6b300-be2a-11e6-94ac-3d324840106c_story. html?utm_term=.bb175270bfde (accessed December 9, 2016).

[33] For a book that documents Russia's hacking of Election 2016, within the context of Russia's foreign policy and use of hacking and "comprising material," written by a former U.S. intel-ligence agent, see Malcolm Nance, *The Plot to Hack America. How Putin's Cyberspies and WikiLeaks Tried to Steal the 2016 Election* (2016). See also Kellner 2017, Corn and Ishikoff 2018; and Hettena 2018. Others have claimed that the alleged Russian hack was an invention of US intelligence services and fake news mass media, a position taken as well by Putin and the Russians and initially by Trump. Presumably, the multiple investigations by Congress, the US government, the global media, and other entities will help document the complex events of the Trump-Russia connections still murky as Trump leaves office in January 2021.

[34] Nina Agrawal, "U.S. disrupts elections too. Like Russia, America has a long history of meddling in foreign votes." *Los Angeles Times*, December 22, 2016: A2.

a number of Latin American countries, and after failing to oust the Nicaraguan Sandinistas through the electoral process funded an illegal Contra war that embarrassed the Reagan administration and destroyed the careers of some of its officials when it was uncovered that an illegal deal selling arms to Iran was funding the Contra war.[35]

Yet given that Republicans had been Cold War super-adversaries of Russian Communism, it was highly bizarre to see so many of Trump's inner circle and Trump himself so enamored of Putin and the Russians.

Further, in late August 2017, Glenn Simpson, a former *Wall Street Journal* reporter who later helped found the private investigation firm Fusion GPS, testified for over ten hours with Senate Judiciary Committee staff behind closed doors about Trump-Russia connections in Election 2016. Simpson reported that his company Fusion GPS had hired a respected former British intelligence officer, Christopher Steele, to compile the dossier, which alleges that Trump had a long-running relationship with Russia and that the Kremlin holds compromising material on him. There were reports that Simpson had turned over 10,000 pages of documents that concerned Trump-Russia connections and there were calls to release the testimony and documents that the Senate Committee was considering.[36]

While Trump-Russia connections were confirmed by the report by Special Counsel Robert Mueller and many Congressional committees that investigated the linkages, the Report did not lead to Trump's impeachment, in part because Mueller did not investigate Trump's financial connections with Russia, which are still a mystery in 2021 as the Trump presidency comes to its ignominious end

---

[35] On the Nicaraguan Sandinistas, the Contra war, and the Reagan administration, see Philip W. Travis, *Reagan's War on Terrorism in Nicaragua: The Outlaw State.* Lexington, Ky.: Lexington Books, 2016.

[36] Ryan Lucas, "Researcher Behind Unverified Trump Dossier Meets Senate Investigators," *NPR*, August 23, 2017 at https://www.npr.org/2017/08/23/545289362/researcher-behind-unv erified-trump-dossier-meets-senate-investigators (accessed July 17, 2017). For a detailed account of the compiling of the Trump-Russia dossier, see Howard Blum, "Ex-Spy Christopher Steele Compiled His Explosive Trump-Russia Dossier. The man behind the infamous dossier that raises the possibility that Donald Trump may be vulnerable to Kremlin blackmail is Russia expert Christopher Steele, formerly of M.I.6. Here's the story of his investigation." *Vanity Fair, March 30, at* https://www.vanityfair.com/news/2017/03/how-the-explosive-russian-dos sier-was-compiled-christopher-steele (accessed August 17, 2017). Sarah Grant, Chuck Rosenberg, "The Steele Dossier: A Retrospective," *Lawfare*, December 14, 2018 at https://www.law fareblog.com/steele-dossier-retrospective (accessed December 8, 2020). Soon after, a wave of books appeared defending and attacking the Steele Dossier making it one of the most explosive and contentious documents in U.S. history in the ongoing saga of Trump's corruption and crimes that are still not fully investigated or documented.

with Trump refusing to concede that he lost the election to Joe Biden, and claiming that the election results were a fraud, an event I discuss below. While more information continues to come out concerning the Trump-Russia connections, and much still remains to be revealed, we can reach some preliminary conclusions and advance some hypotheses concerning the role of Putin and the Russians, and the role of cyberwar and hacking, in the context of the 2016 US presidential election, which helped produce the shocking and surprising victory of Donald Trump, and the even more startling scandals of his administration. In line with the concerns of this book, I will argue that hacking and cyberwar played a decisive role in a U.S. election for the first time, pointing to the increasingly primary role of technopolitics.

Trump, Russia, and Election 2016

From a globalist perspective, the 2016 election and Trump presidency needs to be interpreted in terms of the Cold War and US-Russian relations. The Russian hack and intervention in the US election can be seen as revenge, blowback, for US interventions in Russian elections, as well as what the Russians saw as US interference in elections and political upheaval in Russian satellite countries, and in other elections and countries around the world in which the Russians had interests. During the Cold War, both Russia and the US regularly intervened in elections throughout the world to support, in the case of the US, pro-US candidates while attacking leftist and progressive national candidates who would be conceived as supporting the Soviet Bloc or world socialism and communism. In turn, the former Soviet Union supported parties in its orbit while attacking governments seen as US allies.

While both the US and Russia have intervened in elections throughout the Cold War, yet, as far as we know, Election 2016 is the first time the Russians intervened massively and apparently effectively in a US presidential election and may have influenced the outcome through a campaign of hacking and dissemination of information that makes the campaign the first U.S. presidential election to be decided by cyberwar. It is significant that the revelations of Trump/Russian connections in the election make it clear that the Russians intervened on behalf of the Trump campaign, and that high level members of the Trump team had many contacts with the Russians, although we do not yet know the full nature of the collusion or extent of its coordination. There are many questions to still be raised, yet there are some assertions that we can make as of the time of writing this text in January 2021.

First, we must pose the question: Why did the Russians intervene in favor of Donald Trump in the 2016 election? While there were reports that Trump had secret business connections with Russia and that the Russians had compromising material on Trump (see note 37 above), it is clear that Putin and his crowd hated Hillary Clinton and preferred Donald Trump, although we still don't know why the Russians seemed to like Trump so much and why Trump gushed with such enthusiasm over Putin during the election and then fawned over him at the G-20 conference in Hamburg when they allegedly first met in July 2017. US-Russia-Trump relations have been at the center of many House and Senate investigations, as well as Special Counsel Robert Mueller's investigation of Trump and the election, so presumably we will eventually learn much more about this "special relationship."[37] In addition, the Trump-Russia connection is the focus of US media doing its job investigating shady government actions and it appears that mainstream media like *The New York Times, The Washington Post, The Wall Street Journal* and some of the cable news networks worked feverishly to break new and startling daily revelations of the Trump-Russia saga, which is still unresolved as Trump leaves office in 2021.

There are also books published that help explain why Putin and Russia would take the risk of intervening in the 2016 election. Martha Gesson's *The Man Without a Face. The Unlikely Rise of Vladimir Putin*, initially published in 2012,[38] argues that Putin was devastated by the collapse of the Soviet Union in 1989 when he was a KGB agent assigned to an embassy/Russian government job in Dresden, probably spying for the KGB. Hence, Putin himself was close to seeing

---

[37] The Mueller report was officially titled *Report On The Investigation Into Russian Interference In The 2016 Presidential Election* and documented the findings and conclusions of former Special Counsel Robert Mueller's investigation into Russian efforts to interfere in the 2016 United States presidential election, allegations of conspiracy or coordination between Donald Trump's presidential campaign and Russia, and allegations of obstruction of justice. The report was submitted to Attorney General William Barr on March 22, 2019, and a redacted version of the 448-page report was publicly released by the Department of Justice (DOJ) on April 18, 2019 at https://www.justice.gov/storage/report.pdf (accessed on January 18, 2021). A quite detailed analysis of the Mueller Report, including descriptions of its redactions and the battle over the report is found in the Wikipedia site "Mueller Report" at https://en.wikipedia.org/wiki/Mueller_report#Public_release_of_redacted_report (accessed on January 18, 2021).

[38] Martha Gesson's *The Man Without a Face. The Unlikely Rise of Vladimir Putin*, New York: Riverhead Books, 2012.

the Berlin Wall go down, the Soviet Empire collapse, and George H.W. Bush and others proclaiming that Russia lost the Cold War.[39]

From this optic, Putin was a super sore loser, who wanted revenge and made the choice to disrupt the 2016 election to delegitimize the U.S. electoral system and attempt to help defeat Hillary Clinton, who he blamed for helping stir up anti-Russia revolts in the Ukraine, Crimea, and other parts of Russian when she was Secretary of State under Obama. From this standpoint, the election of Donald J. Trump could be seen as Putin's revenge for the U.S. role in helping take down the Soviet Union and undermining the USSR/Russia during the long Cold War period, and then interfering in Putin's affair during his reign.[40] I'm bracketing the issue, explored in my book *The American Horror Show* whether Trump is Putin's Poodle, a Manchurian candidate, a Russian agent, or someone with deep financial Russian interests subject to blackmail, or something else, some of which we'll hopefully learn more about before it's too late.

The question then arises: how did Putin get his revenge and what are its nature and consequences? My thesis is that Russian Hacking of 2016 election and helping to get Trump elected is rooted in longtime Russian Cold War policies and the psyche of its product Vladimir Putin. Note that I'm saying "helping to get Trump elected" and not causing the outcome, as we obviously lack such evidence. In analyzing Election 2016 and most complex historical events, we need multicausal analysis that explicates the multiple causes and deep layers of economic, political, cultural, media and other factors that help explain as of 2017 why Donald Trump won the 2016 presidential election—which I attempt to do in *The American Horror Show*, published after the election and first 30 days of the Trump administration (Kellner 2017).

We do know, however, that the Democratic National Committee email server, and that of Clinton's campaign manager, John Podesta, among others, were hacked by the Russians, and that Putin and the Russians used WikiLeaks and

---

[39] This account is similar to that in the excellent overview of Putin's career in Steven Lee Myers, *The New Czar. The Rise and Reign of Vladimir Putin*, New York: Random House, 2015.

[40] Putin has long believed that Hillary Clinton was the spearhead of US interference in Russian affairs during her role as Secretary of State under Obama; see David M. Herszenhorn and Ellen Barry, "Putin Contends Clinton Incited Unrest Over Vote," *The New York Times*, December 8, 2011 at https://www.nytimes.com/2011/12/09/world/europe/putin-accuses-clinton-of-ins tigating-russian-protests.html (accessed July 17, 2017).

various global internet networks to circulate fake news, bots, and anti-Hillary stories.[41] From this perspective, key global—digital networks helped Trump win, which along with the anti-globalization discourse with which Trump conned his followers, makes Election 2016 the first US presidential election where global actors and networks, and global politics and cyberwar, played a significant and perhaps decisive role.

In early September 2017, it was revealed that the Russian intervention in election 2016 did not only include hacking the Democrats and releasing embarrassing or divisive emails on WikiLeaks, but that the Russians were actively engaging in an infowar against Hillary Clinton and for Donald Trump on Facebook, Twitter, Google, and various social network sites. While Facebook initially denied that its platform was used to spread anti-Clinton propaganda, in response to Congressional inquiries and media reports, they confessed that they had removed hundreds of fake accounts that attacked Clinton, or spread divisive disinformation, which were linked to Russian intelligence services, and that Facebook had sold at least $100,000 worth of anti-Clinton ads to Russian sources.[42] Fake news stories and fake postings on Facebook and Twitter had weaponized social media and led President Obama to warn Mark Zuckerberg about Russian penetration and use of Facebook before the election, but evidently the warning was not heeded.[43] Twitter was also weaponized by fake individuals linked to Russian intelligence which used bots to send messages to thousands of individuals, constituting en toto one of the most consequential cyberattacks in election history.[44]

---

[41] The hacking is documented in Nance, op. cit., and many mainstream media sources, although it is denied in pro-Trump sources from the swamps and whacko-worlds of "alternative facts" which may be the enduring legacy of the Trump presidency. For a comprehensive analysis of how the Russian hacking interfered in the 2016 election and dangers for the future of US democracy, see Massimo Calabresi, "Hacking Democracy. Inside Russia's Social Media War on America." *Time*, May 29, 2017, pp. 30–35.

[42] Editorial, "Russia's Fake Americans," *The New York Times*, September 9, 2017: A22 and Scott Shane, "The Fake Americans Russia Created to Influence the Election," *The New York Times*, September 2, 2017 at https://www.nytimes.com/2017/09/07/us/politics/russia-facebook-twitter-election.html (accessed October 15, 2017).

[43] Adam Entous, Elizabeth Dwoskin and Craig Timber, "Obama tried to give Zuckerberg a wake-up call over fake news on Facebook," *Washington Post*, September 12, 2017 at (accessed October 15, 2017). https://www.washingtonpost.com/business/economy/obama-tried-to-give-zuckerberg-a-wake-up-call-over-fake-news-on-facebook/2017/09/24/15d19b12-ddac-4ad5-ac6e-ef909e1c1284_story.html (accessed October 15, 2017).

[44] Ibid and Daisuke Wakabayashi and Scott Shane, "Twitter, With Accounts Linked to Russia, to Face Congress Over Role in Election," *New York Times*, September 27, 2017

Social Media Complicity in Trump's Election and Trump's Assault on Democracy

Hence, so-called "social media" clearly bear some major responsibility for Trump's election as President and served his interests as a major force of communication and propaganda throughout his presidency during which he used social media to create chaos, divide the country, put in question the results of the election, spread lies and disinformation, including that he had won the 2020 election against Biden which was "stolen," which enabled Trump to mobilize a violent and thuggish base to harass officials and individuals who Trump sees as their "enemies," and to carry out an insurrection in the U.S. capital on January 6, 2021 as I discuss below.

In particular, Trump was a heavy user of Twitter and tweeted out his messages and rants throughout the day and night. Indeed, Trump may be the first major Twitter presidential candidate, and then president. Indeed, certainly he has been the one using it most aggressively and frequently into his presidency and will no doubt continue until his twitter finger quits tweeting, or he is kicked off the site for violation of rules—which he was in January 2021 as we shall see below.

Twitter was launched in 2006, but I don't recall it being used in a major way in the 2008 election, although Obama used Facebook, and his campaign bragged that he had over a million "Friends." Obama thus used Facebook as part of his daily campaign apparatus, although I do not recall previous Presidential candidates using Twitter in a big way like Donald Trump, although many politicians now have accounts.

Twitter was a perfect vehicle for Trump as you can use its 140 character framework for attack, bragging, and getting out simple messages or posts that engage receivers who feel they are in the know and involved in TrumpWorld when they get pinged and receive his tweets. When asked at an August 26, 2015, Iowa event as to why he uses Twitter so much, Trump replied that it was easy, it only took a couple of seconds, and that he could attack his media critics when he "wasn't treated fairly." Trump has also used Instagram—an online mobile photo-sharing, video-sharing and social networking service that enables its users to take pictures and videos, and share them on a variety of social networking platforms, such as Facebook, Twitter, Tumblr and Flickr.

---

at https://www.nytimes.com/2017/09/27/technology/twitter-russia-election.html (accessed October 13, 2017). As indicted in Notes 35 and 36 above, the U.S. had intervened in many elections over the years against Russia candidates, surrogates in other countries, and allies.

Twitter is perfect for General Trump who can blast out his opinions and order his followers what to think. It enables Businessman and Politician Trump to define his brand and mobilize those who wish to consume or support it. Trump Twitter gratifies the need of Narcissist Trump to be noticed and recognized as a Master of Communication who can bind his warriors into an on-line community. Twitter enables the Pundit-in-Chief to opine, rant, attack, and proclaim on all and sundry subjects, and to subject TrumpWorld to the indoctrination of their Fearless Leader.

Twitter was sharply criticized, as were other social media, for giving Trump unfiltered access to his gigantic following, and eventually, late in his term, they started labelling his lies as information that was not authenticated.[45] Yet Trump continued almost to the end to spew forth daily his litany of lies on television and old media through his orchestration of media events as spectacles and his daily Twitter feed. Trump-friendly TV channels like the Fox Network, Talk Radio shows like Rush Limbaugh and Alex Jones, internet sites like Info Wars and News max, and multiple other sites and news sources to spin Trump's lies and disinformation creating an alternative reality in TrumpWorld and large sectors of the public who followed him religiously, establishing one of the first "cults of personality" in U.S. politics.

After decisively losing to Joe Biden in the 2020 U.S. Presidential election Trump continued his Twitter war, calling the election a fraud, and mobilizing his Republican allies and base, to join him in protesting results of the election.[46] In the early days of January 2021, Trump and his thug army organized an invasion of the Congress, and just after Trump, Giuliani, Don Jr. and others spoke at a demonstration on January 6, 2021, urging Trump's followers to "fight" the allegedly stolen election and to march on the capital, on January 6, 2021 Trump's thug army of Storm Troopers invaded and desecrated the Halls of Congress while terrorizing its members who were forced into lockdown as the marauders trashed offices, broke windows, and beat up guards in a melee in which at least five people were killed, helping to bring the Trump era to its not unexpected implosion,[47] and leading Trump to be impeached for the second time shortly thereafter. Moreover, Trump was permanently banned from Twitter, Facebook, and other social media

[45] Brian Fung, "Social media bet on labels to combat election misinformation. Trump proved it's not enough," *CNN Business* December 8, 2020 at https://www.cnn.com/2020/12/08/tech/facebook-twitter-election-labels-trump/index.html (accessed December 8, 2020).

[46] BBC News, "US election security officials reject Trump's fraud claims," 13 November, 2020 at https://www.bbc.com/news/election-us-2020-54926084 (accessed December 3, 2020).

[47] Jeffrey St. Clair, "Roaming Charges: White Riot, I Wanna Riot of My Own," *Counterpunch*, January 8, 2021 at https://www.counterpunch.org/2021/01/08/roaming-charges-white-riot-they-wanna-riot-of-their-own/ (accessed January 15, 2021).

companies which had fed his insatiable narcissism, mobilized his tribe of cultists, and unleashed the forces of darkness with a twitch of his Twitter finger and the widespread retweeting by his Legions of Darkness, many of which are now fugitives from the law who face decades in prison.

Hence, during the Trump era, elections and presidential politics were fought through cyberwar using digital technology as weapons of mass destruction against the other tribe's political opponents. Cold War tensions heated up between the U.S. and Russia that could be a dangerous challenge in the Biden era. Hence, high-tech war is our future and as we type out our daily emails and work on our computers we should recognize that cyberspace is militarized and that we are at the center of a high-tech war zone. May the Force be with Us.

Kubrick's *2001* and Vision of Techno-Dystopia

9

Abstract

Chapter 9 provides an analysis of filmmaker Stanley "Kubrick's *2001* and the Dangers of Techno-Dystopia." Stanley Kubrick and Arthur C. Clarke's collaboration on their monumental adventure and philosophical masterpiece *2001* (1968) arguably took the science fiction film to new aesthetic and thematic vistas and helped resurrect the SF genre, while paving the way for later SF blockbusters such as the *Star Wars, Star Trek*, and *Alien* film franchises. Indeed, science fiction itself is a potentially visionary genre, putting in question our concepts of the ordinary and conventional sense of reality and presenting us with new ways of seeing, new experiences, and even new worlds. Science fiction projects Otherness, in the forms of civilizations, species, technologies, and ways of life that are other than our own and may present utopias that provide visions of a better life, or dystopias that present dangerous visions of catastrophic futures. Indeed, science fiction has often provided cautionary warnings about where certain current trends could be taking us, or frightening challenges that we could face in the future, as Roland Emmerich's 2004 depiction of extreme world-altering climate change, *The Day After Tomorrow*.

Keyword

2001 • Cinematic visions • Stanley Kubrick • Arthur C. Clarke • Science fiction film • Utopias and dystopias • Roland Emmerich • *The Day After Tomorrow*

Stanley Kubrick's *2001* provides one of the most provocative visions of an alienated, dystopic, and potentially catastrophe technofuture found in contemporary

© The Author(s), under exclusive license to Springer Fachmedien Wiesbaden GmbH, 237
part of Springer Nature 2021
D. Kellner, *Technology and Democracy: Toward A Critical Theory of Digital Technologies, Technopolitics, and Technocapitalism*, Medienkulturen im digitalen Zeitalter, https://doi.org/10.1007/978-3-658-31790-4_9

film. I will argue that *2001* articulates Cold War tensions in the context of the space and arms race, and portrays fears of a future in which technology, presented in the dominant ideology as the solution to social problems and vehicle for a better life, evolves out of human control and depicts a catastrophic future. Following Kubrick's 1964 film *Dr Strangelove or: How I Learned to Stop Worrying and Love the Bomb* (which articulates fears of nuclear war technology), *2001* presents apparent détente between the US and Soviet Union, but articulates mistrust and the dangers of deadly conflict between the superpowers. Yet the villain in the scenario is a computer HAL which takes over control of the spacecraft and propels the ship into a frightening but dazzling alternative space-time continuum. The film dramatizes the endangering nature of computer and space technology, articulating fears of it spiraling out-of-control and taking charge of human destiny, showing how a high-tech utopia can spin into a techno-dystopia in which democracy, individuality, and emotion disappear and humans are lost in a technological apparatus that can be threatening and lethal to their humanity and survival.

Stanley Kubrick's Visionary Cinema

"I don't like to talk about *2001* much because it's essentially a nonverbal experience. Less than half the film has dialogue. It attempts to communicate more to the subconscious and to the feelings than it does to the intellect. I think clearly that there's a basic problem with people who are not paying attention with their eyes. They're listening. And they don't get much from listening to the film. Those who won't believe their eyes won't be able to appreciate this film." Stanley Kubrick (Agel 1970, p. 7)

Where literature gives us moral instruction, philosophical vision, and aesthetic experiences of the highest order, philosophers ranging from Adorno to Derrida to Rorty tell us that movies merely provide templates for contemporary life, as Neil Gabler argues in *Life, The Movie* (2000), or mere mass entertainment and distraction as Adorno and Horkheimer infamously argued (2002). Since the 1960s, however, an increasing number of serious philosophers, theorists, and cultural critics have embraced film as art, or have at least perceived aesthetic potential and value in some cineastes as auteurs who have created a body of exemplary films. While the European art film was for some time considered by many to be models of aesthetic innovation and philosophical depth (Sontag 1966), since the eruption of film into a form of serious culture globally since the 1960s, some critics have found aesthetic value and moral and socio-political critique in Hollywood films of directors such as Alfred Hitchcock, John Ford, and Stanley Kubrick, that latter whom I will be considering in this study.

Indeed, I would argue that Kubrick is an *artiste* of the highest order, that his work provides significant aesthetic pleasures, powerful social and moral critique, and philosophical stimulation and insight. Near his death in 1999, following the release of his last film *Eyes Wide Shut*, many film critics embraced Kubrick as one of the greatest film directors of his era, and his film *2001* is considered by many as one of the masterpieces of world cinema, and even the greatest film ever made.[1]

Stanley Kubrick and Arthur C. Clarke's collaboration on their monumental adventure and philosophical masterpiece *2001* (1968) arguably took the science fiction film to new aesthetic and thematic vistas and helped resurrect the SF genre, while paving the way for later SF blockbusters such as the *Star Wars, Star Trek,* and *Alien* film franchises. Indeed, science fiction itself is a potentially visionary genre, putting in question our concepts of the ordinary and conventional sense of reality and presenting us with new ways of seeing, new experiences, and even new worlds. Science fiction projects Otherness, in the forms of civilizations, species, technologies, and ways of life that are other than our own and may present utopias that provide visions of a better life, or dystopias that present dangerous visions of catastrophic futures. Indeed, science fiction has often provided cautionary warnings about where certain current trends could be taking us, or frightening challenges that we could face in the future, as Roland Emmerich's 2004 depiction of extreme world-altering climate change, *The Day After Tomorrow.*

Visionary science fiction is intimately concerned with science and technology and where they are taking us in the human adventure. It projects visions of utopian futures grounded in science and technological civilization, as well as dystopic catastrophes and even apocalypse produced by science and technology evolving beyond human powers and out of the control of human beings in worlds where technology is more powerful than humans. Kubrick's *2001* is one of the most engrossing and frightening meditations on technology with image, scene, plot line, and the story providing cautionary warnings of the dangers of technologies dominating humans and creating a system where technology ultimately controls human life and destiny.

Indeed, both Kubrick and Clarke's *2001* provides cautionary warnings about the future of humans and technology and dystopian visions of where we could be headed as a human species as we set out to control more of the universe. The Frankfurt School theorized that in the modern era the human project was

---

[1] High evaluations of Stanley Kubrick as auteur are found in Gelmis 1970, Chiohan 2000, Wheat 2000, and Kolker, ed. 2006. Wheat concludes his reading of 2001 by stating outright that "Stanley Kubrick's *2001* is, in my judgment, the grandest motion picture ever filmed" (p. 160).

the domination of nature (1972) through science and technology in which individuality disappeared in mass society, freedom and democracy disappeared in totalitarian political systems, culture degenerated into mass culture, and humans had used science and technology to create mass systems of domination, monstrosities like concentration camps which functioned as systems of mass murder, and nuclear weapons which would destroy humanity and Kubrick and Clarke took the Frankfurt school vision into the future and outer space exploration (although I have no evidence that either Clarke or Kubrick read the Frankfurt School).

Subtitled "A Space Odyssey," the epic and mythic thrust of the narrative of Kubrick's films evokes the wanderings, explorations, and adventure of Homer's *The Odyssey*, which served as an inspiration for the film's scope and complexity. Loosely based on Arthur C. Clarke's 1951 story "The Sentinel", the film was produced in an intense 1964–1968 collaboration between Kubrick and Clarke. Eventually, Clarke's name was placed as sole author of the novel, while Kubrick was listed first in the screenwriter credits.[2]

While Kubrick's collaboration with noted science fiction and science writer Arthur C. Clarke provides helpful technological knowledge and insight that contributed to developing *2001*'s narrative plausibility,[3] Kubrick's film is arguably a much greater work of art and purveyor of aesthetic experience than Clarke's story, and much more provocative philosophically, subverting conventional views of the evolution of humans and technology. Clarke's writing on science and technology is often quite uncritical and conventional, and his fictional texts in general are not particularly original or innovative, compared to Kubrick, or masters of science fiction like Jules Verne, H.G. Welles, or Philip K. Dick.[4] Clarke's prose is highly expository and he explains in dialogue and interior reflections what his characters are thinking and describes their action with explanatory detail. Clarke's science fiction normalizes space travel, often presenting it as an inevitable continuation of the present as in the passage of "The Sentinel" where he writes:

[2] On the genesis and differences between Clarke's short story "The Sentinel," and Kubrick's film, see Clarke (1972), Chiohan (2000), and Wheat (2000). The novelization of the film gives Clarke authorial credit, although he concedes that Kubrick should have been credited as co-author (1972). Wheat draws out in detail the Homeric themes and images in *2001*, and I shall engage Wheat's "triple allegory" reading of *2001* below.

[3] Another technical advisor for *2001*, the late Fred Ordway, should also receive credit for the convincing portrayals of space travel. In an article on Ordway following his death, it was noted that many of the details of space technology in the film were provided by Ordway. See the obituary, Jill Leovy, "Fred Ordway, 1927–2014. Rocket scientist was key advisor on '2001' film." *Los Angeles Times*, July 15, 2014: AA1,5.

[4] For my take on how these giants of science fiction are visionary guides to the "postmodern adventure," see Best and Kellner 2001.

"We had begun our journal early in the slow lunar dawn, and still had almost a week of Earth-time before nightfall. Half a dozen times a day we would leave our vehicle and go outside in the space-suits to hunt for interesting minerals, or to place markers for the guidance of future travelers. It was an uneventful routine. There is nothing hazardous or even particularly exciting about lunar exploration. We could live comfortably for a month in our pressurized tractors, and if we ran into trouble we could always radio for help and sit tight until one of the spaceships came to our rescue" (Clarke, in Agel 1970, p. 16).

Kubrick, by contrast, presents the dangerous dimensions of space travel and portrays encounters with other civilizations which could endanger human sovereignty and even existence. Moreover, Kubrick's characters, unlike those in Clarke's fiction, appear devoid of subjectivity, and we can only glean their thoughts and actions through what we see on the screen. Vast sections of Kubrick's film have little or no dialogue, and the only character expressing emotions or idiosyncratic thoughts and behavior is the computer HAL.

However, Kubrick has mastered his cinematic medium like few other contemporary film artists, combining brilliant innovations in image, sound, and cinematic technique with significant story-telling, social critique and commentary, and philosophical and moral vision. Kubrick's cinematic vision provides spectacles of the awe, wonder, dangers, mysteries, and otherness of space travel, dramatizing the propelling of the human species into another space-time continuum and entering into a rupture with human history as we've known it. Kubrick, like Thomas Pynchon and exemplary writers of his era, is a modernist who combines creative interweavings of aesthetics, technology, narrative and philosophy into original visions of the human adventure via cinematic form and thematic content. As Naremore suggests (2007, p. 24 ff.), Kubrick's style combines cool Brechtian alienation-effects and the aesthetics of the grotesque, with I would suggest, a stunning fusion of music and sound, to produce unique cinematic artifacts with major philosophical themes that approach the sublime, a category that *2001* evokes, as I suggest below.

Kubrick was emerging at the time as one of the American cinema's greatest directors having produced *Dr Strangelove or: How I Learned to Stop Worrying and Love the Bomb* in 1964, preceded by the highly controversial *Lolita* in 1962 and *Spartacus* in 1960. Seeming master of all genres, Kubrick also made an acclaimed anti-war drama *Paths of Glory* in 1957 and a pair of interesting crime dramas (*The Killing* [1956] and *Killer's Kiss* [1955]), preceded by a modernist and existentialist [anti]war film *Fear and Desire* (1953).

Arthur C. Clarke, in turn, was one of the world's most celebrated science fiction writers and popularizer of science, who had published a wide range of stories, novels, and educational science books. Their collaboration helped produce a work that was highly faithful to contemporary science, visionary in its look at the future, and complex and engaging beyond the parameters of any science fiction film that had yet appeared. Indeed, many critics consider *2001* a cinematic masterpiece that appeared on the top of many "best film of the year" lists for 1968 and on many "great film of all time" lists.

Financed by MGM and largely produced in its London Studios, *2001* cost over $10 million, a formidable price tag in 1960s Hollywood for a film belonging to what the big studios then perceived as an offbeat genre. *2001* is one of the most documented films in history concerning its inception, production, and reception, beginning with a book edited and produced by Jerome Agel (1970) and followed by subsequent studies of *2001*'s production and history such as Clarke (1971), Schwam, ed. (2000), Chion 2001, Bizony (2001), and a small library of further studies. The film's completion took much longer than anticipated, as Kubrick and Clarke endlessly reworked the script and the crew devised hundreds of special effects, many of which were technical breakthroughs and which cumulatively generated one of the most impressive array of technological innovations in the history of film, as indicated in a series of documentaries included in the 2012 Kubrick "Triple Feature" Blue-Ray of *2001*, such as "The Legacy of *2001*" and "The Prophecy of *2001*".

Indeed, Kubrick's *2001* demonstrated the potentiality for high-tech cinema of the epoch to present cinematic visions of the future not previously possible which he used as vehicles for the stunning vision of dangers to the human race and the demise of individuality, freedom, and democracy in a techno-future.

The Demise of Democracy and the Individual in 2001

> Wyndam Lewis: "The artist is always engaged in writing a detailed history of the future because he's the only person aware of the nature of the present." H. Marshall McLuhan: "Knowledge of this simple fact is now needed for human survival" (cited in Agel 1970, p. 11).

The screen opens *2001* and remains in black for a three-minute musical overture punctuated by Ligeti's *Atmospheres*; the composer's music would appear also later in the film to help generate an aura of mystery and otherness. A title "The Dawn of Man" introduces the first part of the story which depicts apes who are beginning to communicate through sound, emitting high pitched grunts, and who encounter a fateful monolith which in the film's mythology inspires the creation of technology and helps transform the apes into humanoids. The sequence unfolds in a vast African desert illuminated by the rise of a bright-red sun that portrays drought conditions and bleached bones. The apes/humans are vulnerable to animal attack and are depicted as hovering and frightened, foraging for plants and insects and apparently near starvation. As one group drinks muddy water, another group appears and the two aggressively confront each other, screaming and violently gesturing, an obvious allegory that displays the origins of language and communication in existential need and conflict, and inherent hostility and violence in species destined to struggle for survival. At night, the hoard huddles in fear under a rock, a pathetic species in thrall to fear of predator animals and the horrors of the night.

A new day dawns to the strains of Ligeti's *Requiem* and the man-apes encounter a shimmering monolith, a perfect rectangular shape that appears to be the product of an intelligent being. As they approach the object and touch it, the monolith emanates piercing sounds which seem to transform the man-apes. Shortly thereafter, one of the incipient humanoids takes a bone from a skeleton and begins triumphantly beating the animal skeleton to pieces as the chords of Strauss' *Also Sprach Zarathustra* intensify; the scene cuts with a quick glimpse of a falling animal and henceforth the emerging humanoids have weapons, which they use to kill animals and eat meat, while coming to dominate other groups of less-developed humanoid/animals, emerging as a superior species.

The opening sequence signifies the mystery of human existence and the transformative nature of technology as the ape/men transform sticks and bones into weapons with which they can violently subjugate other species, providing an allegorical scene of how technology helps humans dominate nature and other human beings. The emerging humans stand transfixed before the monolith, a symbol that continues throughout the film as a sign, among other things, of the magical nature of technology and the awe it inspires in human beings. The group of incipient humans who encountered the monolith, master speech and weapons, and subjugate other groups, a sequence portrayed at a watering hole where the group possessing primitive weapons and frightening grunts routs the other group.

Since the film connects the monolith with the discovery of technology, *2001* suggests, in effect, that aliens are the origins of technology and hence human evolution. Whereas Kubrick's *2001* suggests its philosophical points through imagery and cinematic mis-en-scene, Clarke's novel throughout explicitly implies that aliens are the source of the monolith and thus originators of human civilization, while Kubrick's film is agnostic on this point. Initially, Clarke portrays the monolith as an alien monitoring device, writing: "They [i.e. the ape/humans] could never guess that their minds were being probed, their bodies mapped, their reactions students, their potentials evaluated" (1982, p. 21). At every stage that the monolith appears in Clarke's novelization, aliens are presented as determinative forces in human evolution and history. Similar mythologies of aliens would be repeated later in *The X-Files* (1993–2002) mythology (see Kellner 2002), and *2001* helped generate interest in aliens and alien civilizations for the next decades.

Kubrick's *2001*, however, suggests another vision of technology than the more optimistic utopian and technophilic presentation in Clarke's novelization. In one of the cinema's most memorable images, the conquering humanoid throws his bone spiraling up into the air, and in a graceful cut the object becomes an orbiting space satellite, housing nuclear weapons. The "Dawn of Man" sequence of Kubrick's, with the exception of the monolith/alien theme, provides conventional images of humans inventing technology, using it to satisfy their needs, and gain power through the use of tools as instruments of violence and domination. Throughout *2001*, Kubrick will portray dangerous elements and dimensions of technology, and the scenario will progressively depict how domination of humans by technology increasingly threatens humans and could lead to a dystopia of technological domination beyond the control of human beings.

The second part of the film opens in the year 2001 and depicts the confusion sewed by a discovery of a monolith on the moon, which appears to be similar to the monolith encountered by the ape/humans, and suggests to alert viewers an event pointing to the existence of alien civilizations beyond the earth. After passing the nuclear-weapon platforms hovering in space, a PanAm space shuttle, whose sole occupant is Dr. Heywood Floyd (William Sylvester) of NASA, enters a space station that will become the transit point for a trip to the moon. The highly realistic detail in the station depicts a future controlled by big government, military, and corporations, and in which humans and technology, the natural and the artificial, have been gradually imploding into each other in a future technocracy.

Dr. Floyd is served artificial food aboard the ship, falls asleep watching TV, and then passes through a "Voiceprint Identification" device to gain entrance to the space station. After calling his daughter on a Bell "Picturephone," Floyd encounters a group of Soviet scientists who ask him about problems at Clavius Moonbase. The base, apparently, has been cut off from communication and visitation because of an alleged "epidemic" (which turns out to be a cover story for the discovery of the monolith).

The script also plays upon late 1960s Cold War paranoia and mistrust in the images of Soviet and American nuclear weapons in space and apparent detente, but these negative fears will be supplanted, as the story develops, by an even more threatening technology and alien encounter. The scene cuts to Floyd on a shuttle to the moon, enclosed in another technological environment where he eats artificial food and grapples with the complexities of a "zero-gravity toilet." On the moon, Floyd attends what appears to be a corporate board meeting dealing with the need to cover up the monolith's discovery, signaling corporate/state hegemony and the continued proclivities of the corporate state to engage in cover-ups and lies. It also suggests the continuing hegemony into the future of U.S. corporations like PanAm, Bell telephones, and Hilton hotels, all shown with office and logos in the space station (although the parent corporations of the first two U.S. corporations just mentioned have since disappeared in corporate restructuring).

At the closing of the second section, accompanied by the chords of Ligeti's "Choral," Floyd reaches out and touches the monolith. As the group with him assembles for a photograph, a piercing electronic shriek transfixes them, with the cinematic screen displaying a mystical alignment of the monolith and the sun. The electronic sound jolts the viewer into experiencing the endangering nature of technology, symbolized by the monolith and the momentous and unforeseeable effects that technologies can have on human beings, an endangering technology theme that intensifies as the film continues.

2001's third section opens with the graphics "Jupiter Mission: 18 Months Later" and unfolds the sequence in which the HAL-9000 computer comes into conflict with the two mission astronauts, Frank Poole (Gary Lockwood) and Dave Bowman (Keir Dullea). HAL (voiced by Douglas Rain) tells Dave that he is "concerned" and suspects "that there are some extremely odd things about the mission." Dave does not respond and soon after HAL reports that the ship's AE-35 communication unit is about to fail. Dave takes a space pod, and flies outside to detach the unit, but he and Frank find nothing wrong with it. Mission Control in Houston suggests that HAL may have made a mistake, and the crew suddenly finds itself dealing with technological ambiguity and paranoia.

In both the second and third sequences which portray the lives of the astronauts living in space depict a technocracy where democracy, social conviviality, and individuality have all disappeared and the astronauts live as functionaries in a technocratic system, literally cogs in machines. They eat artificial food, sleep in pods, programmed to remain asleep for months, or as long as the mission dictates. Instruments control their daily lives and at most they function to adjust, repair, and carry out the instructions from their computers or the program that is controlling them. There is no democratic debate about how to deal with problems or crises, no individual angst or rebellion, only the cogs in the machine functioning as they have been programmed to.

In the one scene, the astronauts exert agency, the results are disastrous. In a key sequence, Frank and Dave decide to unplug HAL, whose errors may endanger the mission, but HAL reads their lips and moves to defend himself. In a chilling sequence without dialogue, HAL uses the space pod's arms to uncouple Frank, who is repairing the AE-35 communication unit outside in space, and Frank drifts haplessly into space while Dave looks on in horror. Meanwhile, HAL terminates the "life functions" of the three additional sleeping crew members and Dave springs to action, unplugging the computer in one of the film's most poignant sequences.

The fourth and concluding sequence titled "Jupiter and Beyond the Infinite" depicts the space ship arriving at Jupiter and then in conjunction with an alignment of earth, the moon, the spaceship, and Jupiter, the monolith appears once more, signaling an impending apocalyptic event, although this time it is presented as a blue object floating in space, perhaps inviting the space ship through what Clarke and Kubrick called the Star Gate (Clarke 1982, p. 193 f.), which opens into the vast space beyond Jupiter.

In this audacious sequence of formal cinema that proliferates a burst of nonrepresentational images and sounds, Dave is hurtled into another space-time continuum in one of the more astonishing sequences in film history, without dialogue and throwing the viewer into narrative confusion, forcing each viewer to try to make sense of the kaleidoscope of images and sounds. In an audacious and disjointed collage of a rush of mysterious images and sounds, we see a wholly other universe come into being. The rush of non-representational cinema slows down with the character Dave arriving in a mysterious room in which he ages, encounters the monolith, and is reborn as a Star-Child (Clarke 1982, p. 220), who in the final image looks at the audience in a close-up of a young face in a bubble. The screen cuts to black and after the title sequence, there is another three minutes black screen with mysterious music, so that the film has looped to a full return to the beginning, coding it as perhaps a hallucinatory flashback to key episodes within human evolution.

2001's spaceship is titled "Discovery" and invokes the quest to explore the universe, discover new worlds and entities, and to expand the scope of human knowledge (and possibly dominion, as was the case in the modern era of discovery and voyages). The ship itself provides a futuristic environment of technologies and devices to serve all human needs, culminating in the computer that controls the ship. The HAL-9000 computer is programmed to appear to be a techno-human servo-mechanism, although as the narrative unfolds, HAL becomes increasing malevolent and out of control, depicting how development of technology can endanger human control of its destiny, and even human existence. Kubrick's "endangering vision" of a techno-dystopia where malevolent technology is beyond human control subverts techno-optimism and provides a powerful cautionary warning of technology overpowering and potentially destroying human beings in a futuristic techno-dystopia.

As HAL's intelligence is progressively unplugged, he becomes increasingly juvenile, singing an old popular song "Daisy Bell," with its haunting chorus "Daisy, Daisy," and stating repeatedly "I feel afraid" and then "I feel…" as HAL loses his ability to speak and function. The scenes suggest that computers are evolving into both menacing objects that can overpower us and yet are also human appendages that may share our weaknesses and foibles. The HAL-9000 computer, like the monolith, can be read as a potent symbol of endangering technology and the way that technology can control human life and exert negative and unforeseen effects, producing a techno-dystopia, a theme particularly resonant in the contemporary moment as we proceed ever deeper into computer and digital military and weapons culture.

Many critics have seen the name HAL as derived from IBM, since "(i)f you take each letter of IBM and retreat it one notch back down the alphabet (I to H, B to A, M to L), you get HAL" (Wheat 2000, p. 71). Arthur Clarke, however, insists that HAL is derived from "Heuristically programmed ALgorithmic" computer. The IBM story, he claims, is an "annoying and persistent myth" (Clarke 1979, p. 78). IBM at the time stood for big mainframe corporate computers and the control of computers by Big Government and Corporations, an identification that Apple and later generations of PC computers would define themselves against, so associating HAL with big government and business is not an unreasonable connection. HAL also connotes God, as it has godlike control powers over the mission, controlling all of its functions and the mission's trajectory. Yet Dave's rebellion against HAL and eventually unplugging him provides an allegory of how individuals can rebel against and control technologies and their destinies, although *2001* at the same time presents a cautionary warning about computers displacing humans and their sovereignty over the object-world, a theme that would run through 1970s science fiction film, as I note below.

Since the entire space project depends on the reliability of its computers, Kubrick's depiction of a malevolent computer killing crew members and taking control of the mission is an extremely disquieting example of dangerous technology which threaten human beings with destruction. Indeed, it is perhaps the increasing awareness that our technologies can fail, break down, misfire, and destroy us that has curtailed the development of space exploration that was in its Golden Age when *2001* was released and that *Star Trek* and dominant science fiction literature of the era projected was the fate of the human species: to always go further beyond into space and seek new worlds and galaxies for the human adventure. *2001*, however, warns us of the endangering nature of space and military technology, a warning confirmed every time a plane crashes or is shot down, a space mission fails, or war unleashes the dogs of high-tech military weapon systems, an inferno often portrayed in the global mediascape in the Cold War and beyond.

The computer versus human being thematic anticipates the potential conflict between humans and technology and technophobic fears of computers taking over and dominating humans that is a major theme of contemporary philosophy of technology (see Best and Kellner 2001). Moreover, *2001* helped spawn a wave of technophobic Hollywood films in the 1970s, including *Colossus: The Forbin Project* (1970) *THX 1138* (1970), *Logan's Run* (1976), and *Demon Seed* (1977) (on the technophobic film, see Kellner and Ryan 1988). These films, like *2001*, posed individuals against technology and technocratic societies and tended to valorize individual rebellion and revolt, thus replicating the dominant American ideology of individualism.

2001 can also be read as an allegory of the dangers of dehumanization through technology and a technocratic society, in which individuality disappears and democracy dies in darkness. The human characters throughout the film, in fact, seem devoid of passion and engage in common-place and banal dialogue, as if technology has stripped humans of their idiosyncratic and individual characteristics. Encapsulated in techno-environments, dependent on technology for all bodily functions and needs, and enclosed in completely sterile techno-dwellings, humans of the future appear in Kubrick's vision as lacking in personality and uniqueness with no significant social interaction, democratic discussion and consensus, or abilities to collectively control their destiny in align with human needs and desires.

Multiple Allegories of 2001: A Multiperspectivist Reading

"I tried to create a visual experience, one that bypasses verbalized pigeonholing and directly penetrates the subconscious with an emotional and philosophical content... I intended the film to be an intensely subjective experience that reaches the viewer at an inner level of consciousness, just as music does... You're free to speculate as you wish about the philosophical and allegorical meaning of the film," Stanley Kubrick, in Agel (1970, p. 329)

2001 continues to be a cult classic with a bevy of books, articles, Internet sites, and reams of critical commentary in cinema and cultural studies, and an ever-proliferating wealth of interpretations and criticism. Yet it received many hostile notices upon opening, including a pan from Pauline Kael who described it as "a monumentally unimaginative movie," Other reviewers found it as well boring and puzzling (see Agel 1970, p. 246), and to this day the film has its detractors. Nonetheless, *2001* was highly popular with audiences, was ultimately a huge financial success for MGM, and many critics were lavish with praise. With the concluding sequences taken as an icon for a psychedelic explosion of the boundaries of time and space, the film became a classic icon of the counterculture upon its release. Indeed, many viewers watched it repeatedly, and diverse responses appeared from the beginning, generating a battle of interpretations that is still going on.

The mysterious concluding sequences of *2001* subvert narrative cinema and provide some of the most audacious spectacles of pure cinema as a feast of non-representational images and sounds yet to appear in big budget Hollywood film. The special effects, supervised by Douglas Trumball, and fusion of otherworldly sound and music, fragmentation and narrative discontinuity, and explosion of light and color in the audacious concluding sequences provided what many described, in the psychedelic discourse of the time, as a "trip" and for which we could appropriately apply the category of the sublime.[5] Indeed, experiencing *2001* on a full screen with advanced sound systems continues to be an aesthetic delight of the highest order. The narrative as a whole and the mysterious concluding sequences position *2001* as highly modernist and polysemic, and the film has given rise to multiple and competing interpretations—although there are clear narrative guidelines to work with in reading Kubrick's *2001*. Kubrick himself explained the plot in a very straightforward manner in a 1969 interview with Joseph Gelmis:

> You begin with an artifact left on earth four million years ago by extraterrestrial explorers who observed the behavior of the man-apes of the time and decided to influence their evolutionary progression. Then you have a second artifact buried deep on the lunar surface and programmed to signal word of man's first baby steps into the universe—a kind of cosmic burglar alarm. And finally there's a third artifact placed in orbit around Jupiter and waiting for the time when man has reached the outer rim of his own solar system.
>
> When the surviving astronaut, Bowman, ultimately reaches Jupiter, this artifact sweeps him into a force field or star gate that hurls him on a journey through inner and outer space and finally transports him to another part of the galaxy, where he's placed in a human zoo approximating a hospital terrestrial environment drawn out of his own dreams and imagination. In a timeless state, his life passes from middle age to

---

[5] On the sublime, see the classical text by Burke 2008. For Burke, the sublime is connected with terror, vastness and infinity, and that which has the power to compel and destroy us, all themes connected to the techno-dystopia portrayed in Kubrick's *2001*.

senescence to death. He is reborn, an enhanced being, a star child, an angel, a superman, if you like, and returns to earth prepared for the next leap forward of man's evolutionary destiny.

That is what happens on the film's simplest level. Since an encounter with an advanced interstellar intelligence would be incomprehensible within our present earthbound frames of reference, reactions to it will have elements of philosophy and metaphysics that have nothing to do with the bare plot outline itself (Gelmas 1970, pp. 91–92).

Indeed, the rich symbolism and images of *2001* have given birth to a wealth of interpretations. Leonard F. Wheat, for instance, in *Kubrick's 2001: A Triple Allegory*, provides detailed readings of *2001* as an allegory replicating images and themes from Homer's *Odyssey*; the theme of implosion of humans and technology found in the literature of Arthur C. Clark; and the theme of the superman or Ubermensch found in Friedrich Nietzsche. Wheat's readings are provocative and often instructive, although I would argue that *2001* can and has given rise to multiple allegorical readings and has a wealth of meanings requiring multiperspectival readings and eschewing "definitive" interpretations.

Exhibiting a positivist and reductive streak, Wheat, by contrast, insists that his readings are "definitive," and endlessly lists how other critics miss key symbolism or provide wrong interpretations (2000, pp. 11, 14, passim). I would argue that *2001* is such a rich polysemic text that it can provide a wealth of readings and it is unlikely that there will ever be a definitive reading, or readings, although I would not deny that the film has also provided a wealth of misreadings making it one of the most contested films in history.

Interestingly, *Room 237*, a 1980 film about Kubrick's *The Shining* directed by Rodney Ascher, provides multiple readings of Kubrick's version of Steven King's popular novel as an allegory about the genocide of Native Americans, a subversive narrative of the Apollo 11 moon landing as a hoax, a representation of the mythic story of the Minotaur, an allegory about the holocaust, and various other themes which fans of the film document.[6] While none of these readings may convince, like Wheat with *2001*, Ascher finds copious images and themes in *The Shining* and other of Kubrick's films which support the different readings in his documentary.

Kubrick loads his films with symbols and meanings and a wealth of allegorical interpretations and multiple meanings can be found in his work, suggesting that a multiperspectivist approach can help engage key themes in Kubrick's films, although it is probably impossible to provide definitive readings. In alignment with the major theme of my studies of technology collected here, I have focused

---

[6] Kellner and Ryan (1988), pp. 172–178 read *The Shining* as an allegory of the crisis of the patriarchal family in post-60 s America.

on the ways that *2001* endanger dominant ideologies of symbiotic relations between humans and technology, the wonders of space travel and exploration, and the fate of the human in a high tech world. I have presented Kubrick as a filmmaker who subverts dominant ideologies and conventional views of the world and cinema, and as a provocateur who challenges us to delve into his work for multiple meanings, thus challenging the conception that there are definitive readings or meanings to be gleaned from his work.

In addition to symbols like the monolith and forays into pure cinema in the Beyond Jupiter sequence, Kubrick uses color, sound, editing, and other basic elements of film to convey his endangering vision of high tech space travel and a technofuture with dystopic dimensions. To pursue this theme, Kubrick's color symbolism is organized, first, around the oppositions of black and white. The film opens and closes with three minute sequences of pure black screen, an audacious move in a Hollywood studio film, that signals Kubrick's fearless modernism. Throughout the space journey, the black sky symbolizes the infinity and mystery of space, punctuated with the white space vehicle and shining planets and stars. The interiors of the space vehicles are often white, signifying the sterile, hospital-like environment of the human habitants erected in outer space and replicating the white modernist interior design of Corbusier and the Bauhaus. Most of the characters in the post-2001 sequences, and all of the main ones, are ethnic white males, perhaps signifying the continued domination of white men into the future. There are few women in the film, all in subordinate positions, and there is no sexuality, unless the coupling of the space ships with their docking stations is read as sexual union and procreation, as Wheat proposes (2000, p. 66 ff.). Such a reading would suggest that in a technofuture even sexuality is taken over by machines and diminished in humans.

The color red also proliferates multiple symbolic meanings throughout the film. "The Dawn of Man" sunrise is deep red, which cuts from the black three-minute opening to a red dawn perhaps signifying the bloody adventures to come for the emerging humanoids. After the men/apes use bone technology to kill animals and other humans they are shown eating bloody, red meat. As the narrative proceeds into outer space and the future, HAL's "eye" is a menacing red, linked to the murders committed as the humanoid computer spins out of control. The cockpits of the space ships have red flashing lights and technical devices, perhaps signifying the military technology origins of its development and the havoc a space ship can wreak. Red warning signs go off as HAL terminates the life support systems of the three slumbering astronauts signifying their imminent death. Dave Bowman wears a bright red space suit in his clash with HAL and the most dramatic sequences pose HAL's flashing red eye/brain against Bowman's red space suit and his vigorous seizing control of the ship as he unplugs HAL, reducing the malevolent computer to a whimpering child singing and expressing his fear.

Throughout *2001*, the display units flash red in crisis, and the nuclear reactor aboard the ship has a red warning dial, hence red can be seen as a sign signaling endangerment. Many of the space habitant interior scenes also utilize a strong red, white, and blue color schemata, which dominate many scenes featuring HAL, the interior of Discovery, the "Beyond Jupiter" color kaleidoscope, and, of course, American flags. The frequent use of red, white, and blue color schemes perhaps refer to continued U.S. hegemony of the world in the future into outer space, as I suggested earlier in the segment that featured U.S. corporate logos in the moon station. In the same segment, the American astronauts meet Russian colleagues, worried about rumors of an epidemic on the moon planet Clavius where signals from the monolith have been heard and which Heywood Floyd will soon investigate. Clearly, the Americans are dominant over the Russians, depicting both Cold War tensions and suggesting U.S. hegemony in outer space.

Kubrick also constructs a modernist sound collage for the film, mixing popular classic music with electronic and avant garde music and evocative soundscapes that punctuate the various sections of the film. The long "Dawn of Man" sequence has no human dialogue and the silence is punctuated by whispering winds across the plains of Africa, interrupted by grunts and shrieks of the man/apes. The moon sequence features modulated classical music softly played as background, contrasted with the breathing of Heywood Floyd and the search party that encounter the buried monolith on the moon surface, which cuts to a shrill screeching when they touch the monolith. The "Jupiter Mission" section most notably features Bowman's deep breathing as he re-enters the space vehicle and unplugs HAL whose voice unwinds, evoking primitive childhood fears and repetition of his first programming sequence, which included the "Daisy, Daisy" verses. The final "Beyond Jupiter" features electronic sounds and mysterious voices, which could be the aliens, as both images and the soundtrack veer off into a modernist space that breaks with previous codes of cinematic image and sound.

In earlier conceptions of the film and Clarke's novelization (1968, p. 221), *2001* was to end with triggering an explosion of the nuclear weapons shown circling earth on a space satellite depicted earlier in the film, but an explosion of nuclear weapons was deemed by Kubrick to be too close to the ending of *Dr Strangelove*, and he provided a more ambiguous ending with the Star-child in a bubble looking down at earth. This haunting image suggests a wealth of meanings, perhaps suggesting that a new species was born that might bring a new era to earth, but does not really indicate if this would be a form of salvation or damnation. In the film as a whole, a Nietzschean *Ubermensch* theme is suggested by the evolutionary schemata portrayed in the film of ape, ape-man, evolving human, machine-man (i.e. HAL), and its' destruction, which in Nietzschean terms makes man God, and then the rebirth of the new species in the form of the Star Child. This final step in *2001*'s evolutionary continuum is left open, however, so the destiny of the human species and its evolution is left open and mysterious.

Hence, the theology of *2001* is ambiguous and has given rise to multiple inter-pretations. HAL is obviously Godlike in his total control of the space-ship and in his purported perfection, until flaws appear, revealing the super-computer as human-all-too-human. The monolith could signify transcendent forces but it could symbolize malevolent aliens controlling humanity, as well as an omnipotent God. And as noted, in Kubrick's film the Star-child could represent a newly born God, although its' meaning too is highly ambiguous.

One can, of course, over-read Kubrick's symbolism. As Rodney Ascher's film *Room 237* (2012) shows, Kubrick's films are scrutinized for numerology, ana-grams, and other symbologies, as in the example of IBM consisting of the letters following HAL in the alphabet, noted above. *2001*, however, seems to demand multiple hermeneutical readings of its major themes, and its success helped spawn a wave of SF films in the 1970s, including the *Star Wars* series, and a more dystopic mode of space adventure evidenced in the *Alien* series that ran from 1979 to 1997.[7] Yet *2001* also inspired a cycle of films that presented positive visions of aliens ranging from Steven Spielberg's *Close Encounters of the Third Kind* (1978) and *E.T.* (1992) to the Spielberg/Kubrick collaboration of *A.I.* (2001). However, perhaps no science fiction film before or after has achieved the realism, the visionary projection of the future, and the cult status of *2001*.

As the year 2001 arrived, there was a wave of critical articles on the film that attested to its continued cult status and even a book that appraised its science in relation to the development of space travel, computers, and the other themes that were presented in *2001* (see Stork 2001). An article by Brain Libby in *Salon* highlights the film's optimism, noting that "mankind's manic self-destruction is juxtaposed against the species' ultimate survival and evolution." I would suggest,

---

[7] While Chion has a section on science fiction films after 2001 (pp. 156–163), curiously he does not mention the wave of technophobic Hollywood films in the 1970s, including *Colossus: The Forbin Project* (1970), *THX 1170* (1970), and others in the cycle that I mention above, which indicate that Kubrick and Clarke anticipated fear of computers and technology that would continue to be a major theme of many science fiction films to the present; these films are engaged in Kellner-Ryan 1988, Chapter 9.

however, that *2001* goes beyond the dichotomy of optimism and pessimism, providing both hopeful and frightening perspectives on the future and the origins and destiny of human beings.

As I suggest, the film deconstructs human/alien dichotomies by depicting human culture deriving from alien civilizations. It also shows humans and technology as being both highly destructive and creative. Deconstructing human/technology dichotomies, the film might evoke deeply embedded states of fear and aggression within humanity, but finds destructive forces equally embedded in the objects that humans produce.

2001's complex vision of the future thus displays the potential for destruction, as well as creation and higher evolution of the human species, while presenting both destructive and enriching dimensions to technology and human-techno symbiosis. It depicts dangers inherent in a technological civilization and provides dystopic visions of how our technologies can alienate and dehumanize us, and spiral out of control in destructive ways. *2001* thus subverts a dominant ideology of technological optimism and the dangers of faith in technology while presenting disturbing visions of possible dystopias of a high tech society.

Yet the rebirth section near the end of *2001* provides hope of a future destiny for the human species discontinuous from life as we know it, a vision both enthralling and frightening, and deeply ambiguous in Kubrick's cinematic presentation. Ultimately, however, *2001* is enveloped in mystery, requiring many re-viewings and reflections upon its themes and the very origins, nature, and possible destinies of the human adventure. Few other science fiction films had such ambitions and probably none since has reached its plateau of excellence and provocation.

In terms of aesthetic complexity, philosophical and moral vision, and sociopolitical critique, *2001* is both a highly fertile artifact that subverts and undercuts dominant ideologies, views of the future, and the trajectory of the human adventure. It is radically modernist in both its aesthetics and thematics, valorizing complexity, innovation, formal radicality, individual vision, and the sublime, providing a <u>unity of significant form and content</u>, innovative aesthetics and philosophical vision. While presenting dangerous possible futures, it is also illuminating, exciting, and continues to spawn cinematic pleasure of the highest order and hermeneutical contests over its multiple meanings, suggesting that it will remain a cinematic classic and relevant to the human adventure for eons to come.

Thus ultimately Kubrick establishes a democratic relation with his viewer and critic, offering a polysemic vision that challenges the viewer and critic to come up with their own readings or the film, that then can be engaged in with other critical interpretations and readings in what Ricoeur termed "the conflict of interpretations." Kubrick is a modernist who produces complex and multi-layered texts that force the viewer and critic to work to come up with readings and interpretations, to debate meanings of the film with others, and to encourage a wealth of interpretations that can evoke democratic debate over the meaning(s) of the films. Thus there is a democratic potential in modernist films and auteurs, sometimes dismissed as elitist in comparison with popular cinema, but which are properly democratic requiring an engaged spectator and critic who must submit her/his/their reading to democratic discussion and debate of others.

The construction of the history and meanings of a film is thus an exercise in democracy in which the creators collaborate to produce a work of art and spectators and critics work to produce interpretations which they debate in a cinematic public sphere which establishes reputations, classifications, and hierarchies of films and auteurs, themselves subject to democratic debate and challenge and the production of new consensus. Thus to participate in the dance of interpretation and the play of cinematic or literary or philosophical interpretation, one needs to immerse oneself in the pedagogies of critical media and digital literacies which I shall focus on in the final chapter of the book on education.

Digital Technologies, Multi-Literacies, and Democracy: Toward a Reconstruction of Education

10

Abstract

Chapter 10 Engages "Digital Technologies, Multi-Literacies, and Democracy: Toward a Reconstruction of Education," which explores how technology can be used for a democratic reconstruction of education and society in the spirit of John Dewey, Paulo Freire, and Herbert Marcuse. I argue that educators, students, and citizens need to cultivate multiple critical media and digital literacies for contemporary technological and multicultural societies. To meet the challenges of a digital era, teachers, students, and citizens need to develop critical media and digital literacies of diverse sorts, including a more fundamental importance for print literacy, to meet the challenge of restructuring education for a high-tech, multicultural society, and global culture. In a period of dramatic technological and social change, education needs to help produce a variety of types of literacies to make current pedagogy relevant to the demands of the contemporary era.

Keywords

Media and Digital Literacies • Reconstruction of Education • John Dewey • Paulo Freire • And Herbert Marcuse • Critical digital literacies • Radical pedagogy

As I've argued in this book, new digital technologies involve the dramatic multiplication of computer, information, communication, multimedia, and social media technologies which have been changing everything from the ways people work, to the ways they communicate with each other, how they learn, and how they

D. Kellner, *Technology and Democracy: Toward A Critical Theory of Digital Technologies, Technopolitics, and Technocapitalism*, Medienkulturen im digitalen Zeitalter, https://doi.org/10.1007/978-3-658-31790-4_10

consume and spend their leisure time. This technological revolution is often interpreted as the beginnings of a knowledge or information society, and therefore ascribes education a central role in every aspect of life. It poses tremendous challenges to educators to rethink their basic tenets, to deploy digital technologies in creative and productive ways, and to restructure schooling to respond constructively and progressively to the technological and social changes currently underway, in ways that promote democracy and social justice by making digital technologies and literacies accessible to all.

At the same time, important demographic and socio-political changes are taking place throughout the world. Immigration patterns have created the challenge of providing people from diverse races, classes, and backgrounds with the tools and competencies to enable them to succeed and participate in an ever more complex and changing world. In this chapter, I argue that educators, students, and citizens need to cultivate multiple critical digital literacies for contemporary technological and multicultural societies. To meet the challenges of a digital era, teachers, students, and citizens need to develop critical media and digital literacies of diverse sorts, including a more fundamental importance for print literacy, to meet the challenge of restructuring education for a high-tech, multicultural society, and global culture. In a period of dramatic technological and social change, education needs to help produce a variety of types of literacies to make current pedagogy relevant to the demands of the contemporary era.

Radical pedagogy is an important element of a cultural politics aiming at democratic social transformation and in this chapter, I sketch aspects of a critical theory of education, technology, and democracy. I will discuss the fundamental transformations in the world economy, politics, and culture in a dialectical framework that distinguishes between progressive and emancipatory features and oppressive and negative attributes, and how a new critical pedagogy and new technoliteracies are essential for democratic social transformation and justice. Hence, following John Dewey, Paulo Freire, and Herbert Marcuse, I will call for a reconstruction of education to make it more responsive to the challenges of a technological revolution in the context of a democratic and multicultural society.

Digital Technologies and Literacies in a Changing World

The constant development of digital technologies, expansion of global media empires, an explosion of social media, and the unrestricted commercial targeting of children have all contributed to an environment where today's youth are growing up in a mediated world and technoculture far different than any previous generation (Gennaro and Miller 2021). While technological advancements have created novel possibilities for the free flow of information, social networking and global activism, there is also the potential for corporations and governments to increase their control over media, restrict the flow of information, and appropriate these digital technologies and social media for profit and control at the expense of free expression and democracy.

Most children born in the United States in the twenty-first century have never known a time without the Internet, video and computer games, digital devices like the iphones and tablets, with the ability to access previous forms of media like radio, music, television, and movies through these devices.[1] By the early 2000s, over 98% of US households have at least one television set and about one third of young children live in households where the TV is on "always" or "most of the time" (Rideout, Vandewater & Wartella 2003, p. 4). It is also estimated that nearly all young children in the US, "have products—clothes, toys, and the like—based on characters from TV shows or movies" (Rideout et al. 2003, p. 4). Further, already by 2005, it was estimated that before most children are six years of age, they spend about two hours per day with screen media, something that doubles by age eight, and before they are 18 they spend approximately 6½ hours daily with all types of media (Rideout et al. 2005)[2]—numbers that significantly increased by 2020. By 2020, "The Centers for Disease Control and Prevention (CDC) reports

---

[1] While all people born in this millennium have been alive since the invention of the Internet, cellular phones and cable and satellite television, this does not mean that everyone can access key components of digital technology or media culture. Since large numbers of the world's population still lives without electricity and many people do not even have computers (see Note 10 below), billions of people are being left behind the so-called technological revolution. I am writing from the perspective of my situation as an educator since 1968 working in the U.S. and am proposing the transformation of education based on the situation I have experienced and am documenting here, although since many parts of the world are increasingly saturated with digital technologies and social media, the analysis here has relevance elsewhere.

[2] This data is based on random telephone interviews in 2003 with 1,065 parents of children between six months and six years of age. "Screen media" refers to watching TV, watching videos/DVDs, using a computer and playing video games. This research was reported in the Kaiser Family Foundation *Zero to Six* study available on-line at https://www.kff.org/entmedia/ 3378.cfm (accessed October 4, 2009). The following year, a study indicated that the number of hours spent with media increased. See the research based on questionnaires from a 2004 national sample of 2,032 students between 8 and 18 years of age, as well as 694 media-use diaries, as reported in the Kaiser Family Foundation *Generation M* study available on-line at https://www.kff.org/entmedia/entmedia030905pkg.cfm (accessed on October 4, 2009). The figure of 6½ hours per day, includes ¼ of that time spent multitasking with several different media at the same time, thereby increasing media exposure to an estimated 8½ hours per day. Discussions with colleagues suggest that screen-time during the 2020–2021 COVID-19 pandemic that led to shut downs of schools and cities in many parts of the world, including

that children ages eight to 10 spend an average of six hours per day in front of a screen, kids ages 11 to 14 spend an average of nine hours per day in front of a screen, and youth ages 15 to 18 spend an average of seven-and-a-half hours per day in front of a screen."[3]

Since television programs, movies, video and computer games, music, and even toys have become major transmitters of culture, tellers as well as sellers of the stories and culture of our time, it is now, more than ever, that children and all citizens need to learn how to critically question the technologies and media and the messages that surround them and how to use the vast array of new digital technologies available to express their own ideas and participate fully in digital and media culture.

Increasingly, computers and multimedia technologies are becoming a part of everyday life. By 2000, about half of all households (51 percent) in the United States had a computer.[4] In 2015, this percentage had grown to 79 percent, while the American Community Survey, by contrast, indicated that in 2013, 84 percent of households had a computer (desktop or laptop, handheld, or other), with the percentage growing to 87 percent in 2015.[5] At the same time, individuals are becoming increasingly connected to the world of digital information while "on the go" via smartphones and other mobile devices, so digital connections are now established as a primary relation to the world for many.[6]

Los Angeles where I live, increased significantly with kids scheduling recess events, weekend play dates, and other events on-line, as well as their on-line school classes, so it appears that young people throughout the United States spend more time on screen than ever before.

[3] See South Coast Medical Group, "Screen Time: The Effects on Your Child's Health," July 26, 2020 at https://southcoastmedgroup.com/2020/06/screen-time-the-effects-on-your-childs-health/ (accessed December 25, 2020).

[4] U.S. Census Bureau, "Home Computers and Internet Use in the United States: August 2000." *Special Studies,* Issued September 2001 at https://www.census.gov/prod/2001pubs/p23-207.pdf (accessed December 26, 2020).

[5] See Camille Ryan, "Computer and Internet Use in the United States: 2016," *American Community Survey Reports,* August 2018, at https://www.census.gov/content/dam/Census/library/publications/2018/acs/ACS-39.pdf (accessed December 25, 2020).

[6] For documentation, see the PEW Research "Mobile Fact Sheet" which indicates that: "The vast majority of Americans-96%-now own a cellphone of some kind. The share of Americans that own smartphones is now 81%, up from just 35% in Pew Research Center's first survey of smartphone ownership conducted in 2011." Available at https://www.pewresearch.org/internet/fact-sheet/mobile/ (accessed on January 19, 2021).

Likewise, there has been a growing use of computers in the classroom with mixed results.[7] Further, there has emerged a global convergence of media culture, computer culture, games, and social media transmitted on a cornucopia of digital devices that provide information, entertainment, games, and access to culture of all forms, other people, and the world at large. Victoria Carrington (2005) writes that the emergence of digital media texts, "situate contemporary children in global flows of consumption, identity and information in ways unheard of in earlier generations…" (p. 22). In the context of continuously expanding technological and economic transformation, critical media and techno-literacies are an imperative for participatory democracy and citizenship because digital information communication technologies and a market-based media culture have fragmented, connected, converged, diversified, homogenized, flattened, broadened, and reshaped the world.[8] These changes have been reframing the way people think and restructuring societies at local and global levels (Castells 2001; Jenkins 2006).

Put in historical perspective, it is now possible to see modern education as preparation for industrial civilization and minimal citizenship in a passive representative democracy. Modern education, in short, emphasizes submission to authority, rote memorization, and what Freire called the "banking concept" of education in which learned teachers deposit knowledge into passive students, inculcating conformity, subordination, and normalization. Today, these traits are somewhat undercut in certain sectors of the global postindustrial and networked society with its demands for new skills for the workplace, participation in emergent social and political environs, and interaction within novel forms of culture and everyday life.

---

[7] There are growing attempts to assess the impact of computers in class and computer-based learning, studies that will need to be supplemented after the amount of on-line learning forced upon schools by the 2020–2021 global COVID-19 pandemic that closed downs many schools for months and forced moving toward on-line education. This phenomenon intensifies the importance of critical digital literacies for intelligent and productive use of computers in schools and home. For recent studies before the COVID-19 pandemic and shutdown hit, see Anne Boring, "What is the Impact of Students' Use of Computers in Class?," *SciencedPo Learning Lab*, November 21, 2017 at https://www.sciencespo.fr/learning-lab/en/what-is-the-impact-of-students-use-of-computers-in-class/ (accessed December 25, 2020). See also the critical take on on-line learning by Debbie Truong, "More students are learning on laptops and tablets in class. Some parents want to hit the off switch." *Washington Post*, February 1, 2020 at https://www.washingtonpost.com/local/education/more-students-are-learning-on-laptops-and-tablets-in-class-some-parents-want-to-hit-the-off-switch/2020/02/01/d53134d0-db1e-11e9-a688-303693fb4b0b_story.html (accessed December 25, 2020).

[8] For an articulation of the concept of critical media and digital literacies in this paper with practical applications in the classroom, see Kellner and Share 2019, and on techno-literacies see Kahn and Kellner 2006.

A more flexible global economy, based on an ever-evolving technological infrastructure and more multicultural work force demands a more technically literate, interactive, culturally sensitive, and educated work force, while revitalizing democracy requires the participation of informed citizens.[9] Yet, while on one hand, the demands of the expanding global economy, culture, and polity require a more informed, participatory, and active workforce and citizenship, on the other hand, a docile workforce and service industry is still the norm in many sectors of work, society, and culture. Although for a time, ideologues of technocapitalism argued that information and communication technology would of themselves dramatically reorganize and democratize the workplace, schooling, the polity, and everyday life (Gates 1995; Kelly 1995, 1999), it is by now clear the supposed liberating effects of new digital technologies were greatly exaggerated. And while globalization and technological multiplication have highly ambiguous effects (see Chapter 5), they provide educational reformers with the challenge of whether education will be restructured to promote democracy and human needs, or whether education will be transformed primarily to serve the needs of business and the global economy.

To some extent, accelerating technological transformation renders necessary the sort of thorough restructuring of education that radicals have demanded since the Enlightenment of Rousseau and Wollstonecraft through Dewey, Freire, and Marcuse, all of whom saw the progressive reconstruction of education as the key to realizing the promises of democracy. Today, however, intense pressures for change come directly from technology and the economy and not educational reformist ideas, with an expanding global economy and novel technologies demanding innovative skills, competencies, literacies, and practices. It is therefore a burning question as to what sort of restructuring of education and society will take place, in whose interests, and for what ends. More than ever, we need philosophical reflection on the ends and purposes of education, on what we are doing and trying to achieve in our educational practices and institutions.

---

[9] Studies at the turn of the twenty first century reveal that women, minorities, and immigrants now constitute roughly 85 percent of the growth in the labor force, while these groups represent about 60 percent of all workers; see Duderstadt 1999–2000: 38. More recently, Pew Research *News in the Numbers* reveals "The U.S. foreign-born population reached a record 44.8 million in 2018. Since 1965, when U.S. immigration laws replaced a national quota system, the number of immigrants living in the U.S. has more than quadrupled. Immigrants today account for 13.7% of the U.S. population, nearly triple the share (4.8%) in 1970. However, today's immigrant share remains below the record 14.8% share in 1890, when 9.2 million immigrants lived in the U.S." See Abby Budiman, Key findings about U.S. immigrants, *NEWS IN THE NUMBERS*, August 20, 2020 at https://www.pewresearch.org/fact-tank/2020/08/20/key-fin dings-about-u-s-immigrants/ (accessed December 25, 2020).

In this situation, it may be instructive to return to Dewey and see the connections between education and democracy, the need for the reconstruction of education and society, and the value of experimental pedagogy to seek solutions to the problems of education in the present day. A progressive reconstruction of education will urge that it be done in the interests of democratization, ensuring access to information and communication technologies for all, helping to overcome the so-called digital divide and divisions of the haves and have nots, so that education is placed, as Dewey (1997 [1916]) and Freire (1972 and 1998) propose, in the service of democracy and social justice.

Yet we should be more aware than Dewey of the obduracy of divisions of class, gender, and race, and work self-consciously for multicultural democracy and education—a point Marcuse saw.[10] This task suggests that we valorize difference and cultural specificity, as well as equality and shared universal Deweyean values such as freedom, equality, individualism, and participation. Theorizing a democratic and multicultural reconstruction of education forces us to confront the digital divide, that there are divisions between information and technology have and have nots, just as there are class, gender, and race divisions in every sphere of the existing constellations of society and culture. The latest surveys of the digital divide, however, indicate that the key indicators are class and education, as well as race and gender, indicating that education must be transformed across class, gender, racial, and other divides to provide a more equitable education for all, as envisaged by Dewey, Freire, Marcuse, and others.[11]

Rob Shields (2003) has argued that the concept of the "digital divide" serves as a marketing device for the benefit of technology disseminators, so critical theories must seek how to make the development of digital technologies decrease the divide between haves and have nots and provide for a democratization and transformation of education. While no doubt high-tech corporations and affiliated government institutions have promoted the notion of a digital divide, the concept points to some serious problems and challenges. It is clear by now that providing access and computers alone without proper training and pedagogy does not advance education or social justice. Thus, more broadly conceived, the notion of a digital divide points to disparities in terms of access, training, skills, and the actual use of technologies to improve education and promote social justice.

In the global pandemic of 2020–2021, where schools in large sectors of the U.S. and elsewhere were closed, and education took place on-line and at home,

---

[10] See the studies in *Marcuse's Challenge to Education*, co-edited by Cho, Kellner, Lewis, and Pierce, 2009. See also Courts 1998; Weil 1998; and Kellner and Share 2019.

[11] On the "digital divide," see Chapter 2, Note 22.

the digital divide became all the more compelling because without proper computer access and training, students were cut off from education altogether. The pandemic, still raging as I write in January 2021, forced recognition of students, parents, citizens, educators, and other interested parties that it was more essential than ever to get digital technologies that accessed the Internet and social media into all students' hands and develop proper pedagogies that would prepare students and citizens for life in the turbulent twenty-first century.

With the proper resources, policies, pedagogies, and practices, educators can work to reduce the (unfortunately growing) gap between haves and have nots by promoting broad training in information and computer literacies, that embraces a wide range of projects from providing technical skills to engaging students in the production of media projects and innovative uses of the technoculture. Although technology alone will not suffice to democratize and adequately reconstruct education, in a technological society providing proper access and training can improve education if it is taken as an important supplement to basic skills of reading, writing, mathematics, and other key disciplines. That is, technology itself does not necessarily improve teaching and learning, and will certainly not of itself overcome acute socio-economic divisions. Indeed, without proper re-visioning of education and without adequate resources, pedagogy, and educational practices, technology could be an obstacle or burden to genuine learning and will perhaps increase rather than overcome existing divisions of power, cultural capital, and wealth.

In the following reflections, I focus on the role of digital information technology and social media in contemporary education and the need for new pedagogies and an expanded concept of literacy to respond to the importance of information and communication technologies (ICTs) in the field of education. I propose some ways that ICTs and new literacies can serve as efficacious learning tools which will contribute to producing a more democratic and egalitarian society, and not just provide skills and tools to privileged individuals and groups that will improve their cultural capital and social power at the expense of others. How, indeed, can education be re-visioned and reconstructed to provide individuals and groups with the tools, the competencies, the literacies and social practices, to overcome the class, gender, and racial divides that bifurcate our society and at least in terms of economic indicators seem to be growing rather than diminishing?

Many changes in the opening decades of the twenty-first century have contributed greatly to the need for critical media and techno-literacies and the reconstruction of education. New critical pedagogies for digital literacy education is now necessary because of the rapid growth of information communication technology, the expansion of free market global capitalism, and the escalating and always challenged linguistic and cultural diversity that is changing social environments at local as well as global levels.

Looking at the impact of globalization on identity, Manuel Castells (1996, 2001) asserts that people's lives are being shaped by the global forces of the network society. He suggests that the interconnections between technology, economics, culture and identity are challenging, conflicting and impacting upon each other on a global scale.

Already in the 1960s, Marshall McLuhan (1962 and 1964) argued that many of the characteristics of premodern oral culture will again rise up in importance as the complex, verbal, and graphic electronic age proves to be more similar to oral cultures of the ancient past than the last five centuries of typographic literacy. He wrote these ideas before the existence of widespread cellular phones, the Internet, and cable and satellite television defusion, and communications apps such as Skype and Zoom. Yet today as the Internet, social media, and wireless communication become common place in most "First World" countries, as well as in many parts of the "Developing World," his words ring more true today than when he first wrote them a half century ago.

According to McLuhan (1964), before print literacy, humans were hunters and gatherers living in oral societies with tribal cultures that were unified, inclusive, auditory, organic, and had high levels of participation. With the invention of the phonetic alphabet, a new era began. Literacy caused the eye to replace the ear and the cosmic culture became fragmented and separated by a new mechanistic system of repeatability and uniformity. In the fifteenth century these changes exploded with the invention of Gutenberg's printing press. McLuhan calls this the "mechanical age" and attributes the arrival of individualism, rationalism and nationalism to this new literate culture of homogeneity and lineal organization based on the habits of book culture.

The next great change for humanity, according to McLuhan, came with the discovery of electricity and the invention of the telegraph. The new electronic age has and continues to cause an implosion within society that is returning humans to their earlier oral roots. For McLuhan, this latest age of automation and cybernation takes us back to a more participatory, integral, decentralized and inclusive way of living. McLuhan asserts that electricity, with its speed and constancy, is the medium that created simultaneity, it is an extension of our central nervous system, "instantly interrelating every human experience" (p. 358). He suggests that all media are extensions of ourselves; that print is an extension of the eye as the wheel is an extension of the foot.

Now more than ever, we are seeing the transformation of societies into what McLuhan coined, the "global village," and the electronic age that he spoke of is in full force, reshaping societies and identities across the globe. For today's literate society to keep pace with the age of information, education must let go of curriculum that is separated by subjects and "changeover to an interrelation in knowledge," asserts McLuhan (p. 35). He asks, "Would it not seem natural and necessary that the young be provided with at least as much training of perception in this graphic and photographic world as they get in the typographic? In fact, they need more training in graphics, because the art of casting and arranging actors in ads is both complex and forcefully insidious" (p. 230).

Adding economic and technological determinist perspectives to McLuhan's technological determinism, Thomas Friedman (2005) argues that at the turn of this millennium, humans entered the third major shift in globalizing change. He writes that the first great era of globalization began in 1492 when Columbus opened trade between the New World and the Old. During what Friedman calls Globalization 1.0, imperialism and religion drove global integration through brute force as colonizing countries deployed the labor power of exploited peoples until about 1800. The second era, Globalization 2.0, ran from about 1800–2000 and involved multinational companies expanding their markets and labor forces as industrialization reshaped the world. This second era benefited first from the decrease in transportation costs and later from the decrease in telecommunication costs, and was marked by the inventions of new hardware, technology, and industry. Yet in the twenty-first century, Friedman claims that Globalization 3.0 is driven by innovative software and a global fiber-optic network, and asserts that the unique character of this era is "the newfound power for *individuals* to collaborate and compete globally." (p. 10).

Friedman's claim that the world is now less hierarchical with a more level playing field than ever before is overly ideological and optimistic. Friedman is too uncritical of inequalities and injustices of neo-liberal globalization and is often seen as an ideologue for the emergent high-tech society (Klein 2008). However, Friedman's assertion that "the world has been flattened by the convergence of ten major political events, innovations, and companies" (p. 48) is highly provocative and highlights many recent changes in society that are having a global impact. While I do not agree with his utopian conclusion that the world is now flat and there is more equal opportunity, since large numbers of the world's population still lives without electricity, while inequalities are growing in many parts of the world.[12]

---

[12] See the report by "Hannah Ritchie and Max Roser, "Access to Energy." *Our World in Data.* First published in September 2019; last revised in November 2019 at https://ourworldindata. org/energy-access (accessed December 25, 2020). The report summary notes: "940 million (13% of the world) do not have access to electricity. 3 billion (40% of the world) do not have

Yet, Friedman's discussion of the major forces which have changed the world in the last couple of decades in the twenty-first century, makes it clear that the present moment is a different world and will continue to change due to the influences of new ICTs and global economic systems. The examples he describes of transformations in technology, society, and economy provide strong reasons for the need to change education and especially literacy practices. Indeed, I and many colleagues believe that the type of changes that would best accommodate a globalized world perpetually transformed by technology include multiple literacies, of which critical media literacy and critical digital literacies are essential to the reconstruction of education today (Kellner and Share 2019; Gennaro and Miller 2021), as I argue in detail below.

The diversity of ideas and people is increasing in countries, cities, and classrooms, as escalating amounts of information become available and larger numbers of people travel and immigrate across the globe. At the same time there are some reductions of diversity as cultural colonialization and commercial homogenization spreads throughout the global markets with the ease of new information communication technologies. One example of the loss of diversity can be seen in UNESCO's warning that, "Over 50% of the world's 6000 languages are endangered" with one disappearing almost every other week.[13] Joseph Lo Bianco (2000) states that "During this and the next decade there will be the greatest collapse of language diversity in all history." (p. 94). He attributes these changes to an emerging global system being generated by three principal forces: "The first is the almost universal phenomenon of market deregulation; the second is the advanced integration of international financial markets; and the third is the critical facilitating force of instantaneous communications" (p. 93).[14]

One of the common themes running through many analyses of the changes in the relationship between media and society is a high degree of *convergence* that is occurring in numerous ways (Gutiérrez 2003; Luke 2006; Jenkins 2006). Henry Jenkins (2006) insists that we are now living in a *convergence culture*, in which our socio-cultural practices are changing because of the influences of

---

access to clean fuels for cooking. This comes at a high health cost for indoor air pollution. Per capita electricity consumption varies more than 100-fold across the world. Per capita energy consumption varies more than tenfold across the world."

[13] The quote was found on the official UNESCO web site. Retrieved October 23, 2006, from: https://portal.unesco.org/culture/en/ev.php-URL_ID=8270&URL_DO=DO_TOPIC&URL_SECTION=201.html (accessed October 4, 2009).

[14] On the consequences of language loss, see Anthony Woodbury, "Endangered Languages," *Linguistic Society of America*, 2020 at https://www.linguisticsociety.org/content/endangered-languages (accessed December 25, 2020).

technology and economics, and convergence of old and new media. He explains, "Media convergence is more than simply a technological shift. Convergence alters the relationship between existing technologies, industries, markets, genres, and audiences" (p. 15).

Jenkins highlights two major and often contradictory trends, one in which large media corporations threaten democracy by their concentration of ownership, giving less people a greater ability to push and amplify their limited content out to the masses while new media technologies have made it easier on a grassroots level for more people to pull, create and distribute much more diverse media content, thereby offering new opportunities for democracy. This dynamic push and pull of media is a key aspect of convergence and something Jenkins states, "represents a paradigm shift—a move from medium-specific content toward content that flows across multiple media channels, toward the increased interdependence of communications systems, toward multiple ways of accessing media content, and toward ever more complex relations between top-down corporate media and bottom-up participatory culture" (p. 243).

These changes in technology and society are shaping the way people think and relate to media. Jenkins asserts that the larger problem for educators today is not the old notion of a digital divide that separates people based on limited access to the tools of communication, since more people have access today than ever before, but the larger problem today is a *participation gap*, "the unequal access to the opportunities, experiences, skills, and knowledge that will prepare youth for full participation in the world of tomorrow." (Jenkins et al. 2007, p. 3). Jenkins writes, "We need to rethink the goals of media education so that young people can come to think of themselves as cultural producers and participants and not simply as consumers, critical or otherwise" (2006, p. 259).

Framing the changes in technological and social terms, Carmen Luke (2006) argues for an expanded form of media literacy because of three increasingly growing levels of media and technological convergence. One level of convergence is the functional ability of hardware devices to perform multiple tasks, such as a cell phone that can take and send pictures (still and moving images), play music, send and receive text messages, upload and download content online, play games, and can still be used for chatting. A second level of media convergence entails provider convergence which has been greatly enhanced by deregulations of media ownership and the numerous mergers and acquisitions of multinational media corporations. The horizontal and vertical integration of media companies allows fewer corporations the ability to control more different types of services and content (Bagdikian 1997; McChesney 2004). The ability of ICTs to perform more functions and the integration of media providers are creating what Marsha Kinder (1991) labels *transmedia intertextuality*.

According to Luke, the third level of media convergence is a consequence of the two previously mentioned that have had the effect of creating "a much tighter synergy between previously disparate industries, between knowledge and information, consumerism, popular culture, entertainment, communication, and education" (2006, p. 5). As politics, news and entertainment converge into new forms of media, an entire spectator culture is evolving. Spectacle itself is becoming one of the organizing principles of the economy, polity, society, and everyday life (Kellner 2003, 2007), as I argued in Chapter Three. In order to grab larger audiences and increase profit and power, the culture industries aggressively create and promote a synthesized spectacle-centered media culture (Kellner 2020).

James Paul Gee (2000) suggests that technological innovations and hyper-competitive global "fast capitalism" are creating a new type of individual whom he calls the "portfolio person" (p. 43). Gee explains that the idea of 'expertise' has moved "away from 'disciplinary' or academic expertise to a broader notion more compatible with the new capitalist world view" (p. 48). He writes that this business orientation, much like Friedman's flat world perspective, emphasizes "efficient problem solving, productivity, innovation, adaptation, and non-authoritarian distributed systems…In the new capitalism, it is not really important what individuals know on their own, but rather what that they can do with others collaboratively to effectively add 'value' to the enterprise" (p. 49). A problem with this education of the portfolio person is that it is based on a cognitive notion of knowledge workers who have the facility of "higher order thinking," but lack the ability to think *"critiquely."* Gee describes critiquely as the ability "to understand and critique systems of power and injustice" (p. 62). The inability to understand or empathize with marginalized, poor and oppressed people is a major problem of this fast capitalism epistemology and thinking critically from the perspective of class, gender, race, and other forms of oppression provides both insight into systems like education and the need for radical reform.

Another problem with the model that creates the portfolio person is that it advantages most the children from dominant positions in society (i.e. white, male, middle or upper class), who have easier access to this expertise and "school language" based on their lifeworld experiences and privileges. It is much easier to bridge the home culture to the public school domain for students who have been exposed to white middle class values such as reading children's literature from an early age or visiting museums and art galleries, watching educational television and accessing educational web materials.

Yet the common deficit thinking approach that many educators internalize, undervalues the cultural assets that minority and poor students bring to school and often frame those resources as problems to be overcome (Solorzano and Yosso 2001). Gee writes, "We rarely build on their experiences and on their very real distinctive lifeworld knowledge. In fact, they are often asked, in the process of being exposed to specialist domains, to deny the value of their lifeworlds and their communities in reference to those of more advantaged children" (p. 66).

To counteract the problems of inequality and lack of social critique, Gee promotes a Bill of Rights for all students that includes four pedagogical principles: situated practices, overt instruction, critical framing and transformative practice. He writes, "These principles seek to produce people who can function in the new capitalism, but in a much more meta-aware and political fashion than forms of new-capitalist-complicit schooling" (p. 67). The situated practice can help value the different cultural capital (Bourdieu 1986) students bring into the classroom as child-centered experiential practices allow students to discover connections between their lifeworlds and school. A major aspect of these principles is a meta-cognitive awareness about the interconnections of thinking, knowledge and power relations. The need for some overt instruction and critical framing assures that students will engage with texts critically to understand the interconnections and systems of power within which media and digital texts emerge.

The fourth principle of transformed practice suggests that education must involve acting on learning and empower students to use and transform knowledge. When Gee's fourth principle of transformative practice is built on critical framing, then Jenkins' goal of bridging the participation gap can become a reconstruction of education promoting critical media literacy. In a report funded by the MacArthur Foundation, Jenkins and others assert the need for teaching "new media literacies: a set of cultural competencies and social skills that young people need in the new media landscape. Participatory culture shifts the focus of literacy from one of individual expression to community involvement. The new literacies almost all involve social skills developed through collaboration and networking. These skills build on the foundation of traditional literacy, research skills, technical skills, and critical analysis skills taught in the classroom" (Jenkins et al. 2007, p. 4).

Jeff Share and I wrote a *The Critical Media Literacy Guide: Engaging Media and Transforming Education* (2019) in which we attempt to lay out and illustrate key concepts, principles, and practices of what we call "critical media literacy." The "critical" component involves engaging the dimensions of race, gender, class, and sexuality in media analysis, interpretation, and critique, and seeing media as a key locus of power in a global capitalist system. In the next section, I argue that this type of critical media literacy needs to be supplemented by critical digital literacies.

Technology, Democratic Civics Education, and Critical Digital Literacies

Upon first consideration, seeking a suitable definition of "technology" itself appears to be overly technical. Surely, in discussions concerning technology, it is rare indeed that people need to pause so as to ask for a clarification of the term. In a given context, if it is suggested that technology is either causing problems or alleviating them, people generally know what sort of thing is due for blame or praise.

Yet, the popular meaning of "technology" is problematically insufficient in at least two ways. First, it narrowly equates technological artifacts with "high-tech," such as those scientific machines used in medical and biotechnology, modern industrial apparatuses, and digital components like computers, ICTs, and other electronic media. This reductive view fails to recognize, for instance, that indigenous artifacts are themselves technologies in their own right, as well as other cultural objects that may once have represented the leading-edge of technological inventiveness during previous historical eras, such as drums, wall paintings, books, hand tools, or even clothing (McLuhan 1964; Gleick 2011).

Secondly, popular conceptions of technology today make the additional error of construing technology as being merely object-oriented, identifying it as only the sort of machined products that arise through industry. In fact, from the first, technology has always meant far more; and this is reflected in recent definitions of technology as "a seamless web or network combining artifacts, people, organizations, cultural meanings and knowledge" (Wajcman 2004, p. 106), or that which "comprises the entire system of people and organizations, knowledge, processes, and devices that go into creating and operating technological artifacts, as well as the artifacts themselves" (Pearson and Young 2002).

These broader definitions of technology are supported by the important insights of John Dewey. For Dewey, technology is central to humanity and girds human inquiry in its totality (Hickman 2001). In Dewey's view, technology is evidenced in all manner of creative experience and problem-solving. It should extend beyond the sciences proper, as it encompasses not only the arts and humanities, but the professions, and the practices of our everyday lives. In this account, technology is inherently political and historical, and in Dewey's philosophy it is strongly tethered to notions of democracy and education, which are considered technologies that intend social progress and greater freedom for the future.

With the assaults on democracy during the Trump era and rise of global rightwing extremist movements that are authoritarian and anti-democratic, it is more important than ever to develop a civics education that instills the knowledge, values, skills, and competencies necessary to be a democratic citizen. It is telling that one of Trump's last acts as president was to release a "President's Advisory 1776 Report" in which a key paragraph reads: "The declared purpose of the President's Advisory 1776 Commission is to enable a rising generation to understand the history and principles of the founding of the United States in 1776 and to strive to form a more perfect Union. This requires a restoration of American education, which can only be grounded on a history of those principles that is 'accurate, honest, unifying, inspiring, and ennobling.' And a rediscovery of our shared identity rooted in our founding principles is the path to a renewed American unity and a confident American future."[15]

The key phrase "restoration of American education" suggests a rightwing call to a culture war over education and American history like William Bennett and other conservatives fought in Reagan era.[16] Yet a key thrust of the report was a denial of the exceptionalism of American slavery and a belief it was found everywhere and was a normal condition of things. Defining slavery as "the normal everywhere" set off a fierce cultural war and shows the need for a robust concept of civic education on purposes of education derived from Dewey, Freire, Marcuse, and other progressive educators.

John Dewey argued that democracy required an educated citizenry and that education should be geared toward creating citizens and thus required a vigorous civics education. Dewey's views appeared during the Progressive Era when immigrants were flowing into the U.S. and it was urgent to educate all citizens and instill American democratic values. In views of the attacks on democracy by Trump's rightwing extremist hordes and their political and intellectual enablers, the act of creating a vigorous democracy based on civics education is more

---

[15] The report is posted at https://www.whitehouse.gov/wp-content/uploads/2021/01/The-Pre
sidents-Advisory-1776-Commission-Final-Report.pdf (accessed on January 19, 2021).

[16] See William J. Bennett, "The War Over Culture In Education." Heritage Foundation, September 5, 1991 at https://www.heritage.org/education/report/thc-war-over-culture-education (accessed January 19, 2021).

important than ever as future years portend global culture wars over education and democracy.[17]

For Dewey, civics education involved his progressive principles that education involved learning the basic percepts, values, and history of the country, involving political knowledge, social studies, and philosophy and critical thinking to grasp and appraise the key concepts of democracy and its institutional bases, practices, norms, and challenges of a democratic society.[18] For Dewey, learning took place through doing, so civics education involved gaining the information and learning experiences to equip and empower citizens to participate in democratic processes. This process would benefit both the individual and society and produce both individual and social development in a progressive direction.

Civics education thus involved learning to participate in civic life, politics, and progressive social movements for democratic rights and values. It involved grasping the foundations, norms, and core values of the American political system like the right to vote and sanctity of the electoral college, that every citizen has the right to vote, each vote must be counted, and that the citizens should accept the results of the vote. Democratic citizenship involves fighting for a fair vote, defending certified fair and secure election results, and accepting certified results of a fair election. Good citizenship also would involve opposing demagogues like

---

[17] For an important text on the need for revitalizing civics education, see The Association of American Colleges & Universities, "The National Task Force on Civic Learning and Democratic Engagement A Crucible Moment: College Learning & Democracy's Future. A Call to Action and Report from The National Task Force on Civic Learning and Democratic Engagement," at https://www.aacu.org/crucible, (accessed on January 28, 2021). I should note that there are also more boring and less Deweyean concepts of civic education that would reduce it to "Civics can teach you about the rights granted to citizens, as well as their responsibilities, such as serving on juries and engaging in the political process." Or: "Mandatory Duties of U.S. Citizens. 1) Obeying the law. Every U.S. citizen must obey federal, state and local laws, and pay the penalties that can be incurred when a law is broken. 2) Paying taxes. ... 3) Serving on a jury when summoned. ... 4) Registering with the Selective Service." The report is found at = https://www.civics.ks.gov/kansas/citizenship/responsibilities-of-citizens.html (accessed on January 22, 2021). These duties and responsibilities may be part of citizenship, but citizenship for democracy, training individuals to know the fundamental principles of democracy, and to learn to participate in democracy and defend its institutions, principles, and forms of life against authoritarianism and anti-democratic societies and principles is an important part of a stronger Deweyean conception of citizenship.

[18] For a good study of Dewey and citizenship education see Sarah M. Stitzlein, "Habits of Democracy: A Deweyan Approach to Citizenship Education in America Today" at https://docs.lib.purdue.edu/cgi/viewcontent.cgi?article=1514&context=eandc (accessed January 19, 2021).

Trump and his mob who would deny that results of a fair and democratically validated election, as with the 2020 win over Trump by Joe Biden.[19]

Strong civics education involves distinguishing truth from falsity and puncturing Big Lies like "the election was rigged and Trump won by a landslide," as was promoted by Trump and his rampaging hordes in their rampaging in the U.S. Capital on January 6, 2021. This desecration of a Citadel of U.S. democracy dramatizes the need to connect rigorous training in thinking and truth, as well as information literacy, to distinguish truth from falsity, facts from lies, developing critical media and information literacies that can evaluate sources, obtain good information grounded in fact and informed opinion, and to be able to distinguish disinformation and propaganda. Developing critical media and digital information literacies are more important than ever when a large number of U.S. citizens believed the lies of Donald Trump instead of the obvious facts about his presidency and Election 2020 which were disseminated daily by reputable sources in the media and internet.[20]

Civics education involves understanding how the U.S. government established by the Constitution embodies the purposes, values, and principles of American democracy, and how accepting election results and transitioning from one regime to another is a foundation of U.S. democracy—sadly, fundamental tenets denied

---

[19] While the Trump administration challenged voting results in almost all swing states and made over 50 legal challenges, in all cases votes were recounted, often by hand, many of Trump's challenges were thrown out by the court, and experts declared it one of the fairest and most secure elections in U.S. history; see Jen Kirby, "Trump's own officials say 2020 was America's most secure election in history. Homeland Security put out a statement with state and local officials that countered the president's fraud claims." *Vox.com*, November 13, 2020, at https://www.vox.com/2020/11/13/21563825/2020-elections-most-sec ure-dhs-cisa-krebs (accessed on January 29, 2021). See also Pam Fessler, "As States Certify Ballot Totals, An Extraordinary Election Comes To An End," *NPR*, November 25, 2020 at https://www.npr.org/2020/11/25/938617688/as-states-certify-ballot-totals-an-extrao rdinary-election-comes-to-an-end (accessed on January 29, 2021).

[20] The Pro-Trump mobs that invaded the White House carried signs repeating the Trump lie that the 2020 election was stolen and many of the rioters interviewed repeated the Big Lie that Trump won the election, a lie he repeated almost daily in the days after both media announcements of his decisive loss to Biden and the formal certification of Biden's election. According to *The Washington Post* "Fact Checker" team "In four years, President Trump made 30,573 false or misleading claims," Updated Jan. 20, 2021 at https://www.washingtonpost. com/graphics/politics/trump-claims-database/ (accessed on January 22, 2021). The Wikipedia entry on "Veracity of statements by Donald Trump" cites other data bases collecting his lies and offer well-documented examples of Trump's stunning amount of lying throughout his career at https://en.wikipedia.org/wiki/Veracity_of_statements_by_Donald_Trump (accessed on January 22, 2021).

by Trump and great masses of Trump's followers in the 2020 election, demonstrating that civics education must be revitalized by new efforts at developing more robust civic education instruction and teaching the fundamentals of democracy.

Civics education for Dewey also involved technological competencies which today would involve critical media and digital literacies as I am arguing in this chapter. Dewey's progressivist view was unabashedly optimistic and hopeful that it is within the nature of humanity that people may be sufficiently educated so as to be able to understand the society and polity that they live in, the problems which they face, and that people working together can experimentally produce and deploy a wide range of policies and technologies so as to solve those problems accordingly. While I agree strongly with the spirit of Dewey, I also recognize that the present age is potentially beset by the unprecedented problem of globalized media and social dis- and misinformation, inequality in education and access to reliable information, and multiple propaganda sources in both social media and broadcasting media that require more critical views of technology. Hence, critical thinking becomes ever more important to civics education and the ability to distinguish truth from lies, solid information based on facts from disinformation, and democratic consensus from propaganda and demagoguery.[21]

Civics education for democracy thus requires learning how to use information technology to gain solid information and to resist and oppose disinformation and lies. Herbert Marcuse's concept of the "Great Refusal" (1964) is relevant in situations in which students, teachers, and citizens face oppressive or authoritarian forces, as is his calls for resistance to anti-democratic and oppressive forces and the need to create new sensibilities, consciousness, values, and lives (Marcuse 1969; Cho et al. 2009).

Yet, widespread ecological crisis and destruction of the environment by out-of-control technology and economic development, requires more radical critiques of technology and calls for appropriate technologies that serve human life, democracy, and preserving the environment, thus requiring a critical theory of technology that develops its progressive potentialities and possible negative aspects, requiring the creation of technologies in harmony with human beings and their environment.

To this end, I want to highlight the insights of radical social critic and technology theorist Ivan Illich (Kahn and Kellner 2008). Specifically, Illich's notion of "tools" mirrors the broad humanistic understanding of technology outlined so

---

[21] I wish to thank Kimberly Rosenfeld for impressing upon me the importance of civics education.

far, while it additionally distinguishes "rationally designed devices, be they artifacts or rules, codes or operators…from other things such as food or implements, which in a given culture are not deemed to be subject to rationalization" (Illich 1973, p. 22). Consequently, Illich polemicizes for "tools for conviviality," which are technologies mindfully rationed to work within the balances of both cultural and natural limits. In this view, technology so defined will prove useful for a twenty-first century technoliteracy challenged to meet the demands to develop technologies for a sustainable and ecologically secure world.

One of the great insights of Marshall McLuhan is that new media produce new environments in which people live and navigate (1964). For instance, electricity produced entirely new urban and living spaces as well as new sciences that contributed to the development of contemporary physics as well as creating new technologies, including the Internet (see Carr 2010). For McLuhan, a new technology of communication creates a new environment, and he analyzes a progression of stages of society and culture depending on dominant media, moving from oral culture through print culture and electronic media. New media for McLuhan require new literacies and I would argue that he provides an important rationale for reconstructing education and developing multiple technoliteracies in order to properly perceive, navigate, and act in the new technological environment with ever-expanding roles for digital echnologies in information-gathering, education, and many modes of social and political life.

"Literacy" is a concept, often used by educators and policy makers, but in a variety of ways and for a broad array of purposes. In its initial form, basic literacy equated to vocational proficiency with language and numbers, such that individuals could function at work and in society. Thus, even at the start of the twentieth century, literacy largely meant the ability to write one's name and decode popular print-based texts, with the additional goal of written self-expression only emerging over the following decades. Street (1984) identifies these attributes as typical of a model of literacy that is politically conservative in that it is primarily economistic, individualistic, and is driven by a deficit theory of learning. On the other hand, Street characterizes models of literacy as prefiguring positive notions of collective empowerment, understanding social context, learning the encoding and decoding of non-print-based and print-based texts, as well as a progressive commitment to critical thinking-oriented skills.

In this conception, "literacy" is not a singular set of abilities but is multiple and comprises gaining competencies involved in effectively using socially constructed forms of communication and representation that change and evolve with new media, technologies, and social conditions. Learning literacies requires attaining competencies in practices and in contexts that are governed by rules and conventions that are specific to each society and culture. Literacies are socially constructed in different societies in educational and cultural practices involving various institutional discourses and pedagogies. Against the view that posits literacy as static and universal, literacies are continuously evolving and shifting in response to social and cultural changes, as well as the interests of the elites who control hegemonic institutions. Further, it is a crucial part of the literacy process that people come to understand hegemonic codes as "hegemonic." Thus, this conception of literacy follows Freire and Macedo (1987) in conceiving literacy as tethered to issues of power. As they note, literacy is a cultural politics that "promotes democratic and emancipatory change" (viii) and it should be interpreted widely as the ability to engage in a variety of forms of problem-posing and critical analyses of self and society.

Based on these definitions of "technology" and "literacy," it should be obvious that, holistically conceived, literacies are themselves technologies of a sort involving processes that serve to facilitate and regulate human and social interactions. This optic on literacies helps to highlight the constructed and potentially reconstructive nature of literacies, as well as the educative, social, and political nature of technologies. Further, more than ever, we need philosophical reflection on the ends and purposes of education and on what we are doing and trying to achieve in our educational practices and institutions, which itself is a form of technoliteracy in its deepest sense.

Education should produce strong democratic citizens, able, competent, and willing to participate in social and political life in which both the individual and the society develop together. Education must also cultivate critical thinking to resist authoritarian movements that counter the values of democracy. Thus education requires ethical and civics education that enables individuals to learn and appropriate the values and ethos of democracy: listening to other people and views; discussing and debating areas of concern; engaging in consensus on issues where there are differences; and opposing positions that are antithetical to democratic values, norms, and institutions. Of course, in a divided and conflicted society this is not easy and one must accept loss and sometimes struggle in loyal opposition if one's core values conflict with a specific societal consensus or regime that one appraises as anti-democratic or authoritarian.

Less philosophically, contemporary technoliteracies are involved with the need to comprehend and make use of proliferating digital technologies, and the political economy that drives them, but also towards furthering radical democratic understandings and transformations of our worlds through the use of digital technologies as I have been stressing throughout this book. In a world inexorably undergoing processes of globalization and technological transformation, one cannot advocate a policy of clean hands and purity, in which people shield themselves from digital technologies and social media and their transnational proliferation. Instead, technoliteracies must be deployed and promoted that allow for popular interventions into the ongoing (often anti-democratic) economic and political struggles going on, thereby potentially deflecting digital technologies and social media for progressive ends like social justice and ecological well-being.

In this, critical technoliteracies encompass the computer, digital, information, media, and multimedia literacies presently theorized under the concept "multiliteracies" (Cope and Kalantzis 2000; Luke 2000, 1997; Rassool 1999; New London Group 1996). Yet whereas multiliteracies theory often remains focused upon digital technologies, with an implicit thrust towards providing new media job skills for the Internet age, I suggest the need to explicitly highlight the social and cultural appropriateness of technologies and provide a critique of the emergent media economy and society (Best and Kellner 2001; Kellner 1989), and to provide critical literacies that detect its racism, sexism, classism, inequalities, threats to the environment and other social problems while acknowledging progressive potentials. In other words, critical media and digital literacies analyze and critique racism, sexism, homophobia, classism, and other biases in media and digital texts which also providing an understanding of the broader ecologies of media and digital technologies and how they relate to forces of power and domination as well as emancipation (Kellner and Share 2019; Kellner 2020).

Thus, I connect the project of developing a critical theory of society with the project of a reconstruction of education. In the next section, I draw upon the language of "multiple literacies" (Lonsdale and McCurry 2004; Kellner 2000) to augment a critical theory of technoliteracies that I and colleagues have been developing for some years and which this chapter draws upon (Best and Kellner 2001; Kahn and Kellner 2006, 2014; and Kellner and Share 2019). The teaching of critical media and digital literacies will thus be involved in the project of reconstructing education for democracy as in the work of Dewey, Freire, and Marcuse which is discussed earlier in the chapter.

Toward the Reconstruction of Education

Critical technoliteracies in this conception is tied to the project of critical pedagogy and radical democracy and is concerned to develop skills that will enhance democratization, civic participation, and progressive social transformation. It takes a comprehensive approach that teaches critical skills involving how to use media and digital technology as instruments of social communication and transformation. The technologies of communication are becoming more and more accessible to young people and ordinary citizens, and can be used to promote education, democratic self-expression, and social justice. Technologies that could help produce the end of democracy could generate the acceleration of participatory democracy. Critical education today should thus conceive of how to use digital technologies and social media to reconstruct education and help create a more democratic society. Hence, students should learn new forms of critical media and digital literacies that involve both how to use media and computer culture to do research and gather information, as well as to perceive it as a cultural terrain which contains texts, spectacles, games, and new interactive multimedia which requires new modes of literacy, critique, participation, and reconstruction.

Moreover, media and digital culture is a discursive and political location in which students, teachers, and citizens can all intervene, engaging in discussion groups and collaborative research projects, creating their web sites, projects of critique of specific forms of media and digital culture, like doing critical analyses of dominant film, television, popular music and consumer culture (Kellner 2020), of critiquing sites of digital culture like Facebook, Twitter, YouTube, and so on or apps like Zoom or various social media sites. Critical media and digitial literacy projects can also involve producing new media texts for cultural dissemination, and engaging in new modes of social interaction and learning via digital and social media sites, using, for example, digital apps like Moodle or social media sites as places for discussion and learning. Or one can use digital technologies to create one's own texts that express one's views, document your lives, and create your own stories and cultural aesthetic forms. Participating in media and digital culture enables individuals to actively participate in the production of culture, ranging from discussion of public issues to creation of their own individual or group productions. However, to take part in this culture requires not only accelerated forms of traditional modes of print literacy which are often restricted to the growing elite of students who are privileged to attend adequate and superior public and private schools, but new forms of digital literacies as well, thus posing significant challenges to education.

It is indeed a salient fact of the present age that computer and digital culture has now been proliferating for decades and it is increasingly clear that we have to begin teaching computer literacy from an early age on. Computer and digital literacy, however, itself needs to be theorized. Often the term is synonymous with technical ability to use computers, to master existing programs, and maybe engage in some programming oneself. I want, however, to suggest expanding the conception of computer literacy from using computer programs and hardware to a broader concept of critical digital literacies, including information literacy and developing, in addition, more sophisticated abilities in traditional reading and writing, as well as the capability to critically dissect cultural forms taught as part of critical media literacy and new forms of multiple media and digital literacies.

Information literacy involves both the accessing and processing of diverse sorts of information proliferating in our infotainment society.[22] It encompasses learning to find sources of information ranging from traditional sites like libraries and print media to Internet websites and search engines to multi-and social media. Thus, on this conception, genuine computer literacies involves not just technical knowledge and skills, but refined reading, writing, research, and communicating ability that requires heightened capacities for critically accessing, analyzing, interpreting, and processing print, image, sound, and multimedia material, and seeing how these forms are part of a broader culture that includes forms of domination and inequality and that requires reconstruction so that voices of women, people of color, gay and lesbian people, and marginalized groups also have their voices, cultural creations, and forms of culture.

Critical digital information literacies basically involves the ability to discover and access information and intensified abilities to read, to scan texts and computer data bases and websites, and to download or print the information in a form appropriate for further information processing. Utilizing information accessed in an educational context further requires putting it together in meaningful patterns and mosaics, to construct meanings and interpretations, to contextualize and evaluate, and to discuss and articulate one's own views and interpretations.

Within media and digital culture, **visual literacy** takes on increased importance. On the whole, computer screens are more graphic, visual, and interactive

---

[22] In 1991, the Association of Supervision and Curriculum Development (ASCD) concluded: "Information literacy equips individuals to take advantage of the opportunities inherent in the global information society. Information literacy should be a part of every student's educational experience. ASCD urges schools, colleges, and universities to integrate information literacy programs into learning programs for all students" (AASL, 1996). The project has been taken up by the national Forum on Information Literacy (NFIL). Building on these projects, it is thus important to see that computer literacy involves developing a wide range of information literacies, and that the latter also involve developing multi-literacies that access and interpret images, media spectacles, narratives, and multimedia cultural sites in an expanded concept of information that resists its reduction to print paradigms alone. For other conceptions of multimedia literacy, see the discussions of multiple literacies in reading hypertexts in Burbules and Callister 1996; the concept of multiliteracy in the New London Group 1996 and Luke 1997; the concept of hyperreading in Burbules, 1997; the papers in Snyder 1997, and the studies in studies Semali and Watts Pailliotet 1998; Bus and Neuman 2008; and Mills 2015.

than conventional print fields which disconcerted many of us when first confronted with the new cultural environments. Icons, windows, lap tops, smart phones and other digital devices—and the various clicking, linking, and interaction required by computer-mediated hypertext—requires new competencies and a dramatic expansion of literacy. Visuality is obviously crucial, requiring one to quickly scan visual fields, perceive and interact with icons and graphics, and use technical devices like a mouse to access the desired material and field. One must also learn navigational skills of how to proceed from one field and screen to another, how to search for information on the Internet and computer data bases, and how to move from one program to another if one operates, as most now do, in a window-based computer environment.

The expanding multimedia environments require in fact a diversity of multi-semiotic and multimodal interaction, involving interfacing with words and print material and often images, graphics, and audio and video material. The New London Group has produced the concept of "multiliteracy" to describe the types of literacy required to engage new multimedia technology and hypertext fields, while Semali and Watts Pailliotet and their collaborators (1998) propose the concept "Intermediality" to call attention the need to generate literacies that allow inter-action between various media, multimedia, and digital fields, and that promote interdisciplinary and interactive education in an attempt to create education that promotes multicultural understanding and democratic social change. In a similar vein, individuals involved in the University of California at Los Angeles and San Diego with the *la classa magica* project have been using computer and multimedia technology to teach basic reading and writing skills, as well as digital and multimedia literacy and forms of social cooperation and interaction.[23]

As technological convergence develops apace, one needs to combine the skills of critical media literacy with traditional print literacy and new forms of multi-literacy to access the new multimedia hypertext environments.[24] Literacy in my

---

[23] Kris Rodriguez, "'La Clase Magica' program helps children succeed by using technology," *UTSA Today*, June 17, 2010), at https://www.utsa.edu/today/2010/06/magica.html (accessed December 26, 2020).

[24] There are two major modes and concepts of hypertext, one that is primarily literary, that involves new literary/writing strategies and practices and one that is more multimedia, multisemiotic, and multimodal. Hypertext was initially seen as an innovative and exciting new mode of communication that increased potentials for writers to explore new modes of textuality and expression and to expand the field of writing. As multimedia hypertext developed, it was soon theorized as a multisemiotic and multimodal form of communication. Yet some early advocates of hypertext attacked the emergence of the World Wide Web as a debased medium which brought back into play the field of earlier media, like television, forcing the word to renegotiate its power with the image and spectacles of sight and sound, once again

conception involves socially-constructed forms of communication and representation and the corresponding competencies involved in effectively using them. Thus, reading and interpreting print was the appropriate mode of literacy for books, while critical media and technoliteracies require reading and interpreting discourses, images, spectacle, narratives, and the forms and genres of media culture. Forms of multimedia communication involve print, speech, visuality, audio, and multimodal dimensions in a hybrid field which combines these forms, all of which involve skills of interpreting and critique (Kellner and Share 2019; Kellner 2020).

Obviously, here the key root is the multiple, the proliferation of media, technologies, and forms that require a multiplicity of competencies and skills and abilities to navigate and construct a new semiotic terrain of interpretation and critique—hence the term multiliteracy. Multiliteracies involve reading across multiple and hybrid semiotic fields and being able to critically and hermeneutically process print, graphics, images, and perhaps moving images and sounds. The term "hybridity" suggests the combination and interaction of diverse media and the need to synthesize the various forms in an active process of the construction of meaning. Reading a music video, for instance, involves processing images, music, spectacle, and sometimes narrative in a multisemiotic process that simultaneously draws on diverse aesthetic forms. Interacting with a website or social media app involves scanning text, graphics, often moving images, and clicking onto the fields that one seeks to peruse and appropriate. This might involve combining video, audio, print, and graphics in new interactive learning of entertainment environments or internet culture of diverse sorts.

While traditional literacies involve practices in contexts that are governed by rules and conventions, the conventions and rules of multiliteracies are currently evolving so that their pedagogies is a relatively new and developing field. Multimedia fields are not entirely new, however. Multisemiotic textuality was first evident in newspapers (consider the difference between *The New York Times* and *U.S.A. Today* in terms of image, text, color graphics, design, and content), and is now evident in textbooks that are much more visual, graphic, and multimodal than the previously linear and discursive primary verbal texts of old. Yet it is web sites, social media sites and aps, and emergent multimedia that are the most distinctively multimodal and multisemiotic forms. These sites are the new frontier of learning and literacy, and one of the great challenges to education for the millennium. As we proceed into the century, we need to theorize the literacies

---

decentering the written word (see, for instance, Landow 2006; Lankshear and Snyder 2000; Snyder 2002 and the articles in Snyder 1997).

necessary to navigate and interact in these new multimedia environments and to gain the skills that will enable us to learn, work, and create in new cultural spaces and domains.

Parenthetically, I might note that we are soon going to have to rethink SATs and standard tests in relation to digital and media technologies; having the literacy and skills to successfully navigate, communicate, work, and create within digital and multimedia culture is quite different from reading and writing in the mode of print literacy and while this mode of literacy continues to be of utmost importance, it is sublated within multiliteracy, so eventually an entirely different sort of test is going to need to be devised to register individuals' multiliteracy competency and to predict success in an evolving technological and educational environment.[25]

Thus, in my expanded conception, computer literacies involves technical abilities concerning developing basic typing skills, mastering computer programs, accessing information, and using computer technologies for a variety of purposes ranging from verbal communication to artistic expression to political debate, organization, and activism. There are ever more hybrid implosions between media and computer culture as audio and video material becomes part of the Internet, as social and multimedia develop, and as digital technologies and social media become part and parcel of the home, school, and workplace. Therefore, the skills of decoding images, sounds, and spectacle learned in critical media literacy training can also be valuable as part of computer literacy as well.

Furthermore, print literacy takes on increasing importance in computer and technoculture as one needs to critically scrutinize and scroll tremendous amounts of information, putting new emphasis on developing reading and writing abilities. In fact, Internet discussion groups, chat rooms, email, and various forums require writing skills in which a new emphasis on the importance of clarity and precision

---

[25] While I have not myself researched the policy literature on this issue, in the many discussions of SAT tests and their biases which I have read, I have not encountered critiques that indicate the obsolescence of many standardized tests in a new technological environment and the need to come up with new testing procedures based on the new cultural and social fields that we are increasingly immersed in. I would predict that proposals for devising such tests are emerging and that this issue will be hotly debated and contested in the future. I should note that the University of California system has suspended SAT tests until Fall 2024 and that the COVID-19 pandemic required new modes of teaching and learning for which the standardized SAT tests were inappropriate and becoming obsolete. See Sarah Moon, "University of California will suspend SAT and ACT testing admission requirement until 2024" *CNN News*, May 22, 2020 at https://www.cnn.com/2020/05/22/us/uc-suspends-sat-act-for-admissions-until-2024/index.html (accessed January 23, 2021).

is emerging as communications proliferate. In this context of information saturation, it becomes an ethical imperative not to contribute to cultural and information overload, and to concisely communicate one's thoughts and feelings.

In a certain sense, digital technologies are becoming the technological equivalent of Hegel's Absolute Idea, able to absorb everything into its form and medium. Digital technoculture is now not only repositories of text and print-based data, but also contain a wealth of images, multimedia sights and sounds, and interactive environments that, like the media, are themselves a form of education that require a critical pedagogy of electronic, digitized culture and communication. From this conception, computer literacy is something like a Hegelian synthesis of print and visual literacy, technical skills, and media literacies, brought together at a new and higher stage. While Neil Postman (1985) and others produce a simplistic Manichean dichotomy between print and visual literacy, we need to learn to think dialectically, to read together text and image, to decipher sight and sound, and to develop forms of critical computer and multimedia literacies adequate to meet the exigencies of an increasingly high tech society.

Thus, a new critical media and digital pedagogy requires developing interconnected forms of print, media, computer, and multiliteracies, all of which are of crucial importance in the evolving technoculture of the present and fast-approaching postmodern future.[26] Whereas modern pedagogy tended to be specialized, fragmented, and differentiated, and was focused on print culture, a postmodern pedagogy involves developing multiple literacies and critically analyzing, dissecting, and engaging a multiplicity of cultural forms, some of which are the products of digital technologies and social media, requiring developing new critical literacies to engage the emergent cultural forms and media.

In fact, contemporary culture is marked by a proliferation of cultural machines which generate a panoply of diverse aesthetic artifacts within which we wander, trying to make our way through this forest of symbols, images, words, sounds, and spectacle. Critical media and digital literacies require the ability to scan, interact with, traverse, organize, and create pedagogies for new multimedia educational environments. Multimedia literacy thus involves not just reading, but interacting: clicking to move from one field to another if one is involved in a hypertext environment, such as one finds on web and social media sites, capturing, saving, downloading, and perhaps printing material relevant to one's own projects; and maybe responding verbally or adding one's own material if it is a site that invites genuinely interactive participation.

---

[26] For my take on postmodern theory, see Kellner 1989b and 1989c; Best and Kellner 1991, 1997, and 2001. For my earlier sketch of postmodern pedagogy, see Kellner 1989c.

In addition to the linear cognitive skills needed for traditional reading of print material, multimedia literacy thus requires a multisemiotic ability to read hypertexts that are often multidimensional requiring the connecting of images, graphics, texts, and sometimes audio–video material. It also involves new forms of intertextuality and contextualizing multimedia material. Multimedia thus involves making connections between the complex and multilayered cyberworld and its connection with the real world. As Carmen Luke reminds us: "Since all meaning is situated relationally—that is, connected and cross-referenced to other media and genres, and to related meanings in other cultural contexts—a critical literacy relies on broad-based notions of intertextuality" (1997, p. 10). Intertextuality draws attention to the complex ways that language, image, and types of texts are related to various genres, forms, narratives, and modes of meaning such as visual design.

Thus, on one hand, one must learn to read multimedia forms that are themselves overlapping and interrelated, switching from text to graphics to video to audio, decoding in turn sight, sound, and text. In a global information environment, this also may involve switching from sites from one country to another requiring contextual understanding and literacy that is able to read and interact with people and sites from different cultures. As Carmen Luke puts it: "[N]ew [forms of] virtual communication are emerging, which require an intertextual understanding of how meanings shift across media, genres, and cultural frames of reference. Whether one 'visits' the Louvre on-line, joins an international newsgroup of parents of Downs Syndrome children, or visits the www site of an agricultural college in Kenya, cross-cultural understanding and 'netiquette' is increasingly crucial for participating effectively in global communications" (Luke 1997, p. 10).

Crucially, multimedia literacy should be contextual, it requires thematizing the background and power relations of cultural forms (i.e. including analysis of the political economy of the media and technology, of how corporate organizations control production and dissemination, and how oppositional and alternative media and uses are possible; see Kellner 2020). It involves as well engaging the context and power relations of the specific media use in question (i.e. the differences between television watching in the classroom, at home with one's family, with one's friends or alone; or the differences between computer use for research, data organization, email, engaging social media, or playing games, etc.). Multimedia literacy also envisages new modes of collaborative work on research projects or web sites, new forms of student/teacher participation and interaction, and new pedagogical uses for the digital and media technologies which may often appear exotic in the present, but which will become increasingly commonplace in the future and will force a rethinking of education.

Finally, multiliteracy must become *critical*, and in response to excessive hype concerning new technologies and education, it is necessary to maintain the critical dimension. Just as critical media literacy (CML as explicated by Kellner and Share 2019) involves engaging race, class, gender, sexuality and power in media analysis, interpretation and critique to critique racism, sexism, homophobia, classism and other forms of oppression, critical technoliteracies must also critically engage the dimensions of gender, race, class, sexuality, and other determinants of power and domination engaged by critical media and digital literacies.

After an excellent discussion of new modes of literacy and the need to rethink education, Gunther Kress argues that we must move from critique to design, beyond a negative deconstruction to more positive construction (1997). Yet rather than following such modern logic of either/or, we need to pursue the logic of both/and, seeing design and critique, deconstruction and reconstruction, as complementary and supplementary rather than as antithetical choices. Certainly, we need to design new technologies, pedagogies, and curricula for the future, and should attempt to design new social and pedagogical relations as well, but we need to criticize misuse, inappropriate use, overinflated claims, and exclusions and oppressions involved in the introduction of new technologies and media into education.

The critical dimension is needed more than ever as we attempt to develop new critical teaching strategies and pedagogy, as we design new pedagogies and curricula, we must be constantly critical, practicing critique and self-criticism, putting in question our assumptions, discourses, and practices as we experimentally develop new literacies and pedagogy in a democratic transformation of education that produces citizens ready to function competently in democratic social and political life.

Thus, in a democratic transformation of education and literacies, critique is of fundamental importance. From the Deweyean perspective, progressive education involves experiment and the experimental method which involves critique of limitations, failures, and flawed design. In discussing new technologies and multiliteracy, one also needs to constantly raise the question, whose interests are these new technologies and pedagogies serving, are they serving all social groups and individuals, who is being excluded and why? We also need to raise the question both of the extent to which digital technologies and literacies are preparing students and citizens for the present and future and producing conditions for a more vibrant democratic society, or simply reproducing existing inequalities and inequities and teaching students to conform to the existing system.

Critical media and technoliteracies are thus involved in the struggle for a more just and egalitarian future and thus social theory, critique, and democratic transformation. As we battle a fierce pandemic and stunning political ignorance and corruption in the United States and elsewhere as the Trump era passes its last chaotic moments giving rise to a new and highly conflicted Biden era, we have to think anew how viruses, out of control politics, pandemics, and economies, technologies and our human lives are globally interconnected. While digital technologies and social media have been a serious part of the problem in the just ending Trump era, we need to envisage and create solutions and to see that we ourselves and our digital devices can become part of the solution.

References

Adams, J. (1998). *The next world war*. New York: Simon and Schuster.

Agel, J. (1970). *The making of Kubrick's 2001*. New York: Signet.

Alford, C. F. (1985). *Science and the revenge of nature: Marcuse and Habermas*. Gainesville: University Press of Florida.

Altermann, E. (2000). *Sound and fury. The making of the punditocracy*. Ithaca: Cornell University Press.

Altermann, E. (2003). *What liberal media? The truth about BIAS and the news*. New York: Basic Books.

Amann, J. M. (2007). *Fair and balanced, my ass!: An unbridled look at the bizarre reality of fox news*. Washington, DC: Nation Books.

Amoore, L. (Ed.). (2005). *The global resistance reader*. New York: Routledge.

Antonio, R. J., & Kellner, D. (1992). Communication, democratization, and modernity: Critical reflections on Habermas and Dewey. *Habermas, Pragmatism, and Critical Theory, Special Section of Symbolic Interaction,15*(3), 277–298

Appadurai, A. (1996). *Modernity at large: Cultural dimensions of globalization*. Minneapolis: University of Minnesota Press.

Apple, M. (1992). Is new technology part of the solution or part of the problem in education. In J. Beynon & H. Mackay (Eds.), *Technological literacy and the curriculum*. (pp. 105–124). London: Falmer Press.

Armitage, J. (Ed.). (2001). *Virilio live: Selected interviews*. London: Sage.

Arnett, P. (1994). *Live from the battlefield*. New York: Simon and Schuster.

Aronowitz, S. (1985). Why should Johnny read? *The Village Voice Literary Supplement*, (May). 13.

Arquilla, J., & Ronfeldt, D. (1996). *The advent of network*. Santa Monica: Rand Corporation.

Bagdikurai, B. (2004). *The new media monopoly*. Boston: Beacon.

Baker, C. E. (2007). *Media concentration and democracy: Why ownership matters*. New York: Cambridge University Press.

Barber, B. (1984). *Strong democracy. Participatory politics for a new age*. Berkeley: University of California Press.

© The Editor(s) (if applicable) and The Author(s), under exclusive license
to Springer Fachmedien Wiesbaden GmbH, part of Springer Nature 2021
D. Kellner, *Technology and Democracy: Toward A Critical Theory of Digital
Technologies, Technopolitics, and Technocapitalism*, Medienkulturen im
digitalen Zeitalter, https://doi.org/10.1007/978-3-658-31790-4

Bardini, T. (2000). *Bootstrapping: Douglas Engelbart, coevolution, and the origins of personal computing*. Palo Alto: Stanford University Press.

Barnouw, E. (1990). *Tube of plenty: The evolution of American television*. Oxford: Oxford University Press.

Barth, H. (1976). *Truth and ideology*. Berkeley: University of California Press.

Baudrillard, J. (1993). *Symbolic exchange and death*. London: Sage.

Baudrillard, J. (1997). Interview: Jean Baudrillard. *Dazed and Confused*, June, 80–83.

Bauer, A. J. (2012). *Occupying wall street: The inside story of an action that changed America*. Chicago: Haymarket Books.

Bauman, Z. (2000). *Liquid modernity*. Cambridge: Polity.

Baumgartner, J. C., & Morris, J. S. (Eds.). (2007). *Laughing matters: Humor and American politics in the media age*. London: Routledge.

Béja, J.-P. (2013). *The impact of China's 1989 Tiananmen Massacre*. London: Routledge.

Bell, D. (1973). *The Coming of Post-Industrial Society: A Venture in Social Forecasting*. New York: Basic Books, 1973.

Bell, D. (1976). *The coming of post-industrial society*. New York: Basic Books.

Benedikt, M. (Ed.). (1991). *Cyberspace: First steps*. Cambridge: MIT Press.

Benjamin, W. (1999). The artist as producer. In W. Benjamin (Ed.), *Collected Writings* (Vol. II, pp. 134–152). Cambridge: Harvard University Press.

Berlet, C. (1998). Who is mediating the storm? Right-wing alternative information networks. In L. Kintz & J. Lesage (Eds.), *Media, culture and the religious right*. (pp. 249–274). Minneapolis: University of Minnesota Press.

Berman, M. (1982). *All that is solid melts into air*. Baltimore: Penguin Books.

Besser, H. (1993). Education as marketplace. In R. Muffoletto & N. Knupfer (Eds.), *Computers in education: Social, historical, and political perspectives*. Cresskill: Hampton Press.

Best, S., & Kellner, D. (1991). *Postmodern theory: Critical interrogations*. London and New York: MacMillan and Guilford Press.

Best, S., & Kellner, D. (1997). *The postmodern turn*. New York: Guilford Press.

Best, S., & Kellner, D. (2001). *The postmodern adventure*. New York: Guilford Press.

Bizony, P. (2001). *2001 filming the future*. London: Sidgwick and Jackson.

Boggs, C. (2004). *Imperial delusions: American militarism and endless war*. New Jersey: Rowman & Littlefield.

Boggs, C. (2016). *Origins of the warfare state: World war II and the transformation of American politics*. New York: Routledge.

Borgmann, A. (1994). *Across the postmodern divide*. Chicago: University of Chicago Press.

Bowers, C. A. (2000). *Let them eat data: How computers affect education, cultural diversity, and the prospects of ecological sustainability*. Athens: University of Georgia Press.

Braverman, H. (1972). *Labor and monopoly capital*. New York: Monthly Review Press.

Brecher, J., & Costello, T. (1994). *Global village or global pillage: Economic reconstruction from bottom up*. Boston: South End Press.

Brecher, J., Costello, T., & Smith, B. (2000). *Globalization from below*. Boston: South End Press.

Brenner, J. (1994). *Internationalist labor communication by computer network: The United States, Mexico and Nafta*. Unpublished paper.

Breton, A. (1972). *Manifestoes of surrealism*. Ann Arbor: University of Michigan Press.

Brock, D. (2004). *The republican noise machine: Right-wing media and how it corrupts democracy.* New York: Crown.

Brock, D., Rabin-Havt, A., & Matters, M. (2012). *The fox effect: How Roger Ailes turned a network into a propaganda machine.* New York: Anchor Books.

Bromley, H., & Apple, M. (Eds.). (1998). *Education/technology/power: Educational computing as social practice.* Albany: State University of New York Press.

Brook, J., & Boal, I. A. (Eds.). (1995). *Resisting the virtual life.* San Francisco: City Lights.

Brundtland, G. H., et al. (1987). *Our common future: Report of the world commission on environment and development.* Oxford: Oxford University Press.

Bump, P., & Parker, A. (2020, April 27). Trump fills briefings with attacks, boasts, but little empathy. *Washington Post.* https://www.theday.com/article/20200427/NWS21/200 429558. Accessed 26 Jan 2021.

Burbach, R. (2001). *Globalization and postmodern politics. From Zapatistas to high-tech Robber Barons.* London: Pluto Press.

Burbules, N. C. (1997). Rhetorics of the web: Hyperreading and critical literacy. In I. Snyder (Ed.), *Page to screen: Taking literacy into the electronic era.* (pp. 102–122). Sydney: Allen & Unwin.

Burbules, N., & Callister, T. (1996). Knowledge at the crossroads. *Educational Theory., 46*(1), 23–34

Burbules, N., & Callister, T. (2000). *Watch IT: The risks and promises of information technology.* Boulder: Westview.

Burke, E. (2008). *A philosophical enquiry into the origin of our ideas of the sublime and beautiful.* Oxford: Oxford University Press.

Bus, A. G., & Neuman, S. B. (2008). *Multimedia and literacy development.* New York: Routledge.

Carr, N. (2010). *The shallows: What the internet is doing to our brains.* New York: W. W. Norton.

Castells, M. (1996). *The rise of the network society.* Oxford: Blackwell.

Castells, M. (1997). *The power of identity.* Oxford: Blackwell.

Castells, M. (1998). *End of millennium.* Oxford: Blackwell.

Castells, M. (2001). *The internet galaxy: Reflections on the internet, business and society.* Oxford: Oxford University Press.

Chion, M. (2001). *Kubrick's cinematic odyssey.* London: BFI Publishing.

Cho, K. D., Kellner, D., Lewis, T. E., & Pierce, C. (Eds.). (2009). *Marcuse's challenge to education.* Lanham: Rowman and Littlefield.

Chomsky, N. (2012). *Occupy: Reflections on class war, rebellion and solidarity. Occupied media pamphlet.* New York: Zuccotti Park Press.

Clarke, A. C. (1982). *2001: A space odyssey.* New York: Signet. (First published in 1968).

Clarke, A. C. (1972). *The lost worlds of 2001.* New York: Signet.

Cleaver, H. (1994). The Chiapas uprising. *Studies in Political Economy, 44,* 141–157

Cleaver, H. (2000). *Reading capital politically.* (2nd ed.). Leeds: Antithesis.

Cockburn, A., Clair, J. S., & Sekula, A. (2000). *5 days that shook the world.* London: Verso.

Compaine, B., & Gomery, D. (2000). *Who owns the media?: Competition and concentration in the mass media industry.* New York: Routledge.

Conason, J. (2003). *Big lies: The right-wing propaganda machine and how it distorts the truth.* New York: Thomas Crown Books.

Cooley, M. (1987). *Architect or Bee? The Human Price of Technology.* London: Hogarth.

Cope, B., & Kalantzis, M. (Eds.). (2000). *Multiliteracies: Literacy learning and the design of social futures.* New York: Routledge.

Coppola, N. W. (1999). Greening the technological curriculum: A model for environmental literacy. *Journal of Technology Studies,25*(2), 39–46

Corn, D., & Isakoff, M. (2018). *Russian Routlette.* New York: Twelve.

Cornfield, M. (2010). Game-changers: New technologies and the 2008 presidential election. In L. J. Sabato (Ed.), *The year of Obama: How Barack Obama won the white house.* (pp. 205–230). New York: Longman.

Couldry, N., & Curran, J. (Eds.). (2003). *Contesting media power. Alternative media in a networked world.* Lanham: Rowman and Littlefield.

Courts, P. L. (1998). *Multicultural literacies: Dialect, discourses, and diversity.* New York: Lang.

Danaher, K., & Burbach, R. (2000). *Globalize this! Monroe.* Maine: Common Courage Press.

Daniels, J. (1998). *Rubber-necking white supremacist terror: Critical race theory and 'hate' speech in cyberspace.* Paper presented at the 1998 American Sociology Association Meetings, San Francisco.

Davis, M. (1990). *City of quartz: Excavating the future in Los Angeles.* London: Verso.

Debord, G. (2002). *Society of the spectacle.* Detroit: Black and Red. (First published in 1968).

de la Landa, M. (1991). *War in the age of intelligent machines.* New York: Zone Books.

Der Derian, J. (Ed.). (1998). *The Virilio Reader.* Oxford: Blackwell.

Dewey, J. (1916). *Democracy and education: An introduction to the philosophy of education.* Carbondale: Southern Illinois University Press.

Downing, J. (2001). *Radical media.* (2nd ed.). Boston: South End Press. (First published in 1984).

Drew, J. (1998). *Global communications in the post-industrial age: A study of the communications strategies of U.S. labor organizations.* PhD Dissertation, Graduate School, University of Texas.

Dyer-Witheford, N. (1999). *Cyber-Marx. Cycles and circuits of struggle in high-technology capitalism.* Urbana: University of Illinois Press.

Dyrenfurth, M. J. (1991). Technological literacy synthesized. In M. J. Dyrenfurth & M. R. Kozak (Eds.), *Technological literacy.* (pp. 138–186). Peoria: McGraw-Hill.

Edwards, P. N. (1996). *The closed world: Computers and the politics of discourse in cold war America.* Cambridge: MIT Press.

Eisenstein, Z. (1998). *Global obscenities: Patriarchy, capitalism, and the lure of cyberfantasy.* New York: New York University Press.

Ellul, J. (1964). *The technological society.* New York: Vintage Books.

Falk, R. (1999). *Predatory globalization.* London: Blackwell.

Fanon, F. (1967). *For a dying colonialism.* New York: Grove Press.

Feenberg, A. (1991). *Critical theory of technology.* New York: Oxford University Press.

Feenberg, A. (1995). *Alternative modernity.* Berkeley: University of California Press.

Feenberg, A. (1999). *Questioning technology.* New York: Routledge.

Feenberg, A. (2002). *Transforming technology: A critical theory revisited.* Oxford: Oxford University Press.

Feenberg, A. (2017). *Technosystem: The social life of reason.* Cambrdige: Harvard University Press.

Fiske, J. (1994). *Media matters*. Minneapolis: University of Minnesota Press.

Fitzgerald, F. (2001). *Way out there in the blue: Reagan, star wars, and the end of the cold war*. New York: Touchstone Books.

Foucault, M. (1977). *Language, counter-memory, practice*. Ithaca: Cornell University Press.

Fredericks, H. (1994). *North American NGO networking against NAFTA: The use of computer communications in cross-border coalition building*. XVII International Congress of the Latin American Studies Association.

Freire, P. (1972). *Pedagogy of the oppressed*. New York: Herder & Herder.

Freire, P., & Macedo, D. (1987). *Literacy: Reading the word and the world*. Westport: Bergin & Garvey.

Friedman, T. (1999). *The Lexus and the olive tree*. New York: Farrar Straus Giroux.

Friedman, T. (2005). *The world is flat*. New York: Farrar, Straus and Giroux.

Fukuyama, F. (1992). *The end of history and the last man*. New York: Free Press.

Gabler, N. (2000). *Life, the movie: How entertainment conquered reality*. New York: Vintage.

Galloway, S. (2017). The four: The hidden DNA of Amazon, Apple, Facebook, and Google. New York: Portfolio.

Galloway, S. (2018). *The four*. New York: Portfolio Illustrated.

Gardner, H. (1983). *Frames of mind*. New York: Basic Books Inc.

Gates, B. (1995). *The road ahead*. New York: Viking.

Gates, B. (1999). *Business @The speed of thought*. New York: Viking.

Gee, J. P. (2000). Identity as an analytic lens for research in education. *Review of Research in Education,25*, 99–125

Gelernter, D. (1992). *Mirror worlds*. New York: Oxford University Press.

Gelmis, J. (1970). *The film director as superstar*. New York: Doubleday & Company.

Gennaro, S., & Miller, B. (Eds.). (2021). *Young people and social media: Contemporary children's culture in digital space(s)*. Wilmington: Vernon Press.

Gibson, W. (1984). *Neuromancer*. New York: ACE.

Gitlin, T. (1983). *The whole world's watching*. Berkeley: University of California Press.

Gleick, J. (2011). *The information: A history, a theory, a flood*. New York: Pantheon Books.

Goldman, R., & Papson, S. (1999). *Nike culture*. London: Sage.

Goodnow, T. (Ed.). (2011). *The daily show and rhetoric: Arguments, issues, and strategies*. Lexington: Lexington.

Gore, A. (1994). *Remarks prepared for delivery*. Speech at the International Telecommunications Union (Buenos Aires:). https://www.itu.int/itudoc/itu-d/wtdc/wtdc1994/speech/gore_ww2.doc. Accessed 5 Jan 1998.

Gray, C. H. (1989). The cyborg soldiers: The US military and the post-modern warrior. *Levidow and Robins,1989*, 159–178

Gray, C. H. (1997). *Postmodern war*. New York: Guilford Press.

Gray, C. H. (2001). *Cyborg citizen*. London: Routledge.

Gray, J., Thompson, E., & Jones, J. P. (2009). *Satire TV: Politics and comedy in the post-network era*. New York: NYU Press.

Greenwald, G. (2014). *No Place to Hide: Edward Snowden, the NSA, and the U.S. Surveillance State*. New York: Metropolitan Books.

Greider, W. (1998). *Fortress America: The American military and the consequences of peace*. New York: PublicAffairs.

Grossman, K. (2001). *Weapons in space*. New York: Seven Stories Press.

Grossman, E. (2004). *High-tech wasteland*. Orion. July/August. https://www.oriononline.org/pages/om/04-4om/Grossman.html. Accessed 5 Jan 1998.

Grubb, W. N. (1996). The new vocationalism - What it is, what it could be. *Phi Delta Kappan,77*(8), 535–546

Gulati, G. J. (2010). No laughing matter: The role of new media in the 2008 election. *Sabato,2010*, 187–204

Gunther, J., Kantra, S., & Langreth, R. (1994). Digital Warrior. *Popular Science,245*(3), 60–65

Gutiérrez, K. D., & Rogoff, B. (2003) Cultural Ways of Learning: Individual Traits or Repertoires of Practice. *Educational Researcher*, June 1, 2003 at https://doi.org/10.3102/0013189X032005019. Accessed 23 June 2020.

Habermas, J. (1970). *Toward a rational society*. Boston: Beacon Press.

Habermas, J. (1991). *The structural transformation of the public sphere: An inquiry into a category of bourgeois society*. Cambridge: MIT Press.

Hafner, K., & Markoff, J. (1991). *Cyberpunk. Outlaws and hackers on the computer frontier*. New York: Simon and Schuster.

Halberstam, D. (1979). *The powers that be*. New York: Knopf.

Halberstam, D. (2006). Teaching critical media literacies: Theory, praxis and empowerment. *InterActions: UCLA Journal of Education and Information Studies, 2*(1), Article 6. https://repositories.cdlib.org/gseis/interactions/vol2/iss1/6

Hammer, R. (1995). Strategies for media literacy. In P. McLaren, R. Hammer, D. Sholle, & S. Reilly (Eds.), *Rethinking media literacy: A critical pedagogy of representation*. (pp. 225–235). New York: Lang.

Hammer, R., & Kellner, D. (2001). Multimedia pedagogy and multicultural education for the new millennium. *Current Issues in Education, 4*(2). https://cie.ed.asu.edu/volume4/number2/. Accessed 22 Jan 2002.

Haraway, D. (1990). *Simians, cyborgs, and women: The reinvention of nature*. New York: Routledge.

Harding, S. (Ed.). (2004). *The feminist standpoint theory reader: Intellectual and political controversies*. New York: Routledge.

Harding, L. (2014). *The Snowden Files: The Inside Story of the World's Most Wanted Man*. New York: Metropolitan Books.

Hardt, M. (2002). Porto allegre: Today's bandung? *New Left Review, 14*. https://newleftreview.org/issues/ii14/articles/michael-hardt-porto-alegre-today-s-bandung. Accessed 22 January 2021.

Hardt, M., & Negri, A. (2000). *Empire*. Cambridge: Harvard University Press.

Hardt, M., & Negri, A. (2004). *Multitude. War and democracy in the age of empire*. New York: Penquin Press.

Harvey, D. (1989). *The condition of postmodernity*. Oxford: Blackwell.

Harvey, N. (1998). *The Chiapas rebellion and the struggle for land and democracy*. Durham: Duke University Press.

Hayden, M. (1989). What is technological literacy? *Bulletin of Science, Technology and Society,119*, 220–233

Heidegger, M. (2013). *The question concerning technology, and other essays*. New York: Harper Perennial Modern Classics.

Heinonen, S., Jokinen, P., & Kaivo-oja, J. (2001). The ecological transparency of the information society. *Futures,33*, 319–337

Herman, E., & Chomsky, N. (1988). *Manufacturing consent: The political economy of the mass media*. New York: Pantheon.

Hettena, S. (2018). *Trump/Russia: A definitive history*. New York: Melville House.

Hickman, L. (2001). *Philosophical tools for technological culture*. Bloomington: Indiana University Press.

Hill, A. (1998). *Speaking truth to power*. New York: Anchor Books.

Hill, K. A., & Hughes, J. E. (1998). *Cyberpolitics. Citizen activism in the age of the internet*. Lanham: Rowman & Littlefield.

Holbrook, J., Mukherjee, A., & Varma, V. S. (Eds.). (2000S). *Scientific and technological literacy for all. UNESCO and International Council of Associations for Science Education*. Delhi, India: Center for Science Education and Communication.

Holt, J. (Ed.). (2007). *The daily show and philosophy: Moments of Zen in the art of fake news*. New York: Wiley-Blackwell.

Horkheimer, M., & Adorno, T. (1972). *Dialectic of enlightenment*. New York: Continuum.

Huntington, S. (1996). *The clash of civilizations and the remaking of world order*. New York: Simon and Schuster.

Ignatieff, M. (2000). *Virtual war. Kosovo and beyond*. New York: Holt.

Ihde, D. (2010). *Heidegger's technologies: Postphenomenological perspectives*. New York: Fordham University Press.

Illich, I. (1973). *Tools for conviviality*. New York: Harper and Row.

Jameson, F. (1984). Postmodernism–The cultural logic of late capitalism. *New Left Review, 146*, 53–93

Jameson, F. (1991). *Postmodernism, or, the cultural logic of late capitalism*. Durham: Duke University Press.

Jamieson, K. H. (2018). *Cyberwar: How Russian hackers and trolls helped elect a president: What we don't, can't, and do know*. Oxford: Oxford University Press.

Jegede, O. (2002). *An integrated ICT-support for ODL in Nigeria: The vision, the mission and the journey so far."* Paper prepared for the LEARNTEC-UNESCO 2002 Global Forum on Learning Technology. Karlsruhe, Germany.

Jenkins, E. W. (1997). Technological literacy: Concepts and constructs. *Journal of Technology Studies., 23*(1), 2–5

Johnson, S. (1997). *Interface culture: How new technology transforms the way we create and communicate*. New York: Harper and Row.

Joy, B. (2000). Why the future doesn't need us. *Wired, 8*(07), 238–246

Kaczynski, T. (1995). *The Unabomber manifesto: Industrial society & its future*. Berkeley: Jolly Roger Press.

Kahn, R., & Kellner, D. (2005). Oppositional politics and the internet: A critical/reconstructive approach. *Cultural Politics, 1*(1), 75–100

Kahn, R., & Kellner, D. (2006). Reconstructing technoliteracy: A multiple literacies approach. In J. R. Dakers (Ed.), *Defining technological literacy*. (pp. 253–274). New York: Palgrave Macmillan.

Kahn, R., & Kellner, D. (2007). Globalization, technopolitics, and radical democracy. In L. Dahlberg & E. Siapera (Eds.), *Radical democracy and the internet. Interrogating theory and practice*. (pp. 17–36). Houndmills: Palgrave.

Kang, J. (2014). *Walter Benjamin and the media: The spectacle of modernity*. Cambridge: Polity.

Kaufmann, V. (2006). *Guy Debord. Revolution in the service of poetry*. Minneapolis: University Of Minnesota Press.

Kellner, D. (1977). Capitalism and human nature in Adam Smith and Karl Marx. In J. Schwartz (Ed.), *The subtle anatomy of capitalism*. (pp. 66–86). Santa Monica: Goodyear Publishing Company.

Kellner, D. (1989a). *Critical theory, Marxism and modernity*. Cambridge: Polity.

Kellner, D. (1989b). *Jean Baudrillard: From Marxism to postmodernism and beyond*. Cambridge: Polity.

Kellner, D. (1990). "Critical Theory and Ideology Critique," In R. Roblin (Ed.), *Critical Theory and Aesthetics* (pp. 85–123). Lewiston: The Edwin Mellen Press, 1990.

Kellner, D. (1992). *The Persian Gulf TV war*. Boulder: Westview.

Kellner, D. (1995). *Media culture. Cultural studies, identity and politics between the modern and the postmodern* (2nd revised edition 2020). London: Routledge.

Kellner, D. (1997). Intellectuals, the new public spheres, and technopolitics. *New Political Science,41–42* (1997), 169–188.

Kellner, D. (2002). The X-files and conspiracy: A diagnostic critique. In P. Knight (Ed.), *Conspiracy nation. The politics of paranoia in postwar America*. (pp. 205–232). New York: New York University Press.

Kellner, D. (2003a). "The Politics of Spectacle Culture in the Contemporary United States,". In B. Georgi-Findlay & H.-U. Mohr (Eds.), *Millenial Perspectives. Lifeworlds and Utopias* (pp. 99–116). Heidelberg: Universitatsverlag (Winter 2003).

Kellner, D. (2003b). "Globalization, Technopolitics, and Revolution". In J. Foran (Ed.), *The Future of Revolutions. Rethinking Radical Change in the Age of Globalization* (pp. 180–194). London: Zed Books.

Kellner, D. (2004). Technological transformation, multiple literacies, and the re-visioning of education. *E-Learning,1*(1), 9–37

Kellner, D. (2005). "The Conflicts of Globalization and Restructuring of Education,". In M. J. Peters (Ed.), *Education, globalization, and the state in the age of terrorism* (pp. 31–70). Boulder: Paradigm Press, 2005.

Kellner, D. (2009). *Media Spectacle and the 2008 Presidential Election, Cultural Studies<> Critical Methodologies* (Vol. 9, Nr. 6, pp. 707–716) (December 2009).

Kellner, D. (2010a). Barack Obama, the power elite, and media spectacle. *American Study Institute,11*, 25–70

Kellner, D. (2010b). *Cinema wars: Hollywood film and politics in the Bush/Cheney era*. Malden: Blackwell.

Kellner, D. (2012). *Media spectacle and insurrection, 2011: From the Arab uprisings to occupy everywhere*. London: Continuum.

Kellner, D. (2014) "Dialectics of Globalization: From Theory to Practice," In S. Dasgupta & P. Kivisto (Eds.), *Postmodernism in a Global Perspective* (pp. 3–29). London and Delhi: Sage Books.

Kellner, D. (2015). Barack Obama, media spectacle, and celebrity politics. *Marshall and Redmond,2015*, 114–134

Kellner, D. (2017) "Andrew Feenberg, Critical Theory, and the Critique of Technology." In D. P. Arnold & A. Michel (Eds.), *Critical Theory and the Thought of Andrew Feenberg* (pp. 263–284). Switzerland: Palgrave Macmillan, 2017.

Kellner, D., & Kim, G. (2009). YouTube, critical pedagogy, and media activism. *The Review of Education/pedagogy/cultural Studies,32*(1), 3–36

Kellner, D., & Ryan, M. (1988). *Camera politica: The politics and ideology of contemporary Hollywood film.* Bloomington: Indiana University Press.

Kellner, D., & Satchel, R. M. (2020). Resisting youth: From occupy through black lives matter to the trump resistance. In S. Steinberg & B. Down (Eds.), *The SAGE handbook of critical pedagogies.* SAGE. https://doi.org/10.4135/9781526486455. Accessed 12 Jan 2021.

Kellner, D., & Share, J. (2019). *The critical media literacy guide: Engaging media and transforming education.* Rotterdam: Brill-Sense.

Kelly, K. (1995). *Out of control: The new biology of machines, social systems, & the economic world.* New York: Basic Books.

Kinder, M. (1993). *Playing with power in movies, television and video games: From Muppet babies to teenage mutant ninja turtles.* Berkeley: University of California Press.

Klein, N. (2008). *The shock doctrine: The rise of disaster capitalism.* New York: Picador.

Knabb, K. (2007). *Situationist international anthology.* London: Bureau of Public Secrets.

Kolker, R. (Ed.). (2006). *Stanley Kubrick's 2001: A space Odyssey: New essays.* Oxford: Oxford University Press.

Kovalchuk, J. (Ed.). (2021). *Post-industrial society: The choice between innovation and tradition.* London: Palgrave-MacMillan.

Kraut, R., Mukhopadhyay, T., Szczypula, J., Kiesler, S., & Scherlis, W. (1998). Communication and information: Alternative uses of the Internet in households. In *Proceedings of the CHI 98* (pp. 368–383). New York: ACM.

Kress, G. (1997). Visual and verbal modes of representation in electronically mediated communication: The potentials of new forms of text. In I. Snyder (Ed.), *Page to screen: taking literacy into the electronic era.* (pp. 53–79). Sydney: Allen & Unwin.

Kroker, A., & Kroker, M. (1997). *Digital delirium.* London: Palgrave Macmillan.

Kroker, A., & Weinstein, M. (1994). *Data crash.* New York: St. Martin's Press.

Kurzweil, R. (1985). *The singularity is near.* New York: Penguin Group.

Laclau, E., & Mouffe, C. (1985). *Hegemony and socialist strategy.* London: Verso Books.

Lakoff, G. (2003). *Metaphors we live by.* Chicago: University of Chicago Press.

Landow, G. P. (2006). *Hypertext 3.0: Critical theory and new media in an era of globalization.* (3rd ed.). Baltimore: Johns Hopkins University Press.

Lankshear, C., & Knobel, M. (2000). Mapping postmodern literacies: A preliminary chart. *The Journal of Literacy and Technology, 1*(1). Online at: https://www.literacyandtechnology.org/v1n1/lk.html.

Lankshear, C., & Snyder, I. (2000). *Teachers and technoliteracy: Managing literacy, technology and learning in schools.* Sydney: Allen & Unwin.

Lash, S., & Urry, J. (1987). *The end of organized capitalism.* Cambridge: Polity Books.

Lessard, B., & Baldwin, S. (2000). *Net slaves: True tales of working the web.* New York: McGraw Hill.

Levidow, L., & Robins, K. (Eds.). (1989). *Cyborg worlds. The military information society.* London: Free Association Books.

Levinson, P. (1997). *The soft edge.* New York: Taylor & Francis Inc.

Levy, S. (1985). *Hackers.* New York: Dell Books.

Lewis, T., & Gagel, C. (1992). Technological literacy: A critical analysis. *Journal of Curriculum Studies.,24*(2), 117–138

Lievrouw, L. (2011). *Alternative and activist new media*. Cambridge: Polity Press.

Lonsdale, M., & McCurry, D. (2004). *Literacy in the new millennium*. Adelaide: NCVER.

Luke, A., & Luke, C. (2002). Adolescence lost/childhood regained: On early intervention and the emergence of the techno-subject. *Journal of Early Childhood Literacy,1*(1), 91–120

Lyon, D. (1988). *The Information Society: Issues and Illusions*. Polity.

Lyon, D. (2001). *Surveillance society*. Buckingham: Open University Press.

Lyon, D. (2018). *The Culture of Surveillance: Watching as a Way of Life*. Cambridge and Oxford, UK: Polity.

Lyotard, J.-F. (1984). *The postmodern condition: A report on knowledge*. Minneapolis: University Of Minnesota Press.

Mander, J. (1978). *Four arguments for the elimination of television*. New York: William Morrow.

Mander, J., & Goldsmith, E. (1996). *The case against the global economy*. San Francisco: Sierra Club Books.

Marcuse, H. (1964). *One-dimensional man*. Boston: Beacon.

Marcuse, H. (1969). *An essay on liberation*. Boston: Beacon.

Marcuse, H. (1973). The foundations of historical materialism. In *Studies in critical philosophy*. Boston: Beacon (First published in 1932).

Marshall, P. D., & Redmond, S. (Eds.). (2015). *A companion to celebrity*. Edited by P. David. Malden: Wiley-Blackwell, pp. 114–134.

Marx, K. (1973). *Grundrisse*. Baltimore: Penquin Books.

Marx, K. (1975). *Collected works*. (Vol. I)New York: International Publishers.

Marx, K. (1978). *The Marx-Engels reader*. Second edition, Edited by Robert C. Tucker. New York: Norton.

Marx, K. (2007). *Economic and philosophic manuscripts of 1844*. New York: Dover Publications.

Marx, K., & Engels, F. (1975). *The German ideology. Collected works* (Vol. V). New York: International Publishers (First published in 1845).

Marx, L. (2000). *The machine in the garden*. Oxford: Oxford University Press.

Masuda, Y. (1980). The information society as post-industrial society. Washington, DC: World Future Society at Google Books at https://books.google.com/books?id=ynkmIx F1G3AC&printsec=frontcover#v=onepage&q&f=false. Accessed 13 Jan 2021.

McChesney, R. (1993). *Telecommunications, mass media, and democracy: The battle for the control of U.S. broadcasting, 1928–1935*. New York: Oxford University Press.

McChesney, R. (1997). *Corporate media and the threat to democracy*. New York: Seven Stories Press.

McChesney, R. (2000). *Rich media, poor democracy*. New York: New Press.

McChesney, R. (2004). *The problem of the media: U.S. communication politics in the twenty-first century*. New York: Monthly Review Books.

McChesney, R. (2013). *Digital disconnect: How capitalism is turning the internet against democracy*. New York: New Press.

McChesney, R. W., & Nichols, J. (2012). *The death and life of American journalism: The media revolution that will begin the world again*. Washington, DC: Nation Books.

McDonough, T. (Ed.). (2002). *Guy debord and the situationist international*. Cambridge: MIT Press.

McLaren, P., Hammer, R., Sholle, D., & Reilly, S. (1995). *Rethinking media literacy: A critical pedagogy of representation.* New York: Lang.

McLuhan, M. (1964). *Understanding media.* New York: Basic Books.

Melman, S. (1965). *Our depleted society.* New York: Holt.

Melman, S. (1970). *Pentagon capitalism: The political economy of war.* New York: McGraw-Hill.

Miller, J., Engelberg, S., & Broad, W. J. (2001). *Germs: Biological weapons and America's secret war.* New York: Simon & Schuster.

Miller, M. C. (2005). *Fooled again: How the right stole the 2004 election & why they'll steal the next one too (unless we stop them).* New York: Basic Books.

Mills, K. A. (2015). *Literacy theories for the digital age: Social, critical, multimodal, spatial, material and sensory lenses.* Bristol: Multilingual Matters.

Moody, K. (1998). *An injury to no one.* London: Verso.

Mosco, V. (2004). *The digital sublime: Myth, power, and cyberspace.* Cambridge: MIT Press.

Muto, J. (2013). *An atheist in the FOXhole: A liberal's eight-year Odyssey inside the heart of the right-wing media.* New York: Dutton Adult.

Nance, M. (2016). *The plot to hack America. How Putin's cyberspies and wikileaks tried to steal the 2016 election.* New York: Skyhorse Publishing.

Naremore, J. (2007). *On Kubrick.* London: BFI.

National Commission on Excellence in Education. (1983). A nation at risk: The imperative for educational reform. Washington, DC: U.S. Government Printing Office.

National Telecommunications & Information Administration. (2002). A nation online: How Americans are expanding their use of the internet. https://www.ntia.doc.gov/ntiahome/dn/nationonline_020502.htm. Accessed 20 Jan 2009.

Negri, A. (1989). *The politics of subversion. A manifesto for the twenty-first century.* Cambridge: Polity.

Negroponte, N. (1996). *Being Digital.* New York: Vintage.

Nicholas, N. (1996). *Being digital.* New York: Vintage.

New London Group. (1996). A pedagogy of multiliteracies: Designing social futures. *Harvard Educational Review,66,* 60–92.

Noble, D. (1977). *America by design.* New York: Oxford Books.

Noble, D. (1984). *Forces of production.* New York: Knopf.

Noble, D. (1995). *Progress without people.* Toronto: Between the Lines.

Noble, D. (1997). *Religion of technology.* New York: Oxford Books.

Noble, D. (1998). Digital diploma mills: The automation of higher education. *First Monday,3*(1–5). https://firstmonday.org/ojs/index.php/fm/issue/view/88. Accessed 13 Jan 2021.

Noble, S. U. (2018). *Algorithms of oppression: How search engines reinforce racism.* New York: New York University Press.

Offe Claus. (1985). *Disorganized capitalism: Contemporary transformations of work and politics.* London: Polity.

Olson, P. (2013). *We are anonymous: Inside the hacker world of LulzSec, anonymous, and the global cyber insurgency.* Boston: Little, Brown.

Orwell, G. (1950). *1984.* New York: Signet.

Parenti, M. (1986). *Inventing reality.* New York: Saint Martin's.

Park, L.-H., & Pellow, D. N. (2004). Racial formation, environmental racism, and the emergence of Silicon Valley. *Ethnicities,4*(3), 403–424

Pearson, G., & Thomas Young, A. (2002). *Technically speaking: Why all Americans need to know more about technology*. Washington, DC: National Academies Press.

Petrina, S. (2000). The politics of technological literacy. *International Journal of Technology and Design Education,10*(2), 181–206

Plepys, A. (2002). The grey side of ICT. *Environmental Impact Assessment Review,22*, 509–523

Plotnick, E. (1999). *Information literacy*. N.Y.: ERIC Clearinghouse on Information and Technology, New York: Syracuse University.

Poster, M. (1991). *The mode of information: Post-structuralism and social contexts*. Cambridge: Polity.

Poster, M. (1995). *Cyberdemocracy: Internet and the public sphere*. www.hnet.uci.edu/mpo ster/writings/democ.html. Accessed 22 May 1998.

Poster, M. (2006). *Information please: Culture and politics in the age of digital machines*. Durham: Duke University Press.

Postman, N. (1985). *Amusing ourselves to death*. New York: Viking-Penguin.

Postman, N. (1992). *Technopolis: The surrender of culture to technology*. New York: Random House.

Power, C. (1987). Science and technology towards informed citizenship. *Castme Journal.,7*(3), 5–18

Quammen, D. (2013). *Spillover: Animal infections and the next human pandemic*. New York: Norton.

Rash, W. (1997). *Politics on the net. Wiring the political process*. New York: W.H. Freeman.

Rassool, N. (1999). *Literacy for sustainable development in the age of information*. London: Multilingual Matters Ltd.

Ritzer, G. (1996). *The McDonaldization of society*. Thousand Oaks: Pine Forge Press (First published in 1993).

Ritzer, G., & Dean, P. (2015). *Globalization: A basic text* (2nd Edn.). New Jersey: Wiley-Blackwell.

Robertson, R. (1991). *Globalization*. London: Sage.

Robins, K., & Webster, F. (1999). *Times of the technoculture*. London: Routledge.

Rochlin, E. (1997). *Trapped in the net*. Princeton: Princeton University Press.

Rossi, I. (Ed.). (2007). *Frontiers of globalization*. New York: Springer.

Rozack, T. (1969). *The making of a counter culture: Reflections on the technocratic society and its youthful opposition*. Berkeley: University of California Press.

Roszak, T. (1994). *The cult of information: A neo-luddite treatise on high-tech, artificial intelligence, and the true art of thinking*. Berkeley: University of California Press.

Royal Society. (1985). The public understanding of science. London: Royal Society.

Sabato, L. J. (Ed.). (2010). *The year of Obama: How Barack Obama won the White House*. New York: Longman.

Sartre, J.-P. (1965). *What is literature?* New York: Harper and Row.

Sartre, J. -P. (1970, December). Intellectuals and Revolution. *Ramparts*, IX, no. 6.

Sartre, J. -P. (1971, October 17). Interview with John Gerassi. *The New York Times Magazine*.

Sartre, J. -P. (1974). *The writings of Jean-Paul Sartre*. Edited by Michel Contat and Michel Rybalka. Evanston: Northwestern University Press.

Sartre, J.-P. (1975). *Between existentialism and Marxism.* New York: Basic Books.

Schatz, T. (2010). *The genius of the system.* Minneapolis: University Of Minnesota Press.

Schiller, D. (1999). *Digital capitalism.* Cambridge: MIT Press.

Schiller, H. (1990). Culture, Inc. New York: Oxford University Press.

Schumpeter, J. (2009). *Can capitalism survive?: Creative destruction and the future of the global economy.* New York: Harper Perennial.

Schwam, S. (Ed.). (2000). *The making of 2001: A space Odyssey. Introduction by Jay Cocks.* New York: Modern Library.

Schwartau, W. (1996). *Information warfare.* New York: Thunder's Mouth Press.

Selfe, C. L. (1999). *Technology and literacy in the twenty-first century: The importance of paying attention.* Carbondale: Southern Illinois University Press.

Semali, L., & Pailliotet, A. W. (1998). *Intermediality: Teachers' handbook of critical media literacy.* Boulder: Westview.

Shields, R. (2003). *The virtual.* New York: Routledge.

Silberman, M. (2000). *Bertolt Brecht on film and radio.* London: Metheun.

Simpson, L. C. (1995). *Technology, time, and the conversations of modernity.* New York: Routledge.

Singer, P. W., & Friedman, A. (2014). *Cybersecurity: What everyone needs to know.* Oxford: Oxford University Press.

Sklair, L. (2001). *The transnational capitalist class.* Cambridge: Blackwell.

Slouka, M. (1995). *War of the worlds.* New York: Harper and Row.

Solorzano, D. G., & Yosso, T. J. (2001). From racial stereotyping and deficit discourse toward a critical race theory in teacher education. *Multicultural Education,9*(1), 2–8

Steger, M. (2002). *Globalism. The new market ideology.* Lanham: Rowman and Littlefield.

Snowden, E. (2019). *Permanent Record.* New York: Metropolitan Books.

Snyder, I. (Ed.). (1997P). *Page to screen. Taking literacy into the electronic era.* Sydney: Allen and Unwin.

Snyder, I. (2002). *Silicon literacies: Communication, innovation and education in the electronic age.* New York: Routledge.

Sontag, S. (1966). *Against interpretation and other essays.* New York: Farrar, Straus and Giroux.

Stelter, B. (2019). *Hoax: Donald Trump, Fox News, and the Dangerous Distortion of Truth.* New York: Atria/One Signal Publishers.

Stelter, B. (2020). *Hoax: Donald Trump, Fox News, and the dangerous distortion of truth.* New York: Atria/One Signal Publishers.

Sterling, B. (1992). *The hacker crackdown. Law and disorder on the electronic frontier.* New York: Bantam.

Stewart, J., & The Writers of The Daily Show. (2004). *America (The Book): A citizen's guide to democracy inaction.* New York: Grand Central Publishing.

Stiglitz, J. E. (2002). *Globalization and its discontents.* New York: Norton.

Stoll, C. (1995). *Silicon snake oil: Second thoughts on the information highway.* New York: Doubleday.

Stone, S. (1996). *The war of desire and technology at the Close of the mechanical age.* Cambridge: MIT Press.

Stork, D. G. (1998). *Hal's legacy: 2001's computer as dream and reality.* Cambridge: MIT Press.

Street, B. (1984). *Literacy in theory and practice*. Cambridge: Cambridge University Press.

Suppes, P. (1968). Computer technology and the future of education. *Phi Delta Kappan*, April, 420–423.

Tabbi, J. (1996) *Postmodern Sublime: Technology and American Writing from Mailer to Cyberpunk*. New York: Cornell University Press.

Tinker, G. E. (1993). *Missionary conquest: The gospel and native American cultural genocide*. Pennsylvania: Fortress Press of Philadelphia.

Todd, R. D. (1991). The natures and challenges of technological literacy. In M. J. Dyrenfurth & M. R. Kozak (Eds.), *Technological literacy*. (pp. 10–27). Peoria: McGraw-Hill.

Toffler, A. (1993). *The third wave*. New York: Bantham.

Tracey, M. (1998). *Decline and fall of public service broadcasting*. Oxford: Oxford University Press.

Trend, D. (2001). *Welcome to cyberschool: Education at the crossroads in the information age*. Lanham: Rowman & Littlefield.

Turkle, S. (1995). *Life on the screen. Identity in the age of the internet*. New York: Simon & Schuster.

Turner, F. (2008). *From counterculture to cyberculture: Stewart brand, the whole earth network, and the rise of digital utopianism*. Chicago, Ill.: University of Chicago Press.

UNESCO. (1994). The project 2000+ declaration: The way forward. Paris: UNESCO.

UNESCO. (1999). Science and technology education: Philosophy of project 2000+. The association for science education. Paris: UNESCO.

United Nations. (1992). Report of the United Nations conference on environment and development. Rio de Janeiro, Brazil: UNCED.

U.S. Congress. (2001). No child left behind act of 2001. Public Law 107-110. Washington, DC.

U.S. Department of Education. (1996). *Getting America's students ready for the 21st century — Meeting the technology literacy challenge, a report to the nation on technology and education*. Washington, DC: National Education Technology Plan.

U.S. Department of Education. (2004). *Toward a new golden age in American education: How the internet, the law, and today's students are revolutionizing expectations*. Washington, DC: National Education Technology Plan.

Valencia & Solorzano. (2004). *The evolution of deficit thinking: Educational thought and practice*. Edited by Richard R. Valencia.

Vidal, J. (1997). *McLibel: Burger culture on trial*. London: Macmillan.

Virillo, P. (1986). *Speed and politics: An essay on dromology*. New York: Semiotext(e). (First published in 1977).

Virillo, P. (1989). *War and cinema: The logistics of perception*. London: Verso.

Virillo, P. (1990). *Popular defense and ecological struggles*. New York: Semiotext(e).

Virillo, P. (1991a). *The aesthetics of disappearance*. New York: Semiotext(e).

Virillo, P. (1991b). *Lost dimension*. New York: Semiotext(e).

Virillo, P. (1994). *The vision machine*. Bloomington: Indiana University Press.

Virillo, P. (1994). *Bunker archaeology*. New York: Princeton University Press. (First published in 1975).

Virillo, P. (1995). *The art of the motor*. Minneapolis: University of Minnesota Press.

Virillo, P. (1997). *Open sky*. London: Verso.

Virilio, P., & Lotringer, S. (2008). *Pure war*. Cambridge, MA: MIT Press.

Waetjen, W. (1993). Technological literacy reconsidered. *Journal of Technology Education.,4*(2), 5–11

Wagner, D., & Kozma, R. (2003). *New technologies for literacy and adult education: A global perspective.* Paper for NCAL/OECD International Roundtable. Philadelphia, PN. https://www.literacy.org/ICTconf/PhilaRT_wagner_kozma_final.pdf.

Wajcman, J. (2004). *Technofeminism.* Malden: Polity.

Weil, D. K. (1998). *Toward a critical multicultural literacy.* New York: Lang.

Walker, A. (2000). *Stanley Kubrick, director.* New York: Norton.

Wallerstein, I. (2004). *Alternatives: The U.S. confronts the world.* Boulder: Paradigm Press.

Walzer, M. (1983). *Spheres of justice.* New York: Basic Books.

Wark, McKenzie. (2008). *50 years of recuperation of the situationist international.* Princeton: Princeton Architectural Press.

Waterman, P. (1990). Communicating labor internationalism: A review of relevant literature and resources. *Communications: European Journal of Communications,15*(1/2), 85–103

Waterman, P. (1992). *International labour communication by computer: The fifth international?* (Working Paper Series 129). The Hague: Institute of Social Studies.

Webster, F. (1995). *Theories of the information society.* London: Routledge.

Webster, F., & Robins, K. (1986). *Information technology: A luddite analysis.* Norwood: Ablex.

Wells, H. G. (1938). *World brain.* Garden City: Doubleday, Doran.

Wheat, L. F. (2000). *Kubrick's 2001: A triple allegory.* Lanham: Scarecrow Press.

Wiehl, KLis with Lisa Pulitzer. (2020). Hunting the Unabomber: The FBI, Ted Kaczynski, and the Capture of America's most notorious domestic terrorist. *Library Journal.* April 1, 2020 at https://www.libraryjournal.com/?authorName=Lis%20Wiehl%20with%20Lisa%20Pulitzer. Accessed 16 Jan 2021.

Winter, L. (1977). *Autonomous technology: Technics-out-of-control as a theme in political thought.* Cambridge, MA: MIT Press.

Zapatistas Collective. (1994). *Zapatistas: Document of the new Mexican revolution.* New York: Autonomedia.

Zuboff, S. (1988). *In the age of the smart machine: The future of work and power.* New York: Basic Books.

Zuboff, S. (2020). *The age of surveillance capitalism: The fight for a human future at the new frontier of power.* London: Profile Books.

The manufacturer's authorised representative in the EU is Springer
Nature Customer Service Centre GmbH, Europaplatz 3, 69115 Heidelberg,
Germany. If you have any concerns regarding our products, please
contact ProductSafety@springernature.com

Printed and bound by CPI Group (UK) Ltd, Croydon, CR0 4YY

28/04/2026
02098487-0006